# ALTERNATIVE SWEETENERS

# FOOD SCIENCE AND TECHNOLOGY

A Series of Monographs, Textbooks, and Reference Books



1. Flavor Research: Principles and Techniques, *R. Teranishi, I. Hornstein, P. Issenberg, and E. L. Wick (out of print)*
2. Principles of Enzymology for the Food Sciences, *John R. Whitaker*
3. Low-Temperature Preservation of Foods and Living Matter, *Owen R. Fennema, William D. Powrie, and Elmer H. Marth*
4. Principles of Food Science
   Part I: Food Chemistry, *edited by Owen R. Fennema*
   Part II: Physical Methods of Food Preservation, *Marcus Karel, Owen R. Fennema, and Daryl B. Lund*
5. Food Emulsions, *edited by Stig Friberg*
6. Nutritional and Safety Aspects of Food Processing, *edited by Steven R. Tannenbaum*
7. Flavor Research: Recent Advances, *edited by R. Teranishi, Robert A. Flath, and Hiroshi Sugisawa*
8. Computer-Aided Techniques in Food Technology, *edited by Israel Saguy*
9. Handbook of Tropical Foods, *edited by Harvey T. Chan*
10. Antimicrobials in Foods, *edited by Alfred Larry Branen and P. Michael Davidson*
11. Food Constituents and Food Residues: Their Chromatographic Determination, *edited by James F. Lawrence*
12. Aspartame: Physiology and Biochemistry, *edited by Lewis D. Stegink and L. J. Filer, Jr.*
13. Handbook of Vitamins: Nutritional, Biochemical, and Clinical Aspects, *edited by Lawrence J. Machlin*

14. Starch Conversion Technology, *edited by G. M. A. van Beynum and J. A. Roels*
15. Food Chemistry: Second Edition, Revised and Expanded, *edited by Owen R. Fennema*
16. Sensory Evaluation of Food: Statistical Methods and Procedures, *Michael O'Mahony*
17. Alternative Sweeteners, *edited by Lyn O'Brien Nabors and Robert C. Gelardi*
18. Citrus Fruits and Their Products: Analysis and Technology, *S. V. Ting and Russell L. Rouseff*
19. Engineering Properties of Foods, *edited by M. A. Rao and S. S. H. Rizvi*
20. Umami: A Basic Taste, *edited by Yojiro Kawamura and Morley R. Kare*
21. Food Biotechnology, *edited by Dietrich Knorr*
22. Food Texture: Instrumental and Sensory Measurement, *edited by Howard R. Moskowitz*
23. Seafoods and Fish Oils in Human Health and Disease, *John E. Kinsella*
24. Postharvest Physiology of Vegetables, *edited by J. Weichmann*
25. Handbook of Dietary Fiber: An Applied Approach, *Mark L. Dreher*
26. Food Toxicology, Parts A and B, *Jose M. Concon*
27. Modern Carbohydrate Chemistry, *Roger W. Binkley*
28. Trace Minerals in Foods, *edited by Kenneth T. Smith*
29. Protein Quality and the Effects of Processing, *edited by R. Dixon Phillips and John W. Finley*
30. Adulteration of Fruit Juice Beverages, *edited by Steven Nagy, John A. Attaway, and Martha E. Rhodes*
31. Foodborne Bacterial Pathogens, *edited by Michael P. Doyle*
32. Legumes: Chemistry, Technology, and Human Nutrition, *edited by Ruth H. Matthews*
33. Industrialization of Indigenous Fermented Foods, *edited by Keith H. Steinkraus*
34. International Food Regulation Handbook: Policy • Science • Law, *edited by Roger D. Middlekauff and Philippe Shubik*
35. Food Additives, *edited by A. Larry Branen, P. Michael Davidson, and Seppo Salminen*
36. Safety of Irradiated Foods, *J. F. Diehl*
37. Omega-3 Fatty Acids in Health and Disease, *edited by Robert S. Lees and Marcus Karel*
38. Food Emulsions, Second Edition, Revised and Expanded, *edited by Kåre Larsson and Stig E. Friberg*
39. Seafood: Effects of Technology on Nutrition, *George M. Pigott and Barbee W. Tucker*
40. Handbook of Vitamins, Second Edition, Revised and Expanded, *edited by Lawrence J. Machlin*

41. Handbook of Cereal Science and Technology, *Klaus J. Lorenz and Karel Kulp*
42. Food Processing Operations and Scale-Up, *Kenneth J. Valentas, Leon Levine, and J. Peter Clark*
43. Fish Quality Control by Computer Vision, *edited by L. F. Pau and R. Olafsson*
44. Volatile Compounds in Foods and Beverages, *edited by Henk Maarse*
45. Instrumental Methods for Quality Assurance in Foods, *edited by Daniel Y. C. Fung and Richard F. Matthews*
46. *Listeria*, Listeriosis, and Food Safety, *Elliot T. Ryser and Elmer H. Marth*
47. Acesulfame-K, *edited by D. G. Mayer and F. H. Kemper*
48. Alternative Sweeteners. Second Edition, Revised and Expanded, *edited by Lyn O'Brien Nabors and Robert C. Gelardi*

*Other Volumes in Preparation*

Food Extrusion Science and Technology, *edited by Jozef L. Kokini, Chi-Tang Ho, and Mukund Karwe*

# ALTERNATIVE SWEETENERS
## SECOND EDITION, REVISED AND EXPANDED

edited by
Lyn O'Brien Nabors
Robert C. Gelardi

*Calorie Control Council*
*Atlanta, Georgia*

Marcel Dekker, Inc. New York • Basel • Hong Kong

**Library of Congress Cataloging-in-Publication Data**

Alternative Sweeteners / edited by Lyn O'Brien Nabors, Robert C.
Gelardi. -- 2nd ed., rev. and expanded.
p. cm. -- (Food Science and Technology; 48)
Includes bibliographical references and index.
ISBN 0-8247-8475-8 (alk. paper)
1. Nonnutritive sweetners. I. Nabors, Lyn O'Brien.
II. Gelardi, Robert C. III. Series: Food science and
technology (Marcel Dekker, Inc.); 48
[DNLM: 1. Sweetening Agents. W1 FO509P v. 48 / WA 712 A466]
TP422.A48 1991
664.6--dc20
DNLM/DLC
for Library of Congress 91-19959
CIP

This book is printed on acid-free paper.



MARCEL DEKKER, INC.
270 Madison Avenue, New York, New York 10016

Current printing (last digit):
10 9 8 7 6 5 4 3 2 1

PRINTED IN THE UNITED STATES OF AMERICA

# Preface

Alternative sweeteners continue to be a fascinating area of food technology. Food and beverage companies, food technologists and other scientists, and regulators spend considerable effort following sweetener developments. They cannot afford to do otherwise, considering the impact sucrose alternatives are having on low-calorie food production—one of the fastest-growing segments of the food industry.

With a variety of approved sweeteners, manufacturers are able to use the most appropriate sweetener or combination of sweeteners for a given product—the multiple sweetener approach. And recent additional sweetener approvals are making the multiple sweetener approach a reality. An individual sweetener's limitations can be compensated for by combining sweeteners. Product and taste options can be broadened. Additionally, many sweeteners when used in combination have a synergistic effect.

A sweetener by any name is not necessarily just a sweetener. Speciality sweeteners have important properties other than sweetness. For example, some (e.g., isomalt) are being used effectively for bulk in combination with high intensity sweeteners. Others can be used in speciality products designed to inhibit dental caries.

A very significant area for food technologists in the future will be the interaction and combination of alternative sweeteners with low-calorie ingredients such as bulking agents and fat replacers. By using the multiple ingredient approach—combining bulking agents (one or more), fat substitutes, and sweeteners—opportunities for future product development will be limitless.

This book presents the latest information on a wide variety of alternative sweeteners. All chapters have been updated from our first edition and chapters have been added on sweeteners barely mentioned just five years ago. Although there are many additions, there is only one deletion. Developers of thaumatin have found that it cannot be used as an intense sweetener in reduced calorie foods. Since it is more properly a flavor enhancer and palatability improver, thaumatin is not included in this edition. We have once again made a concerted effort to provide the reader with comprehensive, updated information on a wide variety of alternative sweeteners. Substantial reference material is included for those wishing to pursue the topic further.

Lyn O'Brien Nabors
Robert C. Gelardi

# Contributors

Albert Bär, Ph.D. *Bioresco Ltd., Brussels, Belgium*

Abraham I. Bakal, Ph.D. *ABIC International Consultants, Inc., Fairfield, New Jersey*

Iwan H. Blankers *Lactitol Department, CCA biochem b.v., Gorinchem, The Netherlands*

Barbara A. Bopp, Ph.D. *Drug Metabolism Department, Abbott Laboratories, Abbott Park, Illinois*

César M. Compadre, Ph.D. *College of Pharmacy, University of Arkansas, Little Rock, Arkansas*

Ronald C. Deis, Ph.D. *Product Applications, The NutraSweet Company, Deerfield, Illinois*

Basant K. Dwivedi, Ph.D. *NuGen Foods, Ltd., Paterson, New Jersey*

Robert C. Gelardi *Calorie Control Council, Atlanta, Georgia*

Bruno Gentili* *Fruit and Vegetable Chemistry Laboratory, Agricultural Research Service, U.S. Department of Agriculture, Pasadena, California*

Michael E. Hendrick, Ph.D. *Specialty Chemicals Research and Development, Pfizer Inc, Groton, Connecticut*

Barry E. Homler, Ph.D. *Product Applications, The NutraSweet Company, Deerfield, Illinois*

---

**Current Afilliation:* Retired

Robert M. Horowitz, Ph.D. *Fruit and Vegetable Chemistry Laboratory, Agricultural Research Service, U.S. Department of Agriculture, Pasadena, California*

William E. Irwin, M.S., Ph.D. *Palatinit Süßungsmittel GmbH, Elkhart, Indiana*

A. Douglas Kinghorn, Ph.D. *College of Pharmacy, University of Illinois at Chicago, Chicago, Illinois*

Robert G. LaBarge, Ph.D. *New Ventures, Dow Chemical Co., Midland, Michigan*

Gilbert V. Levin, Ph.D. *Biospherics Inc., Beltsville, Maryland*

John E. Long Ph.D. *Archer Daniels Midland Company, Clinton, Iowa*

Gary A. Miller, Ph.D. *Product Development, McNeil Specialty Products Company, New Brunswick, New Jersey*

Mark L. Mitchell, Ph.D. *Research & Development, PMC Specialties Group, A Division of PMC, Inc., Cincinnati, Ohio*

Frances K. Moppett *Technical Service Department, Pfizer Inc, Groton, Connecticut*

Alan H. Moskowitz, Ph.D. *Warner-Lambert Company, Morris Plains, New Jersey*

Lyn O'Brien Nabors *Calorie Control Council, Atlanta, Georgia*

Thomas F. Osberger, MBA *Food Industry Consultant, Upland, California*

Ronald L. Pearson, Ph.D. *Research and Development, PMC Specialties Group, A Division of PMC, Inc., Cincinnati, Ohio*

Paul Price *Chemical and Agricultural Products Division, Abbott Laboratories, Abbott Park, Illinois*

William H. Shazer, Ph.D. *Product Applications, The NutraSweet Company, Deerfield, Illinois*

D. Doel Soejarto, Ph.D. *College of Pharmacy, University of Illinois at Chicago, Chicago, Illinois*

Peter J. Sträter *Palatinit Süßungsmittel GmbH, Mannheim, Germany*

John A. van Velthuijsen *Research & Development, CCA biochem b.v., Gorinchem, The Netherlands*

Gert-Wolfhard von Rymon Lipinski *Hoechst Aktiengesellschaft, Frankfurt, Germany*

Lee R. Zehner, Ph.D. *Biospherics Inc., Beltsville, Maryland*

# Contents

# I
# INTENSE SWEETENERS

# 1

# Alternative Sweeteners: An Overview

**Lyn O'Brien Nabors and Robert C. Gelardi** *Calorie Control Council, Atlanta, Georgia*

Low-calorie and light foods and beverages comprise one of the fastest growing segments of the food and beverage industry. A 1989 low-calorie and dieting consumer survey conducted in the United States found that a majority of adult Americans consume low-calorie products. Since 1986, the number of adult Americans using these products increased from 45% to 51%, or from 78 to 93 million. This is more than double the 42 million consumers using these products in 1978. The 1989 data also show for the first time that a majority of low-calorie consumers, about 60%, are not on a diet but are using low-calorie products as part of an overall healthy lifestyle (1).

A study conducted in 1988 in Great Britain found that over half of the British population consume intense sweeteners regularly. The study data also indicate that individuals are using these products as part of their regular food consumption and not primarily for weight loss or in special dietary regimens (2).

## THE MULTIPLE SWEETENER APPROACH

For nearly a century, low-calorie products were almost entirely dependent on saccharin, the oldest intense sweetener. Now with additional sweeteners available, a multiple sweetener approach can be used. Availability of a variety of sweeteners is important because no sweetener, including sucrose, is perfect for all uses. With several available, each sweetener can be used in the applications for which it is best suited.

Manufacturers also can overcome limitations of individual sweeteners by using them in blends.

Cyclamate and saccharin historically provided sweetness in a number of popular products. Their use in combination was the first practical application of the multiple sweetener approach. The primary advantage of this sweetener blend was that saccharin boosts the sweetening power of cyclamate, while cyclamate masks the aftertaste that some people associate with saccharin. Also, the two sweeteners when combined have a synergistic effect—that is, the sweetness of the combination is greater than the sum of the individual parts. This is true for most sweetener blends. Combinations of other sweeteners, such as (1) aspartame and saccharin and (2) acesulfame-K, aspartame, and saccharin, are being utilized.

Now with the advent of fat substitutes and low-calorie bulking agents, not only a multiple sweetener approach but a multiple ingredient approach is being utilized to provide additional low-calorie product choices. Some specialty sweeteners, for example, many polyols, provide bulk as well as sweetness and combine well with the intense sweeteners. Polydextrose also can provide the bulk needed in many low-calorie products.

Not only is there evidence that humans have an innate desire for sweets (3), but research also indicates that the obese or those who were once obese may have a greater taste preference for fatty liquids mixed with sugar (4). Thus, fat substitution will be increasingly important in the development of new low-calorie products.

## RELATIVE SWEETNESS

Sweetness is subjective and dependent upon a number of factors. The concentration of the sweetener, the temperature at which the product is consumed, pH, the other ingredients in the product, and the sensitivity of the taster are all important. Sucrose is the usual standard. Evaluation of the sweetness of a given substance in relation to sucrose is made on a weight basis. Table 1 provides an approximation of the relative sweetness of many of the alternatives to sucrose discussed in this volume.

## THE IDEAL SWEETENER

Alternative sweeteners are used to (1) provide and expand food and beverage choices to control caloric, carbohydrate, or specific sugar intake; (2) assist in weight maintenance or reduction; (3) aid in the manage-

Table 1 Relative Sweetness of Alternatives to Sucrose

| Alternative sweetener | Approximate sweetness (sucrose = 1) |
|---|---|
| Acesulfame-K | 200 |
| Alitame | 2000 |
| Aspartame | 180 |
| Crystalline fructose | 1.2–1.7 |
| Cyclamate | 30 |
| Dihydrochalcones | 300–2000 |
| Glycyrrhizin | 50–100 |
| Hernandulcin | 1000 |
| High fructose corn syrup, 55% | 1 |
| High fructose corn syrup, 90% | 1+ |
| Isomalt | 0.45–0.64 |
| Isomaltulose | 0.48 |
| L-Sugars | 1 |
| Lactitol | 0.4 |
| Maltitol and hydrogenated starch hydrolysates | 0.7–0.9 |
| Mannitol | 0.7 |
| Monellin | 1500–2000 |
| Saccharin | 300 |
| Sorbitol | 0.54–0.7 |
| Stevioside | 300 |
| Sucralose | 600 |
| Thaumatin | 2000–3000 |
| Xylitol | 1 |

ment of diabetes; (4) assist in the control of dental caries; (5) enhance the usability of pharmaceutical and cosmetics; (6) provide sweetness when sugar is not available; and (7) assist the cost-effective use of limited resources.

The ideal sweetener does not exist. Even sucrose does not fulfill all sweetening needs. Alternative sweeteners are superior to sucrose in some products (e.g., pharmaceuticals and chewing gums).

The ideal sweetener should be at least as sweet as sucrose, colorless, odorless, and noncariogenic. It should have a clean, pleasant, untainted taste with immediate onset and without lingering. The more a sweetener tastes and functions like sucrose, the greater its facility for use in foods and beverages. A sweetener must be compatible with a wide range of food ingredients as sweetness is but one component of complex flavor systems.

The ideal sweetener should be water-soluble and should be stable in both acidic and basic conditions and over a wide range of temperatures. Length of stability is also important.

The safety of an alternative sweetener is essential. The sweetener must be nontoxic and metabolized normally or excreted unchanged.

A sweetener should be competitively priced with sucrose and other comparable sweeteners. It must be easily produced, stored, and transported.

## INTERNATIONAL REGULATORY GROUPS

Food additives are evaluated and/or regulated by numerous international and national bodies. The international groups include the Joint Food and Agriculture Organization/World Health Organization Expert Committee on Food Additives (JECFA), the Codex Alimentarius Commission, and the Scientific Committee for Food of the Commission of the European Economic Communities.

The objective of the joint Food and Agriculture Organization/World Health Organization (FAO/WHO) program on food additives is to make systematic evaluations of food additives and provide advice to member states of FAO and WHO on the control of the additives and related health aspects. The two groups responsible for implementing the program are JECFA and the Committee on Food Additives of the Joint FAO/WHO Codex Alimentarius Commission (5).

JECFA is made up of an international group of experts who serve without remuneration in their personal capacities rather than as representatives of their governments or other bodies. Selection of members is based primarily on their ability and technical experience, with consideration given to adequate geographical distribution. Their reports contain the collective views of the group and do not necessarily represent the decisions or the stated policy of the WHO or FAO. The experts convene to give advice on technical and scientific matters, establishing specifications for identity and purity for food additives, evaluating the toxicological data, and recommending, where appropriate, acceptable daily intakes for humans. The Expert Committee also acts in an advisory capacity for the Codex Committee on Food Additives (5).

The Codex Alimentarius Commission was established in 1962 to implement the Joint FAO/WHO Food Standards Program. Membership is comprised of those member nations and associate members of FAO and WHO that have notified the Director-General of FAO or WHO of their wish to be members. The purpose of the program is:

> to protect the health of consumers and to ensure fair practices of the foodtrade; to promote coordination of all food standards work undertaken by international governmental and non-governmental organizations; to determine priorities and initiate and guide the preparation of draft standards through and with the aid of appropriate organizations; to finalize standards and after acceptance by governments publish them in a Codex Alimentarious either as regional or world-wide standards.

The Codex Alimentarius is intended to guide and promote the elaboration and establishment of definitions and requirements for foods, including food additives, "to assist in their harmonization and in so doing so to facilitate international trade" (6).

The Scientific Committee for Food (SCF) of the Commission of the European Economic Communities was established by the Commission in 1974. The purpose of the Committee is to advise the Commission "on any problem relating to the protection of the health and safety of persons arising from the consumption of food, and in particular the composition of food, processes which are liable to modify food, the use of food additives and other processing aids as well as the presence of contaminants." Committee members are independent persons qualified in medicine, nutrition, toxicology, biology, chemistry, or other similar disciplines. The opinions of the Committee are submitted to the Commission.

As the countries of the European Community (EC) move to establish a single market by 1992, the importance of the SCF has increased. The European Community has agreed to harmonize its member states' laws on food additives, including intense and bulk sweeteners. The SCF's "opinion" is essential to the development of the "Sweeteners Directive," along with the general directive that will regulate all food additives. Under the directive being prepared by the EC, the number of available intense sweeteners is expected to increase and conditions of use for sweeteners may be set. Acesulfame-K, aspartame, cyclamate, saccharin, thaumatin, and neohesperidine are expected to be permitted in the EC, including countries where they are not now allowed (7). A 2.4 calorie per gram value is expected to be assigned to all the bulk sweeteners known as polyols (8).

In the United States, the Committee on Food Chemicals Codex, a full committee of the Food and Nutrition Board, Institute of Medicine, National Research Council, National Academy of Sciences, provides information on the quality and purity of food-grade substances. Specifications and test methods are included in over 800 FCC monographs on substances that are added to or come in contact with foods. The members of the Committee on Food Chemicals Codex are chosen for their special competencies and with regard for appropriate balance.

The *Food Chemicals Codex* is officially recognized by the U.S. Food and Drug Administration (FDA), the principal U.S. food regulatory agency. The fact that an article appears in the *Codex* does not, however, except it from compliance with requirements of acts of Congress or regulations and rulings issued by U.S. government agencies under the authority of congressional acts. *Codex* specifications have been adopted, under certain conditions, by the National Health and Medical Research Council of Australia; the Health Protection Branch of the Department of National Health and Welfare of Canada; The Ministries of Agriculture, Fisheries, and Food of Great Britain; and the Department of Health (Food and Nutrition Branch) of New Zealand. The *Codex* also has been the source of many specifications developed by JECFA (9).

Many countries have special agencies responsible for the regulation of food additives. Some have scientific support. Several of these agencies are mentioned above. The following detailed discussion concerns the United States' regulation of food additives.

## REGULATION OF FOOD ADDITIVES IN THE UNITED STATES

Food additives were first subjected to regulation in the United States under the Food and Drugs Act of 1906. Section 402.a.1 of the Act states that a food shall be deemed adulterated

> if it bears or contains any poisonous or deleterious substance which may render it injurious to health; but in case the substance is not an added substance, such food shall not be considered adulterated under this clause if the quantity of such substance in such food does not ordinarily render it injurious to health. (10, p. 17)

"Added" was not defined, but subsequently has generally been interpreted as a substance not present in the food in its natural state. This section appears to be intended to prohibit any level of added food substance inconsistent with public health. The Federal Food, Drug, and Cosmetic Act of 1938 contains food safety provisions similar to those in the 1906 act.

The basic Food, Drug, and Cosmetic Act was last updated in 1958. Section 201(s) of the act defines the term "food additive" as

> any substance the intended use of which results or may reasonably be expected to result, directly or indirectly, in its becoming a compo-

> nent or otherwise affecting the characteristics of any food (including any substance intended for use in producing, manufacturing, packing, processing, preparing, treating, packaging, transporting, or holding food; and including any source of radiation intended for any such use). (10, p. 5)

Also in 1958, Congress passed the Food Additives Amendment, Section 409 of the Food, Drug, and Cosmetic Act. The law exempts two important groups of substances from this food additive definition, and neither is considered a food additive under U.S. law. Those exempted are (a) substances generally recognized as safe (GRAS), among experts qualified by scientific training and experience to evaluate safety, and (b) substances that either the FDA or the U.S. Department of Agriculture (USDA) had sanctioned for use in food prior to 1958 (so-called "prior-sanction" substances). Additionally, the law does not pertain to pesticide chemicals in or on raw agricultural commodities.

The renowned "Delaney Clause" became part of U.S. food law as part of the 1958 Food Additives Amendment. This much discussed, rarely directly used, clause states that "no additive shall be deemed to be safe if it is found to induce cancer when ingested by man or animal, or if it is found, after tests which are appropriate for the evaluation of the safety of food additives, to induce cancer in man or animal" (10, p. 34). There has been much debate not only over whether the clause is appropriate food safety policy, but also concerning some of its terms and phrases, such as "induce cancer" and "tests which are appropriate." Efforts by the FDA to ban saccharin in 1977 considerably heightened interest in this provision.

The 1958 Food Additives Amendment also forbids the use of any food additive not approved by the FDA, and the agency is forbidden to approve a food additive that has not been shown to be "safe." Section 409 of the act outlines the requirements for requesting approval for a food additive (i.e., "Petition to establish safety") and details the action to be taken by the FDA in dealing with such a petition.

A petitioner, requesting the issuance of a food additive regulation, must provide, in addition to any explanatory or supporting data:

> (1) the name and all pertinent information concerning such food additive, including, where available, its chemical identity and composition; (2) a statement of the conditions of the proposed use of such additive, including all directions, recommendations, and suggestions proposed for the use of such additive, and including samples of its proposed labeling; (3) all relevant data bearing on the physical or other technical effect such additive is intended to pro-

> duce, and the quantity of the additive required to produce such effect; (4) a description of practicable methods for determining the quantity of such additive in or on food, and any substance formed in or on food, because of its use; and (5) full reports of investigations made with respect to the safety of such additive, including full information as to the methods and controls used in conducting such investigations. (10, p. 34)

The federal Food, Drug, and Cosmetic Act does not describe the safety investigations to be conducted on the proposed food additive.

In October 1982, the FDA announced the availability for public comment of a document entitled "Toxicological Principles for the Safety Assessment of Direct Food Additives and Color Additives Used in Food." The document, referred to as the "Redbook," delineates the FDA's stated framework for scientific decision making in evaluating the safety of new direct additives and for maintaining an overview of the safety of approved additives. The document addresses:

> (1) Tiered basic information requirements for direct food additives and color additives used in food based upon population exposure and molecular structure; (2) decision elements based on an evaluation of existing data for determining whether there is a need for more specific toxicological information in making safety judgments; (3) "core" standards for evaluating the sensitivity and rigor of previously conducted toxicological studies; (4) "current" standards for evaluating the sensitivity and rigor of new toxicological studies; (5) guidelines that provide suggested protocols for performance of the most commonly employed toxicological studies; and (6) priority-setting elements that can be applied to all direct food additives and color additives used in food so that the agency can more efficiently monitor the safety of approved additives and focus resources on those issues most likely to benefit the public health. (11)

Redbook II is expected to be available for public comment in late 1990 or early 1991. Important changes are expected in guidelines for metabolic studies, long-term bioassays, and developmental toxicity. It reportedly will include new or expanded sections on pharmacokinetics, immunotoxicity, behavioral toxicity, and unique food additives such as fat substitutes (12).

## ACCEPTABLE DAILY INTAKE

Many bodies evaluating and/or regulating food additives utilize an acceptable daily intake (ADI) in setting conditions of use. The ADI "for man, expressed on a body weight basis, is the amount of a food additive that can be taken daily in the diet, even over a lifetime, without risk" (5).

An ADI is a conservative estimate that incorporates a considerable safety factor. It is established from toxicological testing in animals, and sometimes humans, and is normally estimated by applying an intentionally conservative safety factor (generally a 100-fold safety factor). Animal tests are used to determine the maximum dietary level of an additive demonstrating no toxic effects, a "no observable effect level" or NOEL. The NOEL is then used to determine the ADI. For example, if safety evaluation studies of a given substance demonstrate a NOEL of 1000 mg/kg, using a 100-fold safety factor the ADI would be 10 mg/kg body weight per day for humans.

The ADI does not represent a maximum allowable daily intake level where safety ends and health concerns begin. It is not a precise value, and occasional consumption greater than the ADI should not cause concern.

## CONCLUSION

The development of a new food ingredient is an extremely costly and lengthy process. The toxicological tests alone cost millions of dollars. Nonetheless, the consumer and commercial importance of these products is such that a number of manufacturers continue to make this substantial investment. New alternative sweeteners, therefore, will soon be available. Combining intense sweeteners, bulk sweeteners, low-calorie bulking agents, and fat substitutes provides endless new low-calorie product possibilities. Thus, in addition to alternative sweeteners, polydextrose, fat substitutes, and mixed sweetener functionality also are discussed in this volume.

### REFERENCES


1. Dieting and Low-Calorie Products Survey, Conducted by Booth Research Services, Inc. for the Calorie Control Council, May, 1989.
2. Britain's Sweet Taste, Sweetener Update, the quarterly journal of the International Sweeteners Association, Brussels, Belgium, Summer 1990.
3. J. A. Desor, O. Maller, and L.S. Green, Preference for sweet in humans: Infants, children and adults in *Taste and Development: The Genesis of Sweet Preference* (James M. Weiffenbach, ed.), U.S. Department of Health, Education, and Welfare, Washington, D.C., 1977, p. 171.
4. A. Drewnowski, sweetness and obesity in *Sweetness* (J. Dobbing, ed.), Springer-Uerlag, Berlin, 1987.
5. Joint FAO/WHO Expert Committee on Food Additives, *Toxicological Evalua-*

tion of Certain Food Additives with a Review of General Principles and Specifications, World Health Organization, Geneva, 1974.
6. Codex Alimentarius Commission, Procedural Manual, Food and Agricultural Organization of the United Nations, Rome, 5th ed., 1981, p. 21.
7. Proposal for a Council directive on sweeteners for use in foodstuffs, adopted by the Commission of the European Economic Communities, August 24, 1990.
8. European Economic Community pending directive on nutrition labeling for foodstuffs, Brussels, January 31, 1990.
9. *Food Chemicals Codex*, 3rd ed., National Academy of Sciences, Washington, D.C., 1981.
10. Federal Food, Drug, and Cosmetic Act, as amended, U.S. Government Printing Office, Washington, D.C., January 1979.
11. *Federal Register* [Docket No. 80N-0446], October 15, 1982, p. 46141.
12. K. B. Ekelman, Revising FDA's toxicity testing guidelines for food and color additives, Presented to the American Chemical Society, Division of Agricultural and Food Chemistry, August 26, 1990, Washington, D.C.

# 2

# Acesulfame-K

**Gert-Wolfhard von Rymon Lipinski** *Hoechst Aktiengesellschaft, Frankfurt, Germany*

## INTRODUCTION

Clauss and Jensen (1) in 1967 accidentally discovered a sweet-tasting compound, 5,6-dimethyl-1,2,3-oxathiazine-4(3H)-one-2,2-dioxide, having a new ring system which had previously not been synthesized. Systematic research on dihydrooxathiazinone dioxides revealed quite a number of sweet-tasting compounds in this group of substances. Variations of substituents in positions 5 and 6 of the ring system showed noticeable influence on the intensity and purity of the sweetness. All synthesized dihydrooxathiazinone dioxides, however, exhibited some sweetness, even the ring system without substituents. The maximum sweetness was found in compounds with short-chain alkyl groups. Sensory evaluations of the different dihydrooxathiazinone dioxides showed that the substitutions on the ring system not only influence the intensity but also the purity of the sweetness (Fig. 1).

In addition to variations of substituents on the dihydrooxathiazinone dioxide ring system, structurally similar compounds were synthesized in order to investigate whether variations within the ring system would influence the sweet taste. These evaluations did not reveal any new sweet-tasting compounds; even methylation on the nitrogen in the ring furnishes a compound without sweetness (Fig. 2). An evaluation of the different compounds clearly demonstrated that 6-methyl-1,2,3-oxathiazine-4(3H)-one-2,2-dioxide exhibited the most favorable taste properties. Because production of this compound seemed to be less

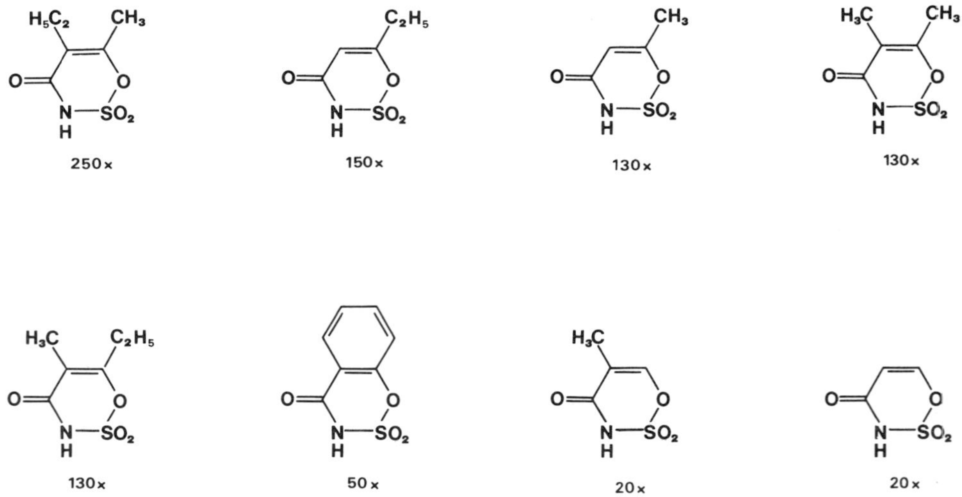


**Figure 1** Dihydrooxathiazinone dioxides with different substituents and sweetness intensities of their sodium salts, compared to a 4% sucrose solution, (tentatively measured by Clauß and Jensen (1)).

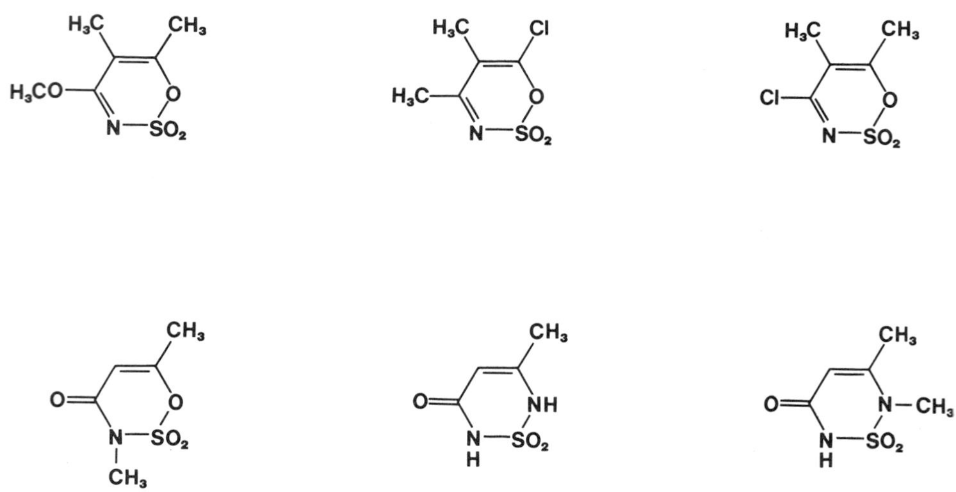

**Figure 2** Heterocyclic compounds without sweet taste.

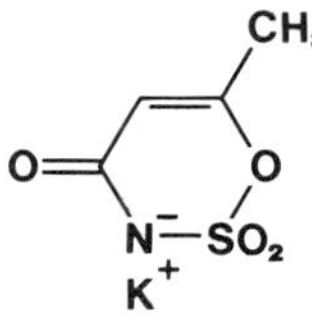

Figure 3 Acesulfame-K.

difficult than that of other dihydrooxathiazinone dioxides, it was chosen for systematic evaluation of its suitability as an intense sweetener for use in foods and beverages (Fig. 3).

In 1978, the World Health Organization registered acesulfame potassium salt as the generic name for this compound. This is often called acesulfame-K.

## SYNTHESIS

Dihydrooxathiazinone dioxides basically can be produced from different raw materials using different production routes. Suitable starting materials are ketones, β-diketones, derivatives of β-oxocarbonic acids, and alkynes, which may be reacted with halogen sulfonyl isocyanates. The compounds formed from such reactions are transformed into *N*-halogen sulfonyl acetoacetic acid amide. In the presence of potassium hydroxide, this compound cyclizes to the dihydrooxathiazinone dioxide ring system by separating out the corresponding potassium salts. Because dihydrooxathiazinone dioxides are highly acidic compounds, salts of the ring system are formed. The production of acesulfame potassium salt requires KOH; however, NaOH or $Ca(OH)_2$ can also be used (1–4).

Acetoacetamide-*N*-sulfonic acid may be another suitable starting material. In the presence of dehydrating agents this compound cyclizes to form the dihydrooxathiazinone dioxide ring system, which may be reacted with KOH to yield acesulfame potassium salt. Again, the production of salts other than the potassium salt seems basically possible (5).

Continuous production of acesulfame-K is possible using the different routes of synthesis. This allows large-scale production.

## PROPERTIES

Acesulfame K-is a white crystalline powder. The crystals are monoclinic, of the P $2_1$/c order. X-ray diffraction demonstrated that the ring system is almost planar, whereas the distances between the single atoms are less than those calculated from the theoretical values. The specific gravity of acesulfame-K is 1.83 g/cm³ (6).

The shelf life of pure crystalline acesulfame-K seems to be almost unlimited at room temperature. Samples kept at room temperature for more than 6 years and either exposed to light or protected from it did not show any signs of decomposition or differences in analytical data, as compared to freshly produced material (4).

Acesulfame-K does not show a definitive melting point. When the product is heated under conditions used for melting point determination decomposition is normally observed at temperatures well above 200°C. The decomposition limit seems to depend on the heating rate. No decomposition of acesulfame-K has been observed under conditions of temperature exposure normally found for food additives. In contrast to acesulfame-K, acesulfame acid has a sharp and definitive melting point at 123°C (2).

Even at room temperature, acesulfame-K dissolves readily in water. The solubility at 20°C is about 270 g/liter water. With rising temperatures, the solubility increases sharply to more than 1000 g/liter at 100°C. In alcohols, however, acesulfame-K is much less soluble. At 20°C, only about 1 g/liter dissolves in anhydrous ethanol. In mixtures of alcohol and water, the solubility increases with rising water content. In 50% ethanol (v/v), about 100 g/liter can be dissolved (2). Aqueous solutions of acesulfame-K are almost neutral (Table 1).

In view of the temperature/solubility ratio of acesulfame-K solutions in water, the product can be easily purified by recrystallization. High-purity acesulfame-K can thus be produced on a technical scale, while meeting the purity requirements for food additives.

**Table 1** Solubility of Acesulfame-K

| Solvent | Temperature (°C) | g/100 ml solvent |
|---|---|---|
| Water | 0 | 15 |
| Water | 20 | 27 |
| Water | 100 | ca. 130 |
| Ethanol | 20 | 0.1 |
| Glacial acetic acid | 20 | 13 |
| DMSO | 20 | 30 |

*Source:* Ref. 2

## SENSORY PROPERTIES

Acesulfame-K exhibits about 200 times the sweetness of a 3% sucrose solution, although slightly higher values have been reported (7). The sweetness intensity depends on the concentrations of the sucrose solution to which it is compared. At the threshold level, the intensity is much higher and decreases with increasing sucrose concentrations to values from 130 to 100 times the sucrose value (Fig. 4). Normally, acesulfame-K can be considered to be about half as sweet as sodium saccharin, similarly sweet as aspartame, and four to five times sweeter than sodium cyclamate. In acid foods and beverages with the same concentrations, a slightly greater sweetness may be perceived, as compared, to neutral solutions.

The sweet taste of acesulfame-K is perceived quickly and without unpleasant delay. The taste is not lingering and does not persist longer than the intrinsic taste of the food. In aqueous solutions with high concentrations of acesulfame-K, a bitter taste can sometimes be detected (8). In foodstuffs with lower concentrations of acesulfame-K, this effect is not of great importance. As for all intense sweeteners, however, assessments of different taste characteristics depend on the product the sweeteners are used in (8). It was further observed that the sweetness of acesulfame-K solutions does not decrease with rising temperatures to the extent of some other intense sweeteners (9).

Although acesulfame-K can be used as an intense sweetener by itself and does not show particular taste problems when used in appropriate concentrations, mixtures with other sweetening agents are of high practical interest (10,11). Studies showed that, in particular, mixtures of acesulfame-K with various other intense sweeteners had especially desirable sensory properties. This qualitative improvement of taste seems to be caused by an addition of the time-intensity profiles of the individual sweeteners. Thus combination balances the characteristics of the sweeteners used in such blends (12). Favorably viewed were mixtures of acesulfame-K and aspartame (ca. 1:1 by weight), acesulfame-K and sodium cyclamate (ca. 1:5 by weight), and sometimes other mixtures, in which both sweetening agents almost equally contributed to the sweetness of the mixture. A strong synergistic taste enhancement was noted in mixtures of acesulfame-K and aspartame or sodium cyclamate, whereas only slight taste enhancement was perceived in mixtures of acesulfame-K and saccharin (13). Acesulfame-K and aspartame each taste approximately 100 times sweeter than a sucrose solution containing 100 g/liter (8). A blend of both sweeteners, however, tastes more than 300 times sweeter than the same sucrose solution.

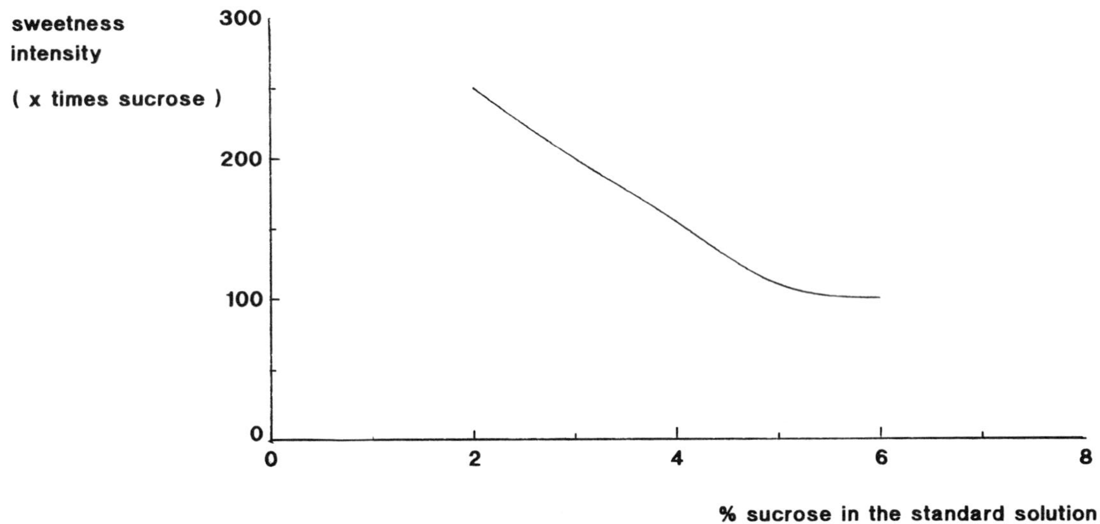


Figure 4 Sweetness intensity of acesulfame-K in aqueous solution.

This effect can apparently not be explained from the individual sweetness intensity curves and should therefore be attributable to synergism (14). Although strongest taste enhancement was found in 1:1 (w/w) blends, pronounced synergism was observed for other blend ratios, too.

Favorable taste reports were given for mixtures of acesulfame-K with sugar alcohols or sugars. Mixtures of acesulfame-K with sugar alcohols have a full and well-balanced sweetness. Therefore, these mixtures are particularly suitable for sugarless confectionary, fruit preparations, and other foods, which require a bulking agent. The rounding effect seems attributable to the fast onset of acesulfame-K's sweetness. Suitable blend ratios are approximately 1:100 in a mixture with xylitol, 1:150 in a mixture with maltitol, 1:150–200 in a mixture with sorbitol, and 1:250 in a mixture with isomalt (12).

## TOXICOLOGY

As toxicological evaluations of intense sweeteners are of crucial importance for the approval and subsequent usage, a full range of toxicological studies was carried out with acesulfame-K.

The acute oral toxicity of acesulfame-K is so low that it can be regarded as practically nontoxic. The $LD_{50}$ was orally determined at 6.9–8.0 g/kg body weight. The intraperitoneal $LD_{50}$ is 2.2 g/kg body weight (15). Subchronic toxic effects were investigated in a 90-day study with rats. The animals were fed concentrations from 0 to 10% acesulfame-K in the diet. Potential carcinogenicity and chronic toxicity were studied in rats that were fed up to 3% acesulfame-K in the diet. A carcinogenicity study was conducted in mice fed with concentrations up to 3% acesulfame-K. In addition, chronic toxicity effects of acesulfame-K were studied in beagle dogs for 2 years (15). Neither indications of carcinogenicity nor other effects excluding the use acesulfame-K as a food additive were found. A no-effect level of 3% of acesulfame-K in the diet was established.

No mutagenicity was found in various studies, including a dominant lethal test; a micronucleus test; bone marrow investigations in hamsters; tests for malignant transformation, DNA binding, and others (15). In reproduction studies with rats fed acesulfame-K, no deviation was seen in fertility, the number of young animals per litter, body weight, growth, or mortality (15).

The toxicological studies on acesulfame-K demonstrated that the compound would be safe for use as an intense sweetener. This view was confirmed by the Joint Expert Committee on Food Additives of the FAO/WHO, according to which the data showed that acesulfame-K did not exhibit mutagenicity or carcinogenicity. Therefore, an acceptable daily intake (ADI) of 0–9 mg/kg of body weight was allocated (16). The Scientific Committee for Foods of the EEC published an assessment stating that long-term studies did not show any dose-related increase in specific tumors nor any treatment-related pathological changes of significance. Therefore, again, an acceptable daily intake of 0–9 mg/kg of body weight was allocated (17). Based on an extensive evaluation of the available animal studies, the Food and Drug Administration allocated an acceptable daily intake of up to 15 mg/kg of body weight (18).

## METABOLISM AND PHYSIOLOGICAL CHARACTERISTICS

Acesulfame-K is not metabolized by the human body. To investigate possible metabolic transformations, $^{14}C$-labeled acesulfame-K was used. Studies were carried out in rats, dogs, and pigs. As animal studies did not show any metabolism, human volunteers were also given labeled acesulfame-K. The different animal species, as well as the human volunteers, excreted the original compound. No activity attributable to metabolites was found. As acesulfame-K is excreted completely unmetabolized, it does not have any caloric value.

In conjunction with the metabolic studies, the pharmacokinetics of acesulfame-K were also investigated. These studies were carried out with rats, dogs, pigs, and later with human volunteers. All animal species, as well as humans, quickly absorbed acesulfame-K, but there was rapid excretion of the compound, mainly in the urine. A multiple-dose study showed an accumulation in tissues (15). Serum determination of acesulfame-K can be carried out by high-performance liquid chromatographic (HPLC) analysis (19). No activity attributable to metabolites was found. After prolonged exposure to acesulfame-K, animals did not show any sign of induced metabolism. Again, after administration of $^{14}C$-labeled acesulfame-K, only the original substance was found in the excreta (15).

Acesulfame-K is not metabolized by bacteria, including *Streptococcus mutans* and other microorganisms which may contribute to the formation of caries. Acesulfame-K was tested in several studies. Although an inhibition of dental plaque microorganisms or *Streptococcus mutans* was not always reported, in other test systems a clear inhibition was demonstrated (20–22). Synergism in the inhibition of bacteria was observed with mixtures of intense sweeteners including acesulfame-K (23) and mixtures of acesulfame-K, saccharin, and fluoride (24). In all studies, however, the concentrations necessary for an inhibitory effect were well above the concentrations used for customary sweetness levels.

## STABILITY AND REACTIONS IN FOODS

Long-term and heat stability are important factors for the use of intense sweeteners in many food products and in beverages. In this regard, various conditions have to be met. pH levels in different foods and beverages vary from neutral to the acid range and may, in extreme cases like certain soft drinks, go down to values around and even below pH 3. In this wide pH range, even after prolonged storage, no decrease of sweetness intensity must be detected.

In aqueous media acesulfame-K is distinguished by very good stability. After several months of storage at room temperature, virtually no change in acesulfame-K concentration was found in the pH range common for beverages. Prolonged continuous exposure at 30°C, conditions that will hardly be found in practice, does not cause losses exceeding 10%, the threshold for recognition of sweetness differences (7). Even at temperatures of 40°C the threshold for detection of sweetness differences is exceeded after several months only for products having pH 3.0 or higher (25).

Extensive studies were performed with buffered aqueous solutions. Results for pH levels and storage conditions commonly found for soft drinks are given in Table 2. After 10 years' storage of a solution buffered to pH 7.5 at room temperature, no significant loss of acesulfame-K was detected.

Acesulfame-K–containing beverages can be pasteurized under normal pasteurizing conditions without loss of sweetness. Pasteurizing for longer periods at lower temperatures is possible, as is short-term pasteurization for a few seconds at high temperatures. Sterilization is possible without losses under the normal conditions, i.e., temperatures around 100°C for products having lower pH levels and 121°C for products around and above pH 4. In a solution of pH 4.0 which was heated to 120°C for 1 hr no loss of acesulfame-K could be measured. Half-life

**Table 2** Stability of Acesulfame-K in Buffered Aqueous Solution

| Storage time (weeks) | pH 3.0 recovery (%) | | pH 3.5 recovery (%) | |
|---|---|---|---|---|
| | 20°C | 30°C | 20°C | 30°C |
| 16 | 98 | 97 | 98 | 100 |
| 30 | 98 | 95 | 99 | 97 |
| 50 | 98 | 91 | 99 | 96 |
| 100 | 95 | — | 98 | — |

values determined at 100°C demonstrate that the common treatments of foods and beverages should not cause any substantial decomposition of acesulfame-K (26).

In baking studies no indication of decomposition of acesulfame-K was found even when biscuits with a low water content were baked at high oven temperatures for short periods. This corresponds to the observation that acesulfame-K decomposes at temperatures well above 200°C.

Potential decomposition products of acesulfame-K can, therefore, be found only under extreme conditions. Under such conditions, compounds of hydrolytic decomposition are mainly acetone, $CO_2$, ammonium salts, sulfate, and amidosulfonate. In the hydrolytic decomposition, the ring system is initially opened, which quickly yields the end products of hydrolysis. Only occasionally and then in very acidic media can traces of derivatives of acetoacetic acid be detected (3).

## APPLICATIONS

Acesulfame-K can be used as a sweetening agent in a wide range of products, for instance in low-calorie products, diabetic foods, sugarless products, oral hygiene preparations, pharmaceuticals, and animal feeds. The product groups acesulfame-K has been used in are shown in Table 3.

**Table 3** Applications of Acesulfame-K

| | |
|---|---|
| Table-top sweetener | Cookies |
| Carbonated soft drinks | Jams |
| Noncarbonated soft drinks | Fruit purees |
| Squashes and dilutables | Hard candy |
| Fruit nectars | Soft candy |
| Powdered beverages | Gum confections |
| Instant tea | Chocolate confections |
| Cider | Marzipan |
| Flavored milk | Chewing gum |
| Cocoa | Pickled vegetables |
| Yogurt | Marinated fish |
| Flavored whey | Tobacco |
| Desserts | Mouthwashes |
| Rice pudding | Toothpaste |
| Ice cream | Drugs |
| Low-calorie diets | Feedstuffs |
| Cake | |

Acesulfame-K is suitable for low-calorie and diabetic beverages because of its good stability in aqueous solutions. If used by itself, acesulfame-K can impart sweetness comparable to 8–10% sucrose. In countries where mixtures of intense sweeteners with other sweetening agents are permitted, mixtures of acesulfame-K with fructose, glucose, high-fructose corn syrup, or sucrose can be used. Generally, beverages containing such mixtures of acesulfame-K with other bulk sweeteners are rated to be fuller-bodied because of the slightly higher viscosity and the different taste profiles of sugars and acesulfame-K. A substantial number of low-calorie beverages, however, are sweetened with mixtures of acesulfame-K and aspartame or other intense sweeteners. These beverages benefit from the synergism and the improved taste characteristics provided by such blends. For example, a sweetness level equivalent to ca. 10% of sucrose is imparted by concentrations in range of 500–600 mg/liter of acesulfame-K or aspartame, whereas the same sweetness level can be achieved by using a blend of only 160 mg/liter of each of the sweeteners. Blend ratios between acesulfame-K and aspartame, in particular, are different in beverages having different flavors in order to match flavor and sweetness profiles (25).

Producing jams and marmalades with only intense sweeteners is difficult. Bulking agents are important to improve the texture and shelf life of such products. In trials for the production of jams with acesulfame-K, highly acceptable preparations were prepared with sorbitol as bulking material. Other sugar alcohols and polydextrose proved to be suitable, too. In combinations with acesulfame-K, the concentration of sorbitol or other bulking agents can be reduced, yielding products with a noticeable reduction in caloric values as compared to sucrose-containing products. These jams and marmalades are less protected from microbial spoilage compared to sucrose-containing products. In view of the low concentration of osmotically active compounds, the addition of preservatives can help to avoid such microbial spoilage. The flavor stability of jams and marmalades containing low levels of dry solids, however, is normally lower than the storage stability of standard products (12).

Confectionary items can be made with acesulfame-K if suitable bulking ingredients or bulk sweeteners are added to give the necessary volume. As the sweetness intensity of most sugar alcohols and bulking ingredients is lower than the sweetness of sucrose, acesulfame-K imparts the desired sweetness, whereas sugar alcohols or low-calorie bulking ingredients provide the necessary bulk and texture. Combinations of acesulfame-K with sorbitol can have a good taste pattern. Gum confections containing sorbitol instead of sucrose, and, in addition, 1000 mg/kg acesulfame-K were produced without difficulties. They showed a good, fruity sweet taste. Hard candies can be manufactured using acesulfame-K as the intense sweetener and isomalt, maltitol, or lactitol as the bulking ingredient. Acesulfame-K rounds the sweetness of these sugar alcohols and brings the taste close to standard, sugar-containing products. Due to the good temperature stability of acesulfame-K it can be added before cooking. Alternatively it can be added together with acids, flavorings, and colorants after cooking. In soft candies acesulfame-K shows a good shelf stability. Again, suitable blends of sugar alcohols and acesulfame-K allow the production of sugarless soft candy coming close to standard products in taste and texture. Sugarless marzipan having good taste, texture, and shelf stability can be manufactured using acesulfame-K as the intense sweetener and isomalt as a bulking ingredient. In chocolate and related products acesulfame-K can be added at the beginning of the production process, e.g., before the rolling. It withstands all treatments including conching without detectable decomposition (27).

Bulking agents with a sweet taste like sorbitol are generally used for the production of sugarless chewing gums. The low sweetness of sugar alcohols is normally enhanced by artificial sweeteners. The fast onset of sweetness of acesulfame-K is beneficial in forming the initial taste. Therefore, acesulfame-K–containing chewing gum has a pleasant sweet taste from the beginning. Due to its good solubility acesulfame-K may be dissolved fairly quickly by the saliva. Prolonged sweetness may be achieved by encapsulation of some acesulfame-K or combinations with other sweetening ingredients having a more lasting sweetness.

In reduced-calorie baked goods, bulking agents like polydextrose have to be substituted for sugar and flour and may help to reduce the level of fats. Acesulfame-K combines well with suitable bulking ingredients and bulk sweeteners and therefore allows production of sweet tasting baked goods having less calories. In diabetic products combinations of acesulfame-K and sugar alcohols like isomalt, lactitol, maltitol, or sorbitol can provide volume and sweetness. Although no reduction in calories is achieved, texture as well as the sweetness intensity can be very similar to sucrose-containing products. Less sweetness for baked in comparison to unbaked goods was reported in one study (28), but extensive work on acesulfame-K in baked goods did not give any indication for losses during baking. Even after prolonged exposure to elevated temperatures, including elevated oven temperatures, no statistically significant deviation from the added concentrations was found. Sensory studies showed good and pleasant sweetness for baked goods containing acesulfame-K alone or in combination with bulking ingredients or bulk sweeteners (29).

Acesulfame-K can easily be used in fruit-flavored dairy products. It withstands pasteurization of fruit preparations and pasteurization of the product itself. For nonfermented products like flavored milk, cocoa beverages, and similar products, the use of acesulfame-K as the single sweetener is advisable, as good heat and shelf-life stability is required. In certain countries approving cyclamate, blends of acesulfame-K and cyclamate are used in this application, too. Whenever temperature stability is not very important, particularly well-rounded taste profiles are obtained from blends of acesulfame-K and aspartame, for example in fruit-flavored yogurt and yogurt analog products. Acesulfame-K is apparently not attacked by lactic acid bacteria and other microorganisms used in fermented milk products. In ice cream and analog products, the modification of freezing and crystallization of water has to be achieved by the use of appropriate levels of sugar substitutes. Again, acesulfame-K can round and enhance the sweetness, as it combines particularly well with the taste of sugar alcohols (30). In addition to the listed food categories, many other products can be sweetened with acesulfame-K, examples of which are shown in Table 3.

The production of table-top preparations containing acesulfame-K is similar to those containing other intense sweeteners. Effervescent tablets, granules, powder, and solutions have been produced. Due to the good solubility of acesulfame-K in water, highly concentrated solutions suitable for household use can be manufactured. No shelf-life problems are anticipated for solutions under normal storage conditions. Similarly, no problems have been reported in the dissolution of tablets or powder. Solid table-top preparations are generally used in hot beverages, and the solubility of acesulfame-K is extremely good at such elevated temperatures.

Outside the food sector, acesulfame-K can be used to sweeten oral hygiene products, pharmaceuticals, tobacco products, and animal feedstuffs. In cosmetics, such as toothpaste and mouthwash preparations, and in similar products, some ingredients, e.g., surfactants, tend to impart a bitter taste. Ordinarily, special flavors and sweeteners are needed to mask such taste and to produce an initial pleasant flavor. Because the sweetness of acesulfame-K is perceived quickly, it is particularly suitable for these oral hygiene products. It is compatible with commonly used flavoring agents in toothpastes and mouthwash preparations. If the flavor concentration is not crucial for the taste character of the cosmetic product, the fast onset of the sweetness of acesulfame-K requires low concentrations of flavoring ingredients. As mentioned before, acesulfame-K is highly compatible with sorbitol, which is often used in toothpaste. If glycerol is used as a humectant and brightener, no solubility problems are incurred. Problems have also not been encountered when acesulfame-K is dissolved in mouthwash formulations containing alcohol, despite the low solubility of acesulfame-K in pure ethanol. In mixtures of ethanol and water, the solubility of acesulfame-K is higher than concentrations normally that are used in mouthwash preparations (31,32). Pharmaceuticals sometimes have unpleasant flavor or taste characteristics; again, acesulfame-K can mask such bitter taste patterns.

## ANALYTICAL METHODS

The thin-layer chromatographic detection of acesulfame-K is simple. Quantitative determinations can be carried out via liquid chromatographic methods and isotachophoresis. As the volatility of acesulfame-K is low, and methylation produces differing ratios of methyl derivatives, the quantitative determination by gas chromatography is impossible. Apart from the direct spectroscopic determination of acesulfame-K at 227 nm, no spectroscopic methods for the determination are available.

The thin-layer chromatographic detection of acesulfame-K can be carried out on polyamide layers. With fluorescent compounds in the layer, the detection can be carried out by extinction of fluorescence. In visible light, detection is possible after spraying with fluorescein and exposing to vapors of $Br_2$. The limit of detection is about $2 \mu g$ (33).

Liquid chromatographic determination of acesulfame-K in foods is simple, as often beverages or aqueous extracts from foods can be injected into the columns immediately after filtration. The methods described are basically reverse-phase separations with detection in ultraviolet (UV) light (34,35). The same liquid chromatographic procedure can be utilized in assaying feeds, which, in contrast to foods, should be clarified by adding zinc sulfate and potassium hexacyanoferrate (36).

Ion chromatography or anion separator columns may be used alternatively. Acesulfame-K and other sweeteners including cyclamate can be detected by a conductivity detector (37).

As conductivity detectors are used in isotachophoresis, too, this method can be used for separation of acesulfame-K from other food ingredients and intense sweeteners. Quantitative determination of acesulfame-K in the presence of other sweeteners has been described (35,38). All methods described allow the separation of acesulfame-K from other intense sweeteners which may be used in combination.

## REGULATORY STATUS

With the completion of a comprehensive safety evaluation program the way for the approval of acesulfame-K as a food additive was paved.

Evaluation of the data by the Joint Expert Committee for Food Additives of the WHO and FAO resulted in the approval of acesulfame-K for food use and the allocation of an ADI of 0–9 mg/kg body weight (16). In addition, specifications were published which have been adopted by many countries (39). The Scientific Committee for Foods of the EEC also evaluated acesulfame-K's safety. Again, acesulfame-K was approved for food use, and an ADI of 0–9 mg/kg body weight was allocated (17).

In the United Kingdom, the first country approving acesulfame-K, the Food Additives and Contaminants Committee published a report on sweetening agents for foods, listing acesulfame-K among "substances that the available evidence suggests are acceptable for use in food" (39). Following this report acesulfame-K was included "The Sweeteners in Food Regulations" approving its use for beverages and most foods without limitations.

In the United States, the approval of acesulfame-K for a number of applications was granted in 1988 (18). The FDA concluded that "acesulfame-K is safe for the intended use." An ADI of up to 15 mg/kg of body weight was granted.

As the ADI values allocated for acesulfame-K may not be easy to compare to use levels, a conversion into the corresponding sucrose levels offers a simple possibility to get an idea of the potential applications. Using the sweetness intensity factor of 200 times in comparison to sucrose the ADI allocated by the WHO and FAO corresponds to 126 g of sucrose for the 70-kg adult. The higher value recently granted by the FDA is equivalent to 210 g of sucrose per day for the 70-kg adult. These values are well above the average daily sucrose consumption. Therefore it can be concluded that acesulfame-K intake will generally be well below the ADI. This view has been confirmed by intake studies and calculations.

Following the favorable assessments published by the WHO and FAO and the Scientific Committee for Foods of the EEC, almost 40 countries have approved the use of acesulfame-K in various products (Table 4). In several countries, approvals according to good manufacturing practice have been granted, e.g., in the United States, the United Kingdom, Ireland, Switzerland, and Cyprus. Generally acesulfame-K is rated favorably, and several countries have expanded the original approvals to further applications.

**Table 4** Some Countries Where Acesulfame-K Has Been Approved

| | |
|---|---|
| Australia | Israel |
| Austria | Italy |
| Belgium | Luxembourg |
| Bulgaria[a] | Netherlands |
| Cyprus | Norway |
| Czechoslovakia | Pakistan |
| Denmark | South Africa |
| Egypt | Sweden |
| Federal Republic of Germany | Switzerland |
| France | United Kingdom |
| Greece | United States |
| Hong Kong | USSR[a] |
| Hungary | Yugoslavia |
| Ireland | |

[a]For oral hygiene products.

Petitions for the approval of acesulfame-K in many more countries are pending, and it may reasonably be expected that acesulfame-K will be approved as a food additive in many more countries in the near future.

In a recent reevaluation the Joint Expert Committe for Food Additives of the WHO and FAO revised the ADI for acesulfame-K and allocated the higher value of 0-15 mg/kg of body weight (41).

## REFERENCES

1. K. Clauss and H. Jensen, Oxathiazinon dioxides—a new group of sweetening agents, *Angew. Chem.* (Engl.), *85:*965 (1973), *Angew. Chem.* (Int. Edit.), *12:* 869 (1973).
2. K. Clauss, E. Lück, and G.-W. von Rymon Lipinski, Acetosulfam, ein neuer Süssstoff, *Z. Lebensm. Unters. Forsch.*, *162:* 37 (1976).
3. H.-J. Arpe, Acesulfame-K, a new noncaloric sweetener. In *Health and Sugar Substitutes*, Proc. ERGOB Conf., Geneva 1978 (B. Guggenheim, ed.), Karger, Basel, 1978, p. 178, and personal communications.
4. G.-W. von Rymon Lipinski and B. E. Huddart, Acesulfame-K, *Chem. Ind.* 427 (1983).
5. European Patent 1 59 516 (March 22, 1984) to Hoechst AG.
6. E. F. Paulus, 6-Methyl-1,2,3-oxathiazin-4(3H)-on 2,2-dioxide, *Acta Cryst.* (B), *31:* 1191 (1975).
7. K. Hoppe and B. Gassmann, Vergleichstabellen zur Süsseintensität von 16 Süßungsmitteln, *Lebensmittelindustrie*, *32:* 227 (1985).

8. K. Paulus and M. Braun, Süsskraft und Geschmacksprofil von Süssstoffen, *Ernährungs-Umschau, 35:* 384 (1988).
9. K. Hoppe and B. Gassmann, Neue Aspekte der Süsskraftbeurteilung und des Einsatzes von Süssungsmitteln, *Nahrung, 24:* 423 (1980).
10. A. I. Bakal, Functionality of combined sweeteners in several food applications, *Chem. Ind.* (18): 700 (1983).
11. R. C. Gelardi, The multiple sweetener approach and new sweeteners on the horizon, *Food Technol., 41* (1): 123 (1987).
12. G.-W. von Rymon Lipinski, The new intense sweetener Acesulfame K, *Food Chem., 16:* 259 (1985).
13. U.S. Patent 4 158 068 (June 12, 1979) to Hoechst AG.
14. G.-W. von Rymon Lipinski and E. Klein, Synergistic effects in blends of Acesulfame K and Aspartame, ACS Conference "Sweeteners—Carbohydrate and Low Calorie," Los Angeles, 1988, No. 27.
15. *Toxicological Evaluation of Certain Food Additives,* WHO Food Additives Series, 1980, No. 16, p. 11.
16. *Evaluation of Certain Food Additives and Contaminants,* Twenty-seventh Report of the Joint FAO/WHO Expert Committee on Food Additives, WHO Technical Report Series No. 696, Geneva, 1983, p. 21.
17. Report of the Scientific Committee for Food No. 16—Sweeteners. Commission of the European Communities, Luxembourg, 1985, p. 8.
18. *Fed. Reg. 53* (145): 28379 (1988).
19. M. Uihlein, Bestimmung von Acesulfam, einem neuen Süssstoff, in Serum mittels Hochdruckflüssigkeitschromatographie, *Dtsch. Lebensm. Rdsch., 74:* 157 (1977).
20. T. H. Grenby and M. G. Saldanha, Studies of the inhibitory action of intense sweeteners on oral microorganisms relating to dental health, *Caries Res., 20:* 17 (1986).
21. H. A. Linke, Adherence of *Steptococcus mutans* to smooth surface in the presence of artificial sweeteners, *Microbios, 36:* 41 (1983).
22. S. C. Ziesenitz and G. Siebert, Nonnutritive sweeteners as inhibitors of acid formation by oral microorganisms, *Caris Res., 20:* 498 (1986).
23. G. Siebert, S. C. Ziesenitz, and J. Lotter, Marked caries inhibition in the sucrose-challenged rat by a mixture of nonnutritive sweeteners, *Caries Res., 21:* 141 (1987).
24. A. T. Brown and G. M. Best, Apparent synergism between the interaction of saccharin, Acesulfame K and fluoride with hexitol metabolism by *Streptococcus mutans, Caries Res., 22:* 2 (1988).
25. G.-W. von Rymon Lipinski, Einsatz von Sunett$^R$ in Erfrischungsgetränken, *Swiss Food, 10*(5): 25 (1988).
26. K. Clauss, unpublished results.
27. G.-W. von Rymon Lipinski and E. Klein, Sunett$^R$—Anwendungen in Süsswaren, *Süsswaren, 32:* 336 (1988).
28. P. E. Redlinger and C. S. Setser, Sensory quality of selected sweeteners: Unbaked and baked flour doughs, *J. Food Sci., 52:* 1391 (1987).

29. G.-W. von Rymon Lipinski, Ch. Klug, and E. Gorstelle, unpublished results.
30. G.-W. von Rymon Lipinski, Einsatz von Sunett[R] in Joghurt und anderen Milcherzeugnissen, *Swiss Food, 10*(6): 29 (1988).
31. G.-W. von Rymon Lipinski, E. Lück, and F.-J. Dany, Acetosulfam—ein neuer Süssstoff für die Verwendung in Mundkosmetika, *Seifen Öle Fette Wachse, 102:* 243 (1976).
32. G.-W. von Rymon Lipinski and E. Lück, Acesulfame K: A new sweetener for oral cosmetics, *Manufacturing Chemist, 52:* 37 (1981).
33. G.-W. von Rymon Lipinski and H.-Ch. Brixius, Dünnschichtchromatographischer Nachweis von Acesulfam, Saccharin und Cyclamat, *Z. Lebensm. Unters. Forsch., 168:* 212 (1979).
34. H. Grosspietsch and H. Hachenberg, Analyse von Acesulfam-K durch Hochdruckflüssigkeits-Chromatographie, *Z. Lebensm. Unters. Forsch., 171:* 41 (1980).
35. U. Zache and H. Gründing, Bestimmung von Acesulfam-K, Aspartam, Cyclamat und Saccharin in fruchtsafthaltigen Erfrischungsgetränken, *Z. Lebensm. Unters. Forsch., 184:* 503 (1987).
36. A. Görner and H. Rückemann, Qualitative Bestimmung von Acesulfam-K in Mischfuttermitteln mittels Hochdruckflüssigkeitschromatographie, *Landwirtsch. Forschung, 33:* 213 (1980).
37. Th. A. Biemer, Analysis of saccharin, acesulfame K and sodium cyclamate by high-performance ion chromatography, *J. Chromatog., 463:* 463 (1989).
38. H. Klein and W. Stoya, Isotachophoretische Bestimmung von Acesulfam-K in Lebensmitteln, *Ernährung/Nutrition, 11:* 322 (1987).
39. *Specifications for Identity and Purity,* FAO Food and Nutrition Paper No. 28, Food and Agriculture Organization of the United Nations, Rome, 1983, p. 3.
40. Food Additives and Contaminants Committee, *Report on the Review of Sweeteners in Food,* London, 1982.
41. Thirty-seventh Meeting of the Joint FAO/WHO Expert Committee of Food Additives, Summary and conclusions document. Document No. TCS190.43, Geneva, 1990.

# 3

# Alitame

**Michael E. Hendrick** *Pfizer Inc, Groton, Connecticut*

## INTRODUCTION

The field of high-potency sweeteners has long been an active area of research. Stimulated by the accidental discovery of the potent sweet taste of aspartame in 1965 (1,2), dipeptides became an area of increasing interest for chemists and food technologists alike. An excellent review has recently been published summarizing the field of dipeptide sweeteners (3). Starting in the early 1970s an intensive, systematic program to develop high-potency sweeteners was carried out at Pfizer Central Research. This program, which involved the synthesis of a very large number of dipeptides of diverse structural types, culminated in the synthesis of alitame, which is the subject of this review. Alitame and structurally related peptide sweeteners are the subject of a U.S. patent (4).

## STRUCTURE

The structure of alitame is shown in Figure 1, which emphasizes the component parts of the molecule. Alitame is formed from the amino acids L-aspartic acid and D-alanine, with a novel C-terminal amide moeity. It is this novel amide moeity (formed from 2,2,4,4-tetramethylthienanylamine) that is the key to the very high sweetness potency of alitame. The structure of alitame was developed by following structural leads from a number of leading structures. Within the series of L-aspartyl-D-alanine amides those structural features which were found to be most conducive to high sweetness potency include small to moderate ring size, presence

**STRUCTURE:**

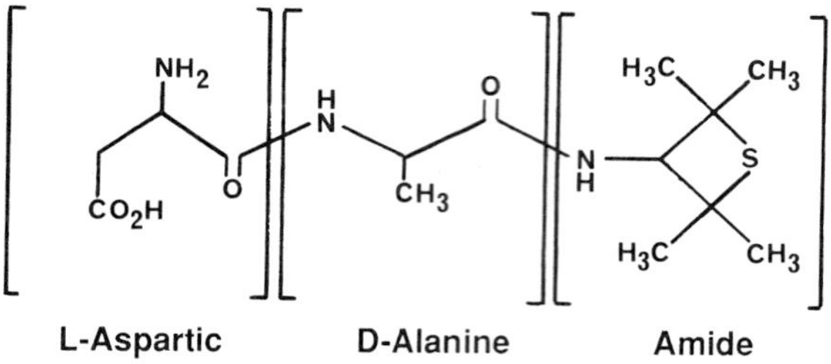


DEVELOPMENT:

Discovered: 1979 Pfizer Central Research
Patented: 1983 (U.S. 4,411,925)
FDA Filing: 1986 Food Additive Petition

**Figure 1** Alitame: structure and development.

of small-chain branching $\alpha$ to the amine-bearing carbon, and the introduction of the sulfur atom into the carbocyclic ring.

## ORGANOLEPTIC PROPERTIES

Alitame is a crystalline, odorless, nonhygroscopic powder. Its sweetness potency, determined by comparison of the sweetness intensity of alitame solutions with concentrations in the range of 50 $\mu$g/ml to a 10% solution of sucrose, is approximately 2000 times that of sucrose. The high sweetness potency of alitame means that this complex molecule can be manufactured and made available to the food industry at a cost competitive with other sweeteners, both carbohydrate and high potency. When compared to threshold concentrations of sucrose (typically 2–3%) the potency of alitame increases to about 2900 times that of sucrose. This phenomenon is typical of high-potency sweeteners. Use of higher concentrations of alitame allows preparation of solutions with sweetness intensity equivalent to sucrose solutions of 40% or greater.

The sweetness of alitame is of a very high quality, sucrose-like, without accompanying bitter or metallic notes often typical of high-potency sweeteners. The sweetness of alitame develops rapidly in the mouth and lingers slightly, in a manner very similar to that of aspartame.

Alitame has been found to exhibit synergy when combined with both acesulfame-K and cyclamate. High-quality blends may be obtained with these and other sweeteners, including saccharin.

**Table 1** Solubility of Alitame in Various Solvents at 25°C

| Solvent | Solubility (%w/v) |
|---|---|
| Water | 13.1 |
| Methanol | 41.9 |
| Ethanol | 61.0 |
| Propylene glycol | 53.7 |
| Chloroform | 0.02 |
| *N*-Heptane | 0.001 |

**Table 2** Solubility[a] of Alitame as a Function of pH and Temperature

| pH | 5°C | 20°C | 30°C | 40°C | 50°C |
|---|---|---|---|---|---|
| 2.0 | 41.7 | 48.7 | 56.4 | 50.3 | 54.0 |
| 3.0 | 32.2 | 39.2 | 46.5 | 50.9 | 53.9 |
| 4.0 | 12.9 | 13.9 | 17.3 | 20.4 | 37.6 |
| 5.0 | 11.7 | 12.8 | 14.9 | 16.8 | 29.2 |
| 6.0 | 11.6 | 13.2 | 14.9 | 19.5 | 32.8 |
| 7.0 | 11.8 | 14.3 | 17.6 | 29.5 | 51.8 |
| 8.0 | 14.8 | 24.9 | 46.8 | 56.2 | 52.1 |

[a]Water solubility (%w/v).

## SOLUBILITY

At the isoelectric pH, alitame is very soluble in water. Excellent solubility is also found in other polar solvents, as is shown in Table 1. As expected from the molecule's polar structure, alitame is virtually insoluble in lipophilic solvents. In aqueous solutions the solubility rapidly increases with temperature and as the pH deviates from the isoelectric pH. This effect is shown in Table 2.

## DECOMPOSITION PATHWAYS

The principal pathways for reaction of alitame with water are relatively simple compared to aspartame (Fig. 2). The major pathway involves hydrolysis of the aspartylalanine dipeptide bond to give aspartic acid and alanyl-2,2,4,4-tetramethylthietane amide ("alanine amide"). The $\alpha,\beta$-aspartic rearrangement, common to all peptides bearing N-terminal aspartic acid (5), also occurs to give the $\beta$-aspartic isomer of alitame. This rearranged dipeptide hydrolyzes at a slower rate than alitame to give the same products as those arising from the parent compound (i.e., aspartic acid and alanine amide). No cyclization to diketopiperazine or hydrolysis of the alanine amide bond is detectable in solutions of alitame that have undergone up to 90% decomposition. All three major decomposition products are completely tasteless at levels that are possible in foods.

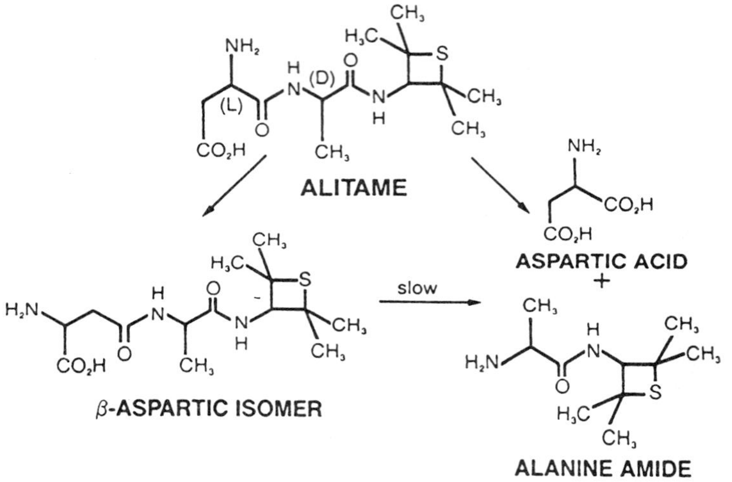


**Figure 2** Main degradation pathways of alitame.

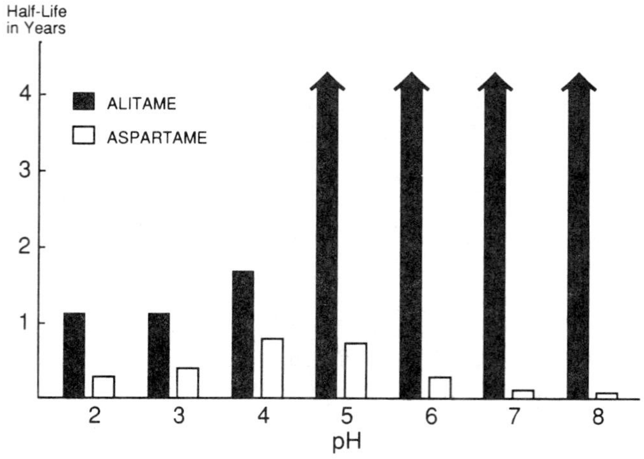


**Figure 3** Alitame and aspartame stability in buffer solutions at 23°C.

## STABILITY

In Figure 3 the half-lives for alitame and aspartame (6) in buffer solutions of various pH are compared. As can be seen, the solution stability of alitame approaches the optimum for aspartic acid dipeptides. At acid pH (2–4), alitame solution half-lives are more than twice those of aspartame. As the pH increases, this stability advantage increases dramatically. In particular in the neutral pH range (5–8) alitame is completely stable for more than one year at room temperature.

Alitame is sufficiently stable for use in hard and soft candies, heat-pasteurized foods, and in neutral pH foods processed at high temperatures, such as sweet baked goods. The dramatic heat stability advantage of alitame over aspartame under simulated baking conditions is illustrated in Figure 4. Actual half-life data for the two sweeteners under these conditions may be found in Table 3. This allows alitame to survive

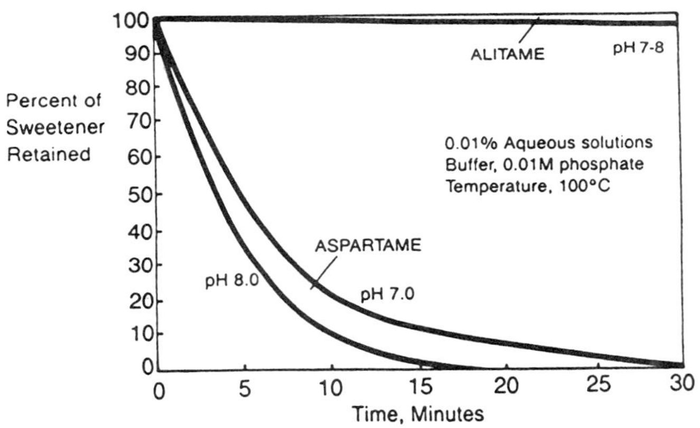


**Figure 4** Thermal stability of alitame and aspartame.

**Table 3** Stability of Alitame and Aspartame at Elevated Temperature in Aqueous Solution

| | | Half-life (hr) | | |
|---|---|---|---|---|
| Sweetener | Temp (°C) | pH 6.0 | pH 7.0 | pH 8.0 |
| Alitame | 100 | 13.5 | 13.4 | 12.6 |
| Alitame | 115 | 2.1 | 2.1 | 2.1 |
| Aspartame | 100 | 0.4 | 0.1 | 0.04 |
| Aspartame | 115 | 0.1 | 0.03 | 0.02 |

the thermal and pH conditions of the baking process with insignificant decomposition.

## COMPATIBILITY

Alitame exhibits excellent functionality and is compatible with a wide variety of freshly prepared foods. In accordance with its chemical structure, it can undergo chemical reaction with certain food components. In particular, high levels of reducing sugars, such as glucose and lactose, may react with alitame in heated liquid or semi-liquid systems, such as baked goods, to form Maillard reaction products. High levels of flavor aldehydes can behave similarly. Such reactions have been reported for aspartame and other aspartyl dipeptides (7–10). Therefore, the compatibility of alitame with a given recipe will depend on the actual ingredients present and the thermal and pH exposure involved in the manufacturing process.

Prolonged storage of alitame in a few standard acidic liquid beverage recipes may result in an incompatibility as measured organoleptically (off-flavors). This is not reflected in storage stability as measured by chemical assay for alitame and its degradation products. Levels of off-flavorant(s) are below the limits of modern analytical detection, and adjustment of these formulations to eliminate or modify incompatible ingredients may resolve such problems. Substances which may produce off-flavors on storage with alitame in liquid products are hydrogen peroxide and sodium bisulfite.

## METABOLISM

Following oral administration to the mouse, rat, dog, or man, alitame is well absorbed. Most of an oral dose (77–96%) is excreted in urine as a mixture of metabolites (Fig. 5). The remainder (7–22%) is excreted in the feces, primarily as unchanged alitame. Radiochemical balances of 97–105% were obtained in the four species.

In all four species the metabolism of alitame is characterized by the loss of aspartic acid followed by conjugation and/or oxidation at the sulfur atom of the alanine amide fragment yielding the corresponding sulfoxide isomers and the sulfone. Further hydrolysis of the alanine amide fragments does not take place. In the rat and dog the alanine amide is partially acetylated, and in man it is partially conjugated with glucuronic acid. No observable cleavage of the alanine amide bond or rupture of the thietane ring takes place.

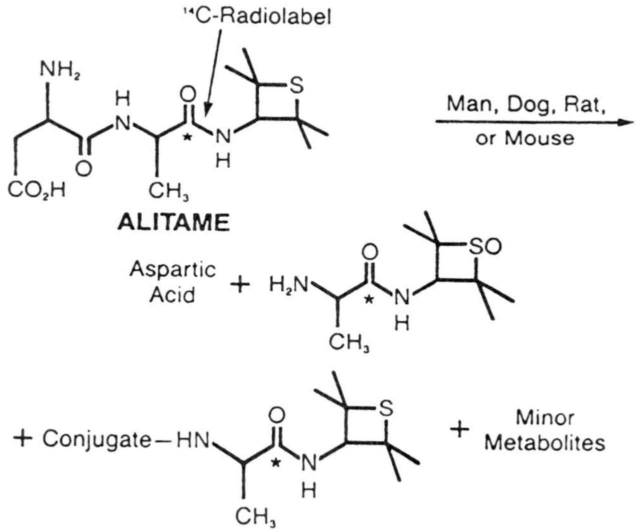


**Figure 5** Metabolism of alitame.

Since the aspartic acid portion of the molecule is available for normal amino acid metabolism, alitame is partially caloric. The maximum caloric contribution of 1.4 calories per gram is clearly insignificant at the use levels in the diet.

## SAFETY

Alitame has been evaluated in the extensive series of studies designed to establish its safety as a sweetener in the human diet. A list of the major studies completed is shown in Table 4. Based upon substantial safety factors determined in these studies, it is concluded that alitame is safe for its intended use as a component of the diet of man.

The appropriateness of the animal studies is supported by the fact that the major metabolites of the compound are common to laboratory animals and man. The cumulative data amassed during the safety evaluation demonstrate that this sweetener has an extremely low order of toxicity that supports its estimated mean chronic daily human dietary intake of 0.34 mg/kg body weight, assuming that alitame is the sole sweetener in all food categories requested in the Food Additive Petition. In all of the studies the no-effect level (NOEL) was consistently above 100 mg/kg, or >300 times the mean chronic human exposure.

**Table 4** Major Safety Studies Completed

| |
|---|
| Genetic toxicology |
| Three-month dog (0.5, 1, 2% of the diet) |
| Three-month rat (0.5, 1, 2% of the diet) |
| Rat teratology (100, 300, 1000 mg/kg/day) |
| Rabbit teratology (100, 300, 1000 mg/kg/day) |
| One-month mouse (1, 2, 5% of the diet) |
| Twelve-month rat (0.01, 0.03, 0.1, 0.3, 1% of the diet) |
| Eighteen-month dog (10, 30, 100, 500 mg/Kg/day) |
| Rat reproduction (0.1, 0.3, 1% of the diet) |
| Twenty-four-mouth mouse oncogenicity (0.1, 0.3, 0.7% of the diet) |
| Twenty-four-month rat oncogenicity (0.1, 0.3, 1% of the diet) |
| Man, metabolism |
| Man, no effects (15 mg/kg/day, 14 days) |
| Man, no effects (10 mg/kg/day, 90 days) |

The primary effect of alitame in animals treated with levels above the NOEL (>100 mg/kg) for periods ranging from 5 days to 2 years was the dose-related increase in liver weight secondary to the induction of hepatic microsomal metabolizing enzymes, a common adaptive response of the liver to xenobiotics. This high-dose effect abated during chronic studies and after cessation of treatment. At a dose to man of 15 mg/kg/day (44 times the mean chronic intake estimate and equivalent to consumption of 18 liters of alitame-sweetened carbonated beverage per day by a 60-kg individual), no enzyme induction was observed over a 14-day period.

Furthermore, neither enzyme induction nor any other effects were observed over a 90-day period in a double-blind toleration study of 130 subjects of both sexes of three races (Caucasian, black, and Asian) receiving 10 mg alitame/kg/day, a dose equivalent to consumption of 12 liters carbonated beverage per day.

There was no evidence of carcinogenic potential in rats and mice treated for 2 years at doses as high as 564 and 1055 mg/kg/day, respectively. Alitame, in doses as high as 1000 mg/kg/day during organogenesis, was devoid of embryotoxic or teratogenic potential in rats and rabbits. In two-generation reproduction studies there were no compound effects on mating behavior, pregnancy rate, the course of gestation, litter size, or upon maternal or offspring survival.

Mutagenicity assays, both in vitro and in vivo, revealed no genotoxicity at gene or chromosomal levels in microbial and mammalian test systems. Alitame had no detectable activity in a battery of pharmacological systems employed to assess autonomic, gastrointestinal, renal, and central nervous system functions. There were no effects on fasting blood glucose or upon the disposition of an oral glucose load.

## REGULATORY STATUS

In 1986 a Food Additive Petition was submitted to the U.S. Food and Drug Administration, requesting broad clearance for alitame (11). The petition requests approval of alitame as a sweetener and flavoring in specified foods, in amounts necessary to achieve the intended effect, and in accordance with good manufacturing practice. Table 5 presents the 16 food product categories requested in this petition.

In addition to the United States, permission for use of alitame in food has been requested from a number of foreign countries and regulatory agencies. These include the Scientific Committee for Food of the EEC, JECFA, the United Kingdom, The Netherlands, France, Switzerland, Sweden, Canada, and Australia.

**Table 5** Food Product Categories Requested in 1986 Food Additive Petition to U.S. FDA

| |
|---|
| Baked goods and baking mixes (restricted to fruit-, custard- and pudding-filled pies; cakes; cookies, and similar baked products) |
| Presweetened, ready-to-eat breakfast cereals |
| Milk products (restricted to flavored milk- or dairy-based beverages and mixes for their preparations, yogurt, and dietetic milk products) |
| Frozen desserts and mixes |
| Fruit and water ices and mixes |
| Fruit drinks, ades, and mixes, including diluted juice beverages and concentrates (frozen and non-frozen) for dilute juice beverages |
| Confections and frosting |
| Jams, jellies, preserves, sweet spreads |
| Sweet sauces, toppings, and mixes |
| Gelatins, puddings, custards, fillings, and mixes |
| Beverages, nonalcoholic and mixes |
| Dairy product analogs (restricted to toppings and topping mixes) |
| Sugar substitutes |
| Sweetened coffee and tea beverages, including mixes and concentrates |
| Candy (including soft and hard candies and cough drops) |
| Chewing gum |

## REFERENCES

1. A. Ripper, B. E. Homler and G. A. Miller, Aspartame. In *Alternative Sweeteners* (L. O. Nabors and R. C. Gelardi, eds.), Marcel Dekker, Inc., New York, 1986, pp. 43–70.
2. R. H. Mazur, J. M. Schlatter, and A. H. Goldkamp, *J. Am. Chem. Soc., 91:* 2684 (1969).
3. J. M. Janusz, Peptide sweeteners beyond aspartame. In *Progress in Sweeteners* (T. H. Brenby, ed.), Elsevier Science Publishing Co., Inc., New York, 1989, pp. 1–46.
4. T. M. Brennan and M. E. Hendrick, U.S. Patent No. 4,411,925 (October 25, 1983).
5. M. Bodanszky and J. Martinez, Side reactions in peptide synthesis. In *The Peptides,* Vol. 5, (E. Gross and J. Meienhofer, eds.), Academic Press, Inc., New York, 1983, pp. 143–148.
6. B. M. Homler, *Food Technology* (July): 50–55 (1984).
7. A. S. Cha and C.-T. Ho, *J. Food Science, 53:* 562 (1988).
8. Y. Ariyoshi and N. Sato, *Bull. Chem. Soc. Japan, 45:* 2015 (1972).
9. M. M. Hussein, R. P. D'Amelia, A. L. Manz, H. Jacin, and W.-T. C. Chen, *J. Food Science, 49:* 520 (1984).
10. J. A. Stamp and T. P. Labuza, *J. Food Science, 48:* 543 (1983).
11. *Fed. Reg., 51* (188): 34503 (1986).

# 4

# Aspartame

**Barry E. Homler, Ronald C. Deis, and William H. Shazer** *The NutraSweet Company, Deerfield, Illinois*

## INTRODUCTION

The sweet taste of the compound *N*-L-$\alpha$-aspartyl-L-phenylalaline-1-methyl ester (APM-aspartame) was discovered in December 1965 by G.D. Searle & Co. chemist James Schlatter during the synthesis of gastrin tetrapeptide to be used in a bioassay (69,71). Mr. Schlatter noted the following in his affidavit of discovery:

> In December 1965 I was working with Dr. Mazur on the synthesis of the C terminal tetrapeptide of gastrin. We were making intermediates and trying to purify them. In particular, on an occasion of December, 1965, I was recrystallizing aspartylphenylalanine methyl ester (aspartame) which had been prepared by the hydrogenolysis of the protected dipeptide ester prepared and given to me by Dr. Mazur. I was heating the aspartame in a flask with a methanol when the mixture bumped onto the outside of the flask. As a result, some of the powder got onto my fingers. At a slightly later stage, when licking my finger to pick up a piece of paper, I noticed a very strong, sweet taste. Initially, I thought I must have still had some sugar on my hand from earlier in the day. However, I quickly realized this could not be so, since I had washed my hands in the meantime. I therefore, traced the powder on my hands back to the container into which I had placed the crystallized aspartyl-phenylalanine methyl ester. I felt that this dipeptide ester was not likely to be toxic and I

therefore tasted a little of it and found that it was the substance which I had previously tasted on my finger.

What Schlatter had discovered was what years later would be named NutraSweet brand sweetener—a nutritive sweetener with the taste of sugar.

Following the discovery of the sweetness of the compound, Searle examined about 200 analogs of APM, but decided to commercialize the original discovery (72), which became known by the generic term *aspartame*. The major reasons for the choice of the original compound were (20):

1. None of the analogs could be manufactured any more economically than APM.
2. The potency was satisfactory, and the taste quality was excellent.
3. The components L-aspartic acid and L-phenylalanine and the methyl ester are dietary components, and the body metabolizes them normally.
4. Searle speculated that APM would have an excellent chance of surviving the most severe and searching toxicity testing.

## PHYSICAL CHARACTERISTICS AND CHEMISTRY

The structure of aspartame is shown in Figure 1. Aspartame is a dipeptide of two amino acids: L-phenylalanine as the methyl ester and L-aspartic acid. It is slightly soluble in water (about 1.0% at 25°C) and is sparingly soluble in alcohol (14,15). Aspartame is not soluble in fats and oils. Aspartame as sold for commercial use meets all requirements of the Food Chemical Codex (46,99).

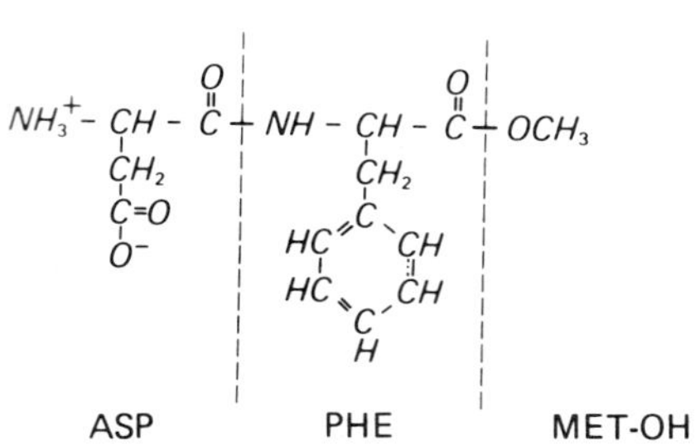


**Figure 1** Chemical structure of aspartame.

Being a peptide, the compound is amphoteric, the negative logs of its dissociation constants are 3.1 and 7.9 at 25°C, and its isoelectric point is 5.2 (71). Aspartame is a white, odorless crystalline powder having a clean, sweet taste.

## Stability

Aspartame is an O-methyl ester. In liquids and under certain conditions of moisture, temperature, and pH, the ester bond is hydrolyzed, forming the dipeptide aspartylphenylalanine and methanol. Alternatively, methanol may be eliminated by the cyclization of aspartame to form its diketopiperazine (DKP). Ultimately, aspartylphenylalanine can be hydrolyzed to its individual amino acids—aspartate and phenylalanine (69–72).

Decomposition appears to follow first-order kinetics, and occurs under moist storage and manufacturing conditions. The primary reactions that occur are shown in Figure 2. The stability of dry APM is quite good, and is affected only by extremely high temperatures, which are not typical for the production of dry food products (14,15). The rate of decomposition, as measured by the amount of APM converted to DKP, is shown in Figure 3.

The combined effect of pH, temperature, and time on the stability of aspartame in solutions is shown in Figure 4. At 25°C, the maximum stability is observed at pH ~4.3 (71). Aspartame is most stable in the

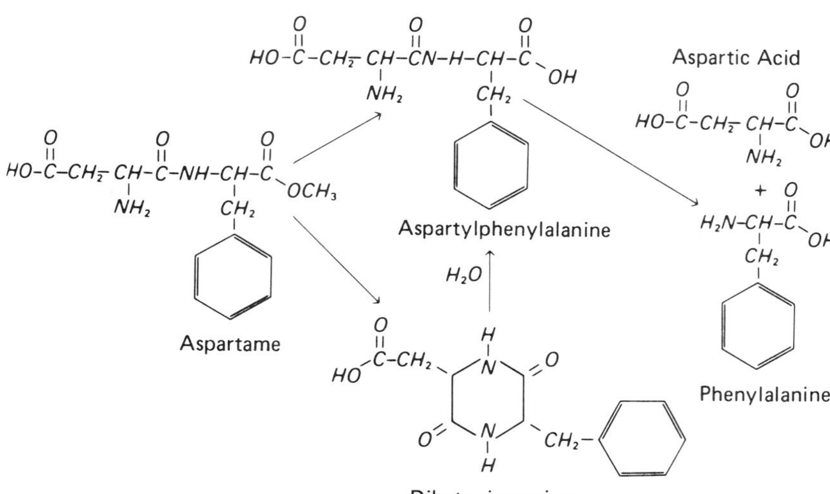


**Figure 2** Principal conversion products of aspartame.

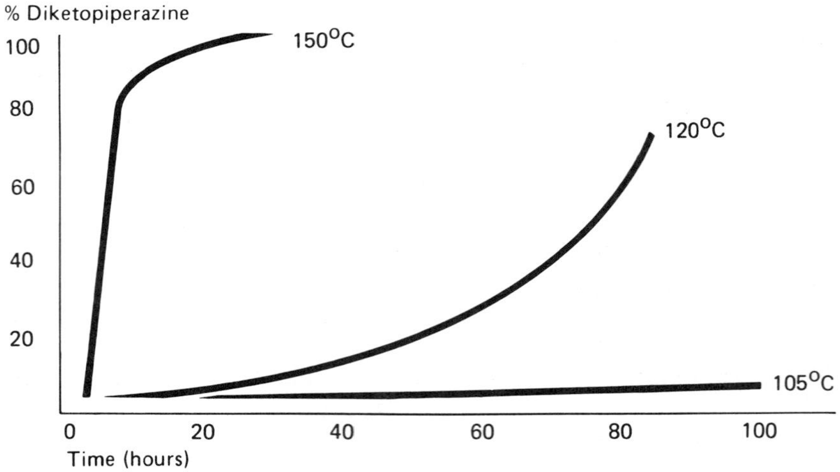


**Figure 3** Stability of dry aspartame: rate of decomposition to diketopiperazine at 105°, 120°, and 150°C.

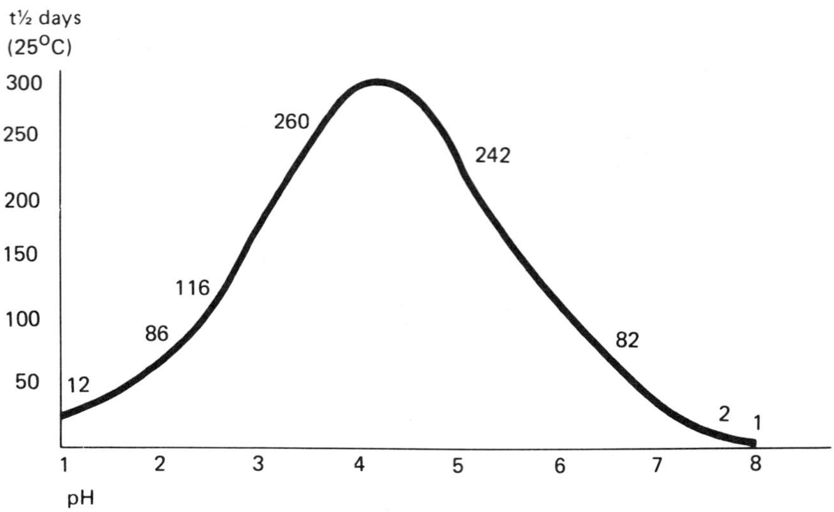


**Figure 4** Stability of aspartame in aqueous buffers at 25°C.

weak acidic range in which most foods exist—between pH 3 and 5—but functions very well over a broad range of pH conditions.

For example, a frozen dairy dessert may have a pH ranging from 6.5 to over 7.0, but, due to the frozen state, the rate of reaction is dramatically reduced. And, because of the lower free moisture, the shelf-life stability of aspartame exceeds the textural shelf-life stability of these products.

Stabilities at 40°C (104°F) and 80°C (175°F) at various pHs are shown in Figure 5 and 6. High temperatures are used to demonstrate changes observed over reasonable time periods. The horizontal lines of these figures represent the almost perfect stability of aspartame, whereas the more vertical lines indicate almost instant conversion and loss of sweetness. The 80°C curves demonstrate that APM can withstand high-temperature–short-time (HTST) or ultra-high-temperature (UHT) processing with minimal loss of APM.

## Sweetness Potency

In food and beverage systems, the potency of APM has been judged to be 160–220 times the sweetness of sucrose (14,15) (Table 1). In general, there is an inverse relationship in the potency of APM compared to the concentration of sucrose being replaced. The relative sweetness varies

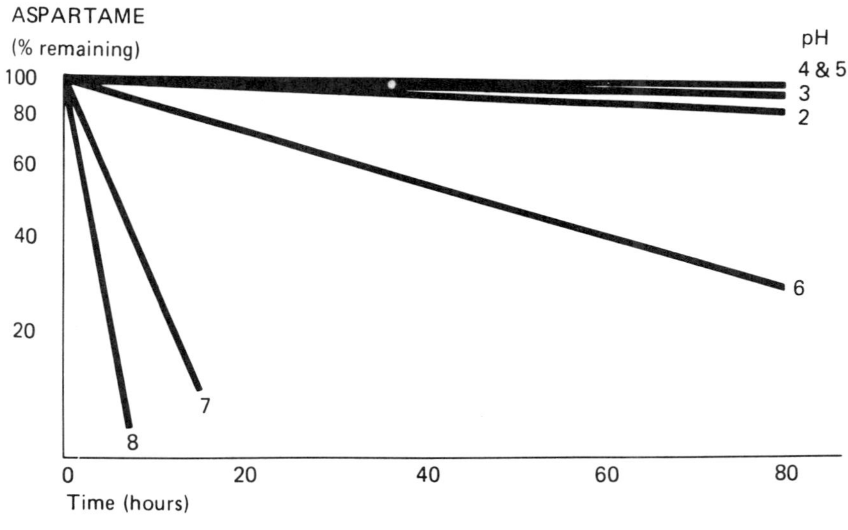


**Figure 5** Stability of aspartame in aqueous buffers at 40°C.

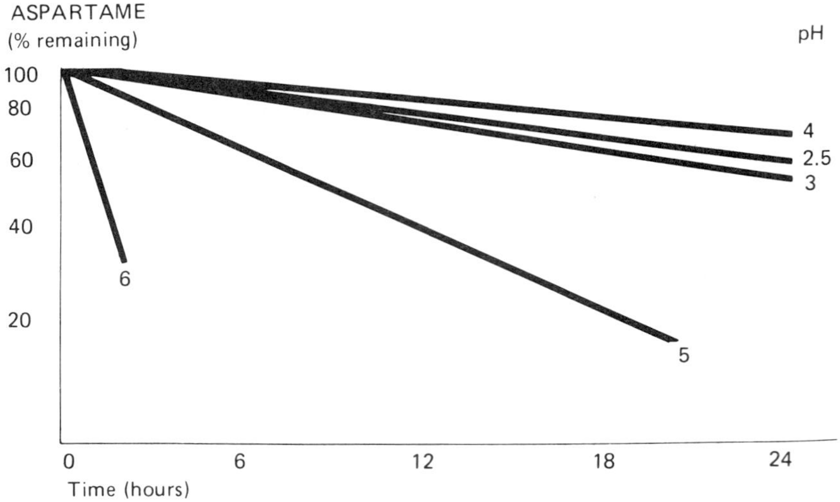


**Figure 6** Stability of aspartame in aqueous buffers at 80°C.

with the flavor system, pH, tasting temperature, and the amount of sucrose or other sugars being replaced (14,15).

## TASTE

APM has a clean, sweet taste like sugar without the bitter chemical or metallic aftertaste often associated with artificial sweeteners. The sugar-like taste of APM is well documented in studies by food industry and university researchers (22,23,65,66,72,74,75,82,94–96,123).

Further internal studies comparing aspartame and sucrose using quantitative descriptive analyses (QDA) showed that aspartame follows very closely the profile of sucrose (Fig. 7).

### Flavor-Enhancing Property

Some food and beverage flavors are enhanced or extended by APM, particularly acid fruit flavors. Sensory evaluation studies have shown that this flavor enhancement or flavor extension is best with flavors naturally derived rather than with artificial flavors (11). In certain food applications, this flavor-enhancing property may allow for lower use levels of APM or can be an attribute allowing a product claim, as in chewing gum, that the sweetness and flavor of gum with APM lasts four times longer than sugar-sweetened gum (9). This property should be considered in product formulation. Any lingering sweetness from aspar-

**Table 1** Reported Potency Values for Aspartame in Various Systems

| | Potency[a] | Sucrose in the control medium (%) | Tasting temperature (°C) | Tasting temperature (°F) |
|---|---|---|---|---|
| Water | 400 | 0.34 | 22 | 72 |
| | 215 | 4.3 | 22 | 72 |
| | 133 | 10 | 22 | 72 |
| | 100 | 15 | 22 | 72 |
| | 182 | 2 | 8 | 46 |
| | 160 | 4 | 8 | 46 |
| | 100 | 12 | 8 | 46 |
| | 43 | 30 | 8 | 46 |
| | 61 | 10 | 3 | 37 |
| | 54 | 10 | 22 | 72 |
| Buffer (citrate) | 160 | 4 | 8 | 46 |
| Phosphate at pH 3.2 | 100 | 12 | 8 | 46 |
| Powdered soft drink mixes | 133 | 10 | 8 | 46 |
| | 143 | 10 | 3 | 37 |
| | 180 | 11 | 8 | 46 |
| Chewing gum | 100–199 | 69 | 22 | 72 |
| Gelatin | 160–172 | 18 | 3 | 37 |
| Pudding mix | 200 | 17 | 8 | 46 |
| Presweetened cereal | 180 | 37 | — | — |
| Soft drinks | 180 | 10 | 8 | 46 |
| Coffee/tea | 180 | 4–6 | 70 | 158 |
| Flavored yogurt | 175–220 | 17 (13 added) | 8 | 46 |
| Canned fruit | 175 | 20 | Chilled | |

[a]The potency is equal to the concentration of sucrose divided by the concentration of aspartame at equal sweetness.

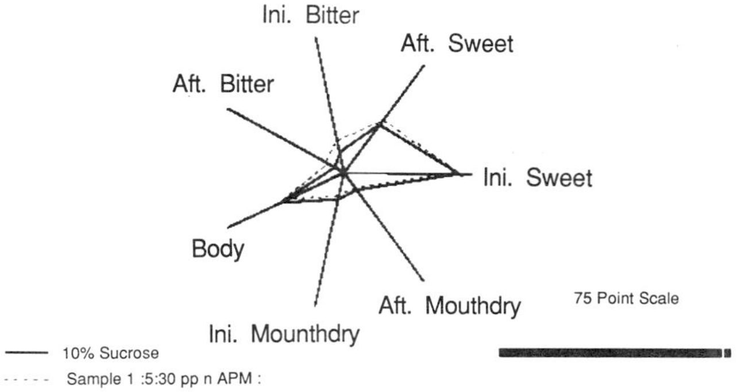


**Figure 7** QDA Analysis.

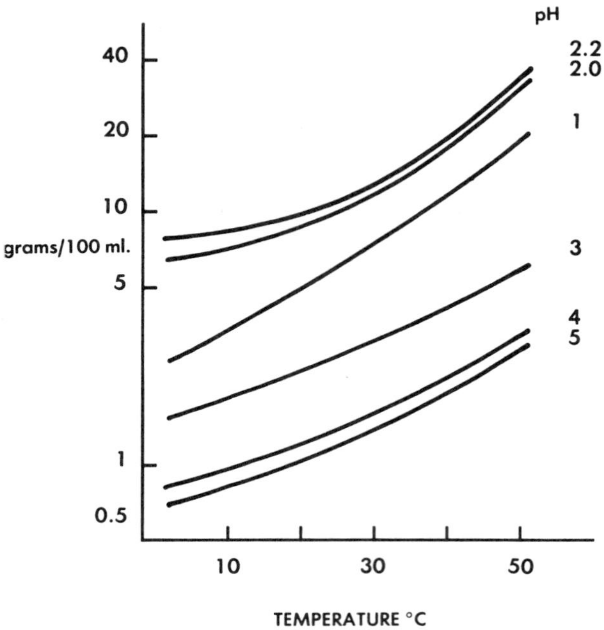


**Figure 8** Effect of pH and temperature on aspartame solubility in water.

tame can be modified by using slightly less sweetener or certain salts (93).

### Solubility in Food and Beverage Applications

The rate of dry mix dissolution is an important consideration, particularly for the many consumer products prepared by mixing aspartame-sweetened products with water or milk at various temperatures. For aspartame only, solubility is a function of pH and temperature, as shown in Figure 8. The formulation of table-top sweetener products and other presweetened beverage and dessert mixes must take into account this combination of factors. Improvements in solution rate by improving dispersion for even wetting are described in the patent literature (16,21,48–50,53,103,104).

## UTILITY IN FOOD AND BEVERAGE SYSTEMS

Aspartame is a nutritive sweetener and flavor enhancer that has a taste like sugar when used in food and beverages. Because aspartame is a

**Table 2** Sugar Function in Food Products

| |
|---|
| Humectant |
| Freeze point depression |
| Ice crystal control |
| Control of starch gelation |
| Texturizer |
| Browning/Caramelization |
| Flavor |
| Fermentable |
| Dispersion of other ingredients |
| Viscosity development |
| Volume control |
| Glass formation |
| Preservative |

dipeptide, it yields 4 kcal/g upon digestion and metabolism, as do proteins and carbohydrates. It is for this reason, the yield of calories, that aspartame is classified by the Food and Drug Administration (FDA) as a nutritive sweetener. Aspartame is the only nutritive high-potency sweetener approved for use in foods and beverages by the FDA.

It is important to realize and remember that, while aspartame does yield caloric value, its intense sweetness (180–200 times sugar) and subsequent low level of use to produce the desired sweetness in a finished product results in essentially no caloric impact.

Unlike nutritive carbohydrate sweeteners, aspartame (and other high potency sweeteners) provides *only* sweetness and cannot impart the other physical properties to food that are expected of bulk carbohydrate sweeteners. Thus a bulking agent or macronutrient substitute will be needed in products where water is not the replacement for removed sugar and the functional properties of sugar are required. In a cake, sugar will have to be replaced by another low-calorie or sugar-free bulking agent. There are other applications where this need will apply, including frozen desserts, frozen novelties, confections, and hard candy.

Table 2 shows some of the functions that a bulking agent will have to provide when used to replace sugar.

Currently there are products on the market or in development that will act as bulking agents to allow the above products to be produced with aspartame.

In food and beverage applications, APM can:

1. Sweeten foods to taste virtually the same as sugar-sweetened counterparts

2. Reduce calories as much as 95%
3. Intensify and extend flavors, particularly fruit flavors
4. Avoid nutrient dilution
5. Be combined with sugars and artificial sweeteners

## FOOD AND BEVERAGE APPLICATIONS

Table 3 lists the currently approved uses in the United States for aspartame. In dry-product applications, such as powdered drink and dessert mixes, the stability of APM is quite good; the stability is similar to that of the pure compound (see Fig. 3).

The potential for aspartame to hydrolyze or cyclize may, under some conditions of excessive heat, limit some applications of aspartame. The newly developed heat-protected aspartame provides for use of aspartame in areas once thought not to be feasible. One such area is that of baked goods. The new heat-protected form has removed this limitation, and the use of aspartame for baking has been petitioned to the FDA (126).

As previously noted, aspartame can withstand the heat processing used for dairy products, juices, aseptics, and others where HTST and UHT conditions are utilized. It has been demonstrated that aseptic juice products—even when heated to abusive conditions of 275°F—showed little APM loss through processing (103).

Table 4 gives a summary of recoveries from a series of processing tests conducted at The NutraSweet Company to support to the FDA the

**Table 3** Currently Approved Uses for Aspartame in the United States

| | |
|---|---|
| Table-top sweetener | Chewing gum |
| Carbonated beverages | Cocoa mixes |
| Powdered soft drinks | Refrigerated or frozen fruit/juice drinks |
| Tea beverages | Frozen desserts |
| Ready-to-eat cereal | Yogurt products |
| Dry mix gelatins, puddings, and fillings | Refrigerated flavored milk |
| Refrigerated gelatins | Wine coolers |
| Chewable multivitamins | Aseptic juice/fruit drinks |
| Creme fillings of prebaked cookies | Breath mints |
| Frozen cheesecake | Instant coffee |
| Frozen yogurt | Frozen or refrigerated fruit/toppings |
| Pharmaceuticals | Dry mix dairy product |
| Fruit spreads | Analog toppings |
| Frozen dairy/nondairy frostings | |
| Toppings and fillings | |

Table 4 Effect of Temperature and Hold Time on APM Recovery

| | pH | Temperature (°F) | Hold time (sec) | APM recovery (%) |
|---|---|---|---|---|
| 10% Orange | 3.51 | 276 | 42.7 | 92.8 |
| | 3.51 | 232 | 30.5 | 100.5 |
| | 3.39 | 205 | 34.1 | 96.8 |
| | 3.45 | 206 | 66.4 | 98.9 |
| | 3.43 | 185 | 65.6 | 99.7 |
| | 3.46 | 170 | 83.3 | 100.7 |

use of aspartame in aseptically processed beverages. The loss of APM due to a certain combination of moisture, pH, and temperature results in a gradual loss of perceived sweetness, but without the development of off-flavors because the conversion products are tasteless (14,15). It has also been demonstrated in several flavors of carbonated soft drinks that the proportion of aspartame conversion and the perceived sweetness and "overall liking" of products is not a direct relationship. There is a wide range over which aspartame sweetness levels are acceptable.

### Carbonated Beverages

Carbonated beverages are a major market for aspartame. Previous studies (14,53,76) demonstrated that aspartame could function as the sole sweetener in all carbonated beverages. Results also showed that aspartame stability in carbonated beverages could be very acceptable with slight modification of pH and some control of the distribution system. Since the first edition of this book in 1986, the use of aspartame in carbonated beverages has grown considerably and the market for diet beverages has grown rapidly across a wide range of climatic and distribution environments.

We understand the role of temperature in the rate of aspartame conversion in beverages of varying pH very well and have developed models that can predict conversion at different constant temperatures. In reality, however, temperature fluctuates during the distribution of product. Therefore, a constant temperature model does not truly predict how a product will exhibit aspartame loss or quality loss.

To assess the "real world," studies have been conducted to monitor distribution temperatures and times (57). Over 1000 temperature points were recorded for a cola beverage at hourly intervals over a 6-week period. The effect of temperature on aspartame concentration was predicted using degradation models.

Comparison of real-world data with the predicted data of constant temperature showed that degradation was nearly 45% less than constant temperature storage would have shown. Thus the stability of aspartame-sweetened beverages, with reasonable distribution control, becomes significantly less of an issue.

### Dairy Products

Sweetened dairy products are available in many different forms. Each product poses a number of challenges for the food technologist when formulating with aspartame in place of sugar or other carbohydrate-based sweeteners. Solids replacement, processing methodology, and stability considerations are three important factors in manufacturing aspartame-sweetened dairy products.

### Refrigerated Dairy Products

Flavored milks, puddings, and coffee whiteners are examples of sweetened, fresh, fluid dairy products. In formulating these products to contain aspartame, added sugars are removed. The primary function of sugar in these products is to provide sweetness. The volume lost due to the removal of sugar from the product can be replaced by milk. Since sugar provides slight viscosity, its absence can be compensated through incorporation of low levels of common thickening agents and stabilizers. In puddings, higher levels of starch can be used.

These products should be pasteurized under HTST or UHT processing conditions to maximize the recoveries of aspartame. Due to the neutral pH of flavored milks and coffee whiteners, long exposure to pasteurization temperatures should be minimized. Batch pasteurization (prolonged exposure times) should be avoided unless the aspartame is to be added during the last 1–2 minutes of the process.

Refrigerated product stability is similar for aspartame-sweetened products and sugar-sweetened products. It has been shown that the more effective the pasteurization in eliminating microbial counts, the greater the stability of aspartame (58). While true extended shelf life (ESL) or aseptic products are not yet possible for aspartame-sweetened fresh dairy products (due to the neutral pH), it is possible to extend aspartame-sweetened dairy products' shelf lives through frozen distribution. Products such as frozen coffee whiteners and frozen soft serve mix maintain shelf lives similar to sugar-sweetened counterparts under these conditions.

### Frozen Dairy Products

Frozen desserts and soft serve products (including frozen yogurt) are the primary dairy products considered for sweetening with aspartame. Sugar replacement by bulking agents is necessary in aspartame-sweetened frozen desserts. Several "nonsugar" bulking agents are available for use in frozen desserts (52). By blending various bulking agents the functional properties such as freeze-point depression, ice crystal control, and structure/bulk contribution that sugar normally provides to frozen desserts can be maintained with sweetness provided by aspartame.

Frozen dessert processing and stability are virtually the same for sugar-sweetened and aspartame-sweetened products. However, the use of batch pasteurization should be avoided for aspartame-sweetened products. Exposure of aspartame to prolonged heating in neutral pH products results in unnecessary aspartame degradation. Typical frozen dessert/ice cream shelf-life patterns result for aspartame-sweetened frozen desserts in frozen distribution.

### Fermented Dairy Products

By far the most significant sweetened fermented dairy product is yogurt. Aspartame-sweetened yogurt is prepared by adding an aspartame-sweetened fruit prep to plain yogurt or by adding aspartame to plain yogurt. The procedure for making yogurt is unchanged. In these aspartame-sweetened products, sugar is replaced in the fruit prep by water or juice and stabilizers, or in the yogurt by milk and milk solids. Both stirred and fruit on the bottom (FOB) forms of yogurt may be prepared. In FOB yogurt, aspartame is segregated from the milk as part of the fruit preparation.

The pH of yogurt/fermented daily products (4.0–4.5) is ideal for APM stability. At the finished product pH, it has been shown that the active cultures in these products do not affect APM stability (59) and product shelf lives similar to sugar-sweetened products result. Fermented dairy drinks have also shown similar results.

Table 5 lists pending approvals for petitions currently filed with the FDA.

### Baked Goods

A petition for the approval of aspartame in baked goods and baking mixes was filed with the Food and Drug Administration in October 1987. This petition requested use of aspartame in combination with safe and suitable ingredients when exposed to high temperatures for a prolonged

**Table 5** Pending Approvals for Aspartame Application

| |
|---|
| Bulk packaging table-top powder |
| Ready-to-serve puddings |
| Hard candy—Cough drops |
| Malt beverages containing fruit |
| Hot breakfast cereals |
| Confections—soft candy |
| Baked goods—baking mixes |
| Nonrefrigerated, noncarbonated beverages |

period of time in the presence of moisture. The category requested for clearance is as follows:

> 21 CFR 170.3(1), baked goods and baking mixes, including all ready-to-eat and ready-to-bake products, flours and mixes requiring preparation before serving. These products include, but are not limited to cakes, cookies, pies, pastries, coffee cakes, and sweet rolls which may be:
> 1) Commercially baked and are ready-to-serve,
> 2) Commercially prepared (frozen or refrigerated) and must be baked or heated prior to serving,
> 3) Commercially prepared (frozen or refrigerated) and must be thawed or warmed prior to serving, and
> 4) Commercially assembled mixes which must be prepared in the home for baking.

Baked goods provide a wide range of challenges to the aspartame molecule. Moisture leves range from approximately 2 to 60%, pH ranges from 3 to 8, baking temperatures range from 200 to 450°F, and baking times range from 4 to 60 minutes. Aspartame can be stable for long periods of time under dry conditions or at low pH (see Figs. 3,4). For other conditions, aspartame has been encapsulated to take advantage of its dry heat stability, protecting the molecule from the moist environment during batch preparation and the early stages of baking, but releasing it in time to provide uniform dispersion. NutraSweet was granted U.S. Patent #4,704,288 on aspartame encapsulation, and several patents are pending.

Prototypes of cookies, yellow cake, chocolate cake, fruit pie, and cheesecake have been developed to serve as the basis for the petition and to demonstrate functionality of the heat-stable aspartame. A similar petition has also been filed in Canada with the Health Protection Branch (HPB).

The opportunity to replace sucrose in baked goods is significant—the typical sucrose content of standard baked goods is 0–45% (this is much higher if frostings and fillings are considered). From a caloric standpoint, this may represent up to 30% of the baked product. Baking stability of encapsulated aspartame is very good, and shelf stability has been shown to be excellent to good for a wide range of baked products, dependent on percent moisture, pH, and temperature. Stability in cookies is excellent over a 9-month shelf life, and stability in cakes is typical of the industry norms. Common macronutrient substitutes for replacement of sugar bulk are sorbitol, mannitol, polydextrose, isomalt, lactitol, and maltodextrins.

### Confections

A petition has been filed with FDA for clearance of aspartame for use in a broad spectrum of confectionary products. This category includes soft candy such as compound coatings, gelled candies, nougat, caramel, marshmallow, and about 30 other products that make up confectionery by FDA regulations.

Sugar is the primary ingredient in nearly all confections, representing 20–70% of typical formulations. During processing, temperature/time effects may be avoided by adding aspartame at later stages along with colors, flavors, etc. Aspartame stability has been demonstrated for a wide range of confections, including hard candies, compound coatings, nougats, caramels, starch jellies, and fondants. Macronutrient substitutes for sugar (bulk replacements) are commonly known to the industry—malitol, hydrogenated starch hydrolysates, sorbitol, mannitol, isomalt, and lactitol.

### Hard Candy

A boiled candy with aspartame is prepared by heating an appropriate sugar substitute bulking agent to reduce moisture. This mix is then cooled to allow for the addition of acids, colors and flavors as per industry standards and practices. It is at this point of low moisture that aspartame is incorporated into the candy mass, followed by panning and further cooling. With this procedure aspartame recovery in the process is over 98%.

## PATENTS

A scientist at G.D. Searle & Co. discovered that *N*-L-$\alpha$-aspartyl-L-phenylalanine-1-methyl ester (APM) had a sweet taste and could be used as a food and beverage sweetener. As a result of this discovery, Searle was granted U.S. Patent 3,492,131 and numerous foreign patents relating to the use of APM as a sweetening agent (97). The U.S. patent expires in December 1992.

In addition, Searle has received patents in the United States and several other countries relating to APM. An example is U.S. Patent 3,780,189, which covers compositions comprising aspartame and saccharin (98). This U.S. patent expires in November 1996.

## SAFETY

In terms of types of studies and the amounts given to humans in controlled systems, aspartame is one of the most thoroughly tested food additives. It has been the subject of over 100 scientific studies (4,73,101).

In assessing the safety of a food ingredient, it is necessary to take into account the probable consumption levels of the ingredient and to compare those levels to the consumption levels potentially injurious to human health. This has been done with aspartame and its principal metabolites—phenylalanine, aspartate, and methanol. The data demonstrate that there is a substantial margin of safety for aspartame and its metabolites beyond any reasonable consumption level anticipated to be achievable by normal human beings.

### Metabolic Studies

Metabolic studies were performed to describe the absorption, distribution, metabolism, and excretion of APM, by means of radiolabeled techniques. The results uniformly showed that APM is rapidly digested to its three moities—phenylalanine, aspartic acid, and the methyl ester. These are then absorbed, metabolized, and excreted by normal pathways, as are the identical moieties when they occur naturally in foods. These studies were performed in mice, rats, dogs, and rhesus monkeys (8,17,18,51,79,80,86,101).

### Pharmacology Studies

Pharmacology studies performed on all major physiological systems have demonstrated a total lack of pharmacological activity associated with APM (19,42,84,92,101).

### Toxicology and Carcinogenicity

APM has been administered to a range of animal species in acute, subchronic, and chronic tests. Mice, hamsters, rats, dogs, and rhesus monkeys were given APM in doses up to 13 g/kg in subchronic studies and up to 8 g/kg in chronic studies (mice and rats only). No significant effects or indications of toxicity or carcinogenicity attributable to APM administration have been demonstrated (55,56,87,88,101,124).

### Mutagenicity

These tests were designed to evaluate the potential of APM to produce mutation through evaluation in host-mediated, dominant lethal, and in vivo cytogenic assays. In all this work, APM was shown not to alter the mutagenic frequency, and therefore, it does not cause mutations (18, 101,125).

### Teratology

In all teratology studies, dietary administration of APM was without significant effect on fertility, the development of fetal abnormalities, mean gestational period, maternal survival, or litter size or sex distribution (67,85,101).

### Diketopiperazine Toxicology Overview

A similar toxicological evaluation of the major decomposition product, DKP, has been performed. Because APM may contain very small amounts of DKP and because APM may convert to DKP under certain storage conditions, the human population may consume small but varying amounts of DKP. The thorough examination of DKP through toxicology, teratology, mutagenicity, and pharmacology tests has shown no adverse effects directly attributable to DKP up to doses of 3 g/kg (56,101).

### Clinical Studies Overview

The clinical trials and series of human studies with APM are unprecedented for a food additive prior to its approval (4). Studies were performed to demonstrate that APM is well tolerated by normal adults, children, and adolescents. This included both short-term (6 weeks) and long-term (12 weeks) clinical studies designed to demonstrate the ability of normal human beings to tolerate APM in their diet, without untoward effects, in doses up to 1.8 g per day (24,47). Some studies used a single administration or less-than-24-hr administration of APM in doses up to 200 mg/kg body weight (12,24,43,45,106,107,112,115,117,119,121).

Studies were also performed on special groups. Diabetic and obese populations potentially may consume larger than average quantities of aspartame-containing products because of their unique dietary and nutritional situations. Therefore, an evaluation of the ability of these groups to tolerate aspartame was done. Both diabetic (13,54,61,101,107,121) and obese persons (60,101,107) tolerate aspartame well, even when fed doses of 1.8 g/day for up to 21 weeks. Lactating women were evaluated using a 50 mg/kg dose to determine if aspartame would affect the composition of their milk. No significant effect was noted (69,113). Glutamate-sensitive adults were evaluated because of the metabolic relationship of glutamate and aspartame (44,107,112,115).

Phenylketonurics metabolize the amino acid phenylalanine very poorly due to a genetically transmitted metabolic disease. Because aspartame contributes phenylalanine to the diet, its impact on this population was studied. These studies showed that phenylalanine from aspartame was metabolized in an identical manner to any other phenylalanine-containing product (62–64,101,118–120). If phenylketonuria (PKU) patients are controlling the amount of phenylalanine in their diets, the phenylalanine from aspartame must be included in the daily calculation.

There are few sweeteners or food additives that have had the kind of regulatory agency review that aspartame has had prior to its use in foods and beverages. Prior to final approval in the United States in July 1981, data on aspartame and its decomposition product, diketopiperazine, were reviewed by the Joint Expert Committee on Food Additives of the Food and Agricultural Organization/World Health Organization (1); pivotal studies and data were subjected to review by Universities Associated for Research and Education in Pathology (UAREP) (6,7); specific scientific issues were scrutinized before a board of independent scientists at an open hearing (4); and the aspartame regulations have been the subject of public comment on many occasions since mid-1974.

In addition to review by the FDA before approval for use was granted in 1981, aspartame was reviewed and approved for use by the regulatory agencies of over 70 countries, the Joint Expert Committee on Food Additives (JECFA) of the Codex Alimentarius (Food and Agriculture Organization/World Health Organization) and the Scientific Committee for Foods of the European Economic Communities (20). The FDA established an acceptable daily intake (ADI) for aspartame of 50 mg/kg body weight/day (20).

The result of all of these reviews of aspartame has been a determination that aspartame is a safe food additive for the general population. An extensive review of the scientific data, including both animal and human studies conducted prior to and since the marketing of aspartame is presented by Butchko and Kotsonis (20). It is recommended that the overview presented in that paper be reviewed, as the wealth of data presented is too detailed to be presented here.

## HISTORY IN THE UNITED STATES

Aspartame was first approved as a sweetener and flavor enhancer by the U.S. FDA for table use, and as an ingredient in certain dry food applications on July 26, 1974 (5). The approval responded to a petition from G.D. Searle & Co. covering the safety studies and application for use in table-top sweetener products, and a companion petition from General Foods Corporation covering the other dry foods applications. The regulation became codified in 21 CFR 172.804.

Immediately following this approval, formal objections to the regulation were filed; they were based on safety concerns arising from toxic effects of certain amino acids in animals (77,78). Separately and for unrelated reasons, a comprehensive audit of the authenticity of selected toxicological studies on aspartame was undertaken, and the regulation authorizing the use of aspartame was stayed on December 5, 1975, pending the findings (7).

Following the authentication of pivotal studies and acceptance of the report by the FDA, a Public Board of Inquiry (PBOI) was established to examine the original objections to the regulation and other safety questions (6). The PBOI was finally convened for a three-day hearing on January 30, 1980. The FDA hearing record contains all prehearing documents, transcripts, initial decision of the Board, hearing participants' comments on the initial decision, and final responses to comments.

On July 24, 1981, the Commissioner of the Food and Drug Administration reviewed in detail the contested aspects of aspartame's safety and issued his final decision on Searle's petition for the food additive use of aspartame (4). The Commissioner's decision concluded that aspartame had been shown to be safe under Section 409 of the Federal Food, Drug, and Cosmetic Act (21 USC 348), and vacated the stay of the original regulation, effective October 22, 1981.

The chronology of aspartame approval in the United States is as follows. Aspartame may be used as a sweetener in:

1. Dry, free-flowing sugar substitutes for table use (not to include use in cooking) in package units not to exceed the sweetening equivalent of two teaspoons of sugar (October 1981) (25).
2. Sugar substitute tablets for sweetening hot beverages, including coffee and tea (October 1981) (25).
3. Cold breakfast cereals (October 1981) (25).
4. Chewing gum—sweetener and flavor enhancer (October 1981) (25).
5. Dry bases for
   beverages,
   instant coffee,

gelatins, puddings, pie fillings, and
dairy product analog toppings (October 1981) (25).
6. Carbonated beverages and carbonated beverage syrup bases (July 1983) (26).
7. Chewable multivitamin food supplements (May 1984) (28).
8. Refrigerated single strength and frozen concentrates of fruit juice type drinks (November 1986) (29).
9. Frozen confections and novelties (November 1986) (29).
10. Breath mints (November 1986) (30).
11. Tea beverages (November 1986) (31).
12. Aseptically packaged diluted fruit juice beverages (June 1988) (32).
13. Refrigerated ready-to-use gelatin desserts (June 1988) (33).
14. Fruit (including grape) wine beverages with ethanol contents below 7% percent v/v (June 1988) (34).
15. Yogurt-type products (June 1988) (35).
16. Refrigerated flavored milk beverages (June 1988) (36).
17. Frozen dairy/nondairy desserts (June 1988) (37).
18. Fillings for prebaked cookies (October 1988) (38).
19. Frozen cheesecakes, fruit, and fruit toppings (June 1988) (39).
20. Frozen dairy and nondairy frostings, toppings, and fillings (June 1988) (40).
21. Fruit spreads, fruit toppings, fruit syrups (July 1989) (41).

## WORLDWIDE STATUS

In addition to the United States, aspartame has been approved for food and beverage and/or table-top sweetener use in over 90 countries. The countries whose government agencies have approved the marketing of products containing APM are shown in Table 6, along with an indication of the general categories of approval.

## REGULATIONS UNIQUE TO ASPARTAME

Regulatory conditions unique to aspartame have been implemented in several countries. A condition of the U.S. approval in 1981 was that the manufacturer monitor the consumption of aspartame in food products. A method was agreed upon between G.D. Searle & Co. and the U.S. Food and Drug Administration and has been implemented. The detailed tracking required is possible because of data gathering systems already in place in the United States. However, other countries with the same requirement may lack such a method.

In addition, because approximately 50% of the weight of aspartame yields phenylalanine upon hydrolysis, the presence of phenylalanine is

**Table 6** Worldwide Regulatory Status of Aspartame as of October 1990

| Country | Food | Beverage | Table-top |
|---|---|---|---|
| Africa/India | | | |
| Algeria | + | + | + |
| India | − | − | + |
| Kenya | − | + | + |
| Morocco | + | + | + |
| Nigeria | − | − | + |
| Pakistan | + | + | + |
| Sri Lanka | + | + | + |
| Tunisia | + | + | + |
| Zimbabwe | − | + | − |
| Australia/New Zealand | | | |
| Australia | + | + | + |
| New Zealand | + | + | + |
| Eastern Europe | | | |
| Bulgaria | + | + | + |
| Czechoslovakia | + | + | + |
| Hungary | + | + | + |
| Poland | − | − | + |
| Romania | + | + | + |
| U.S.S.R. | + | + | + |
| Yugoslavia | + | + | + |
| Far East/Asia | | | |
| Burma | − | − | + |
| China | + | + | + |
| Hong Kong | + | + | + |
| Indonesia | + | + | + |
| Japan | + | + | + |
| Korea | + | + | + |
| Malaysia | + | + | + |
| Philippines | + | + | + |
| Singapore | + | − | + |
| Tahiti | − | + | − |
| Taiwan | + | − | + |
| Thailand | + | + | + |
| Latin America | | | |
| Argentina | + | + | + |
| Aruba | − | + | − |
| Bahamas | − | + | − |
| Barbados | − | + | − |
| Bermuda | − | + | − |
| Brazil | + | + | + |
| Chile | + | + | + |

**Table 6** Continued

| Country | Food | Beverage | Table-top |
|---|---|---|---|
| Colombia | + | + | + |
| Costa Rica | − | + | + |
| Curacao | − | + | − |
| Dominican Republic | − | − | + |
| Ecuador | − | + | + |
| El Salvador | − | − | + |
| Guatemala | − | + | + |
| Honduras | + | + | + |
| Jamaica | − | + | + |
| Panama | − | + | + |
| Paraguay | + | + | + |
| Peru | + | + | + |
| Trinidad | − | + | + |
| Uruguay | − | + | − |
| Venezuela | + | + | + |
| Middle East | | | |
| Bahrain | − | − | + |
| Egypt | + | + | + |
| Gulf States | − | + | − |
| Iraq | − | − | + |
| Israel | + | + | + |
| Jordan | − | + | − |
| Kuwait | − | − | + |
| Lebanon | − | + | − |
| Oman | − | − | + |
| Qatar | − | − | + |
| Saudi Arabia | − | + | + |
| Turkey | + | + | + |
| U.A.E. | − | − | + |
| Yemen | − | − | + |
| North America | | | |
| Canada | + | + | + |
| Mexico | + | + | + |
| U.S.A. | + | + | + |
| South Africa | | | |
| South Africa | + | + | + |
| Western Europe/EEC | | | |
| Austria | + | + | + |
| Belgium | + | + | + |
| Cyprus | − | + | + |
| Denmark | + | + | + |
| Finland | + | + | + |

**Table 6** Continued

| Country | Food | Beverage | Table-top |
|---|---|---|---|
| France | + | + | + |
| Germany | + | + | + |
| Gibraltar | − | − | + |
| Greece | + | + | + |
| Iceland | + | + | + |
| Ireland | + | + | + |
| Italy | − | + | + |
| Liechtenstein | + | + | + |
| Luxembourg | + | + | + |
| Malta | − | + | + |
| Netherlands | + | + | + |
| Norway | + | + | + |
| Portugal | + | + | + |
| Sweden | + | + | + |
| Switzerland | + | + | + |
| Spain | + | + | + |
| United Kingdom | + | + | + |

noted for the benefit of persons with phenylketonuria (PKU). Label statements and notations in ingredient listings vary by country.

Depending on local regulations, aspartame may be regulated as a "sweetener," an "artificial sweetener," a "food additive," or by a combination of regulatory classes. These various regulations have resulted in requirements such as (1) the name aspartame must be shown on the primary display panel of consumer products; and (2) the amount of aspartame per serving unit or other specified unit must be stated.

## PRODUCTION OF ASPARTAME

There are many patents that cover the manufacture of aspartame. In general, the process requires as starting material the free amino acids phenylalanine and aspartic acid. The phenylalanine is methylated and then coupled with aspartic acid through a series of chemical reactions, and the final sweetener product is crystallized and dried.

## CONCLUSION

The formulation of products with aspartame is not a simple substitution of one sweetener for another, but requires reformulation and objective evaluation. However, technological ingenuity in reformulation coupled with creative new marketing strategies can result in many new and acceptable food and beverage products.

Aspartame can help food producers exploit one of the major trends in the food market today: the trend toward "light foods." Although not a blanket substitute for sugar and other carbohydrate sweeteners, aspartame does provide many opportunities for formulating new products. These opportunities reflect aspartame's ability to be blended with other sweeteners, its unique flavor-enhancing properties, and its clean sugar like taste. When new compatible bulking agents become available, even more opportunities will arise.

The availability of aspartame to food manufacturers worldwide has been one of the major factors responsible for the growth of the "light" and "low-calorie" segments of the food industry. The profile of the sweetener, as one made from components of food (amino acids) coupled with good taste, has also contributed to acceptance by consumers and health professionals of aspartame as a safe sweetening ingredient (107).

## REFERENCES

1. WHO, Aspartame, in *Toxicological Evaluation of Certain Food Additives,* WHO Food Additive Series, 1980, No. 15 Twenty-fourth Report of the Joint FAO-WHO Expert Committee on Food Additives, Rome, World Health Organiz. Rep. Ser., No. 653, 1980, pp. 18–86.
2. Aspartame: Amendment to food additive regulations for aspartame in chewable multivitamin food supplements, *Fed. Reg., 49:* 22468–22469 (1984).
3. Aspartame: Amendment to food additive regulations for aspartame in carbonated beverages, *Fed. Reg., 48:* 31376–31382 (1983).
4. Aspartame: Commissioner's final decision, *Fed. Reg., 46:* 38284–38308 (1981).
5. Aspartame, *Fed. Reg., 39:* 27317–27319 (1974).
6. Aspartame: Ruling on objections and notice of hearing before a public board of inquiry, *Fed Reg., 44:* 31716–31718 (1979).
7. Aspartame: Stay of effectiveness of food additive regulation, *Fed. Reg., 40:* 56907 (1975).
8. R. L. Aspinall, R. N. Saunders, W. F. Pautsch, and E. F. Nutting, The biological properties of aspartame. V. Effects on a variety of physiological parameters related to inflammation and metabolism, *J. Envir. Path. Toxicol., 3:* 387–395 (1980).
9. B. J. Bahoshy, R. E. Klose, and H. A. Nordstrom, Chewing gums of longer lasting sweetness and flavor, U.S. Patent No. 3,943,258 to General Foods Corp., White Plains, NY (March 9, 1976).
10. A. I. Bakal, Functionality of combined sweeteners in several food applications, *Chem. Ind.:* 700–708 (Sept. 19, 1983).

11. R. E. Baldwin and B. M. Korschgen, Intensification of fruit flavors by aspartame, *J. Food Sci., 44:* 938–939 (1979).
12. G. L. Baker, L. J. Filer, Jr., and L. D. Stegink, Plasma and erythrocyte amino acid levels in normal adults fed high protein meals: Effect of added monosodium glutamate (MSG) or monosodium glutamate plus aspartame (APM), *Fed. Proc., 36:* 1154 (1977).
13. B. K. Bauer-Nehrling, M. P. McLane, P. J. Kobe, S. K. Kamath, and D. L. Horowitz, Evaluation of aspartame in diabetes control, paper presented at the 1983 Annual Meeting of the American Diabetes Association [abstract in *Diabetes, 32* (Suppl. 1): 64A (1983).]
14. C. I. Beck, Application potential for aspartame in low calorie and dietetic foods, in *Low Calorie and Special Dietary Foods* (B. K. Dwivedi, ed.), CRC Press, West Palm Beach, FL, 1978, pp. 59–114.
15. C. I. Beck, Sweetness, character, and applications of aspartic acid-based sweeteners, in *ACS Sweetener Symposium* (G. E. Inglett, ed.), AVI Publishing, Westport, CT, 1974, pp. 164–181.
16. G. H. Berg and J. F. Trumbetas, Method of producing fast-dissolving dipeptide sweetener for sweet-sour systems, U.S. Patent 3,868,472, to General Foods Corp., White Plains, NY (Feb. 25, 1975).
17. R. G. Bianchi, E. T. Muir, D. L. Cook, and E. F. Nutting, The biological properties of aspartame: II. Actions involving the gastrointestinal system, *J. Envir. Pathol. Toxicol., 3:* 355–362 (1980).
18. R. G. Bost and A. Ripper, A commercially feasible aspartic acid based sweetener. Special supplement to feeding, weight and obesity abstracts, in *Proceedings of Sweeteners and Dental Caries* (J. H. Shaw and G. G. Roussos, eds.), Information Retrieval Inc., Washington, DC, 1978, pp. 269–288.
19. R. L. Brunner, C. V. Voorhees, L. Kinney, and R. E. Butcher, Aspartame: Assessment of developmental psychotoxicity of a new artificial sweetener, *Neurobehav. Toxicol., 1:* 79–86 (1979).
20. A. Butchko, H. H., F. N. Kotsonis, Aspartame: Review of recent research, *Comments Toxicology, 3*(4): 253–278 (1989).
21. J. A. Cella and W. H. Schmitt, Low calorie dry sweetener composition, U.S. Patent 3,753,739, to Alberto Culver Co., Melrose Park, IL (August 21, 1973).
22. M. R. Cloninger and R. E. Baldwin, L-Aspartyl-L-phenylalanine methyl ester (aspartame) as a sweetener, *J. Food Sci., 39:* 347–349 (1974).
23. M. R. Cloninger and R. E. Baldwin, Aspartylphenylalanine methyl ester: A low-calorie sweetener, *Science, 170:* 81–82 (1970).
24. Effect of aspartame and aspartate loading upon plasma and erythrocyte free amino acids in normal adult volunteers, *Nutr. Rev., 36:* 110–12 (April 1978).
25. *Fed. Reg., 46*(200): 50947–50948 (Oct. 1981).
26. *Fed. Reg., 48*(132): 31376–31382 (July 8, 1983).
27. *Fed. Reg., 49:* 6672 (1984).
28. *Fed. Reg., 49* (105): 22468–22469 (May 30, 1984).
29. *Fed. Reg., 51:* 42999–43000 (1986).
30. *Fed. Reg., 51:* 43000–43001 (1986).

31. *Fed. Reg., 51:* 43001–43003 (1986).
32. *Fed. Reg., 53:* 20838–20839 (1988).
33. *Fed. Reg., 53:* 20837–20838 (1988).
34. *Fed. Reg., 53:* 20839–20840 (1988).
35. *Fed. Reg., 53:* 20840–20841 (1988).
36. *Fed. Reg., 53:* 20841–20842 (1988).
37. *Fed. Reg., 53:* 20842 (1988).
38. *Fed. Reg., 53:* 40878–40879 (1988).
39. *Fed. Reg., 54:* 23646–23647 (1988).
40. *Fed. Reg., 54:* 23647 (1988).
41. *Fed. Reg., 54:* 31332–31333 (1989).
42. J. D. Fernstrom, M. H. Fernstrom, and M. A. Gillis, Acute effects of aspartame on large neutral amino acids and monoamines in rat brain, *Life Sci., 32:* 1651–1658 (1983).
43. L. J. Filer, Jr., G. L. Baker, and L. D. Stegink, Effect of aspartame loading on plasma and erythrocyte free amino acid concentrations in one-year-old infants, *J. Nutr., 113:* 1591–1599 (1983).
44. L. J. Filer, Jr., G. L. Baker, and L. D. Stegink, Effect of aspartame loading in individuals subject to Chinese restaurant syndrome, *Am. Clin. Nutr., 33* (Suppl., April): 933 (1980).
45. L. J. Filer, Jr., M. C. Brummel, G. L. Baker, T. R. Tephly, and L. D. Stegink, Methanol (MEOH) metabolism in humans, *Fed. Proc., 37:* 319 (1978).
46. *Food Chemical Codex*, third edition, National Academy Press, 1981, pp. 28–29.
47. G. H. Frey, Use of aspartame by apparently healthy children and adolescents, *J. Toxicol. Envir. Health, 2:* 401–415 (1976).
48. I. Furda and J. F. Trumbetas, Multifunctional sweetening composition, U.S. Patent 3,934,048, to General Foods Corp., White Plains, NY (Jan. 20, 1976).
49. M. Glicksman and B. N. Wankier, Low calorie sweetening composition and method for making sure, U.S. Patent 3,922,369, to General Foods Corp., White Plains, NY (Nov. 25, 1975).
50. M. Glicksman and B. N. Wankier, Method for making a low calorie sweetening composition, U.S. Patent 3,761,288, to General Foods Corp., White Plains, NY (Sept. 25, 1973).
51. A. E. Harper, Aspartame, in *Sweeteners: Issues and Uncertainties,* Academy Forum Series, National Academy of Sciences, Washington, DC, 1975, pp. 182–188.
52. B. E. Homler, A. Kedo, L. Hill, and W. H. Shazer, FDA approves four new aspartame uses, *Food Technology 41:* 41–44 (1987).
53. J. Hornyak and H. D. Stahl, Dry carbonation source for a beverage concentrate and a method of preparing same, U.S. Patent 3,939,289, to General Foods Corp., White Plains, NY (Feb. 17, 1976).
54. D. L. Horwitz, Aspartame use by diabetics, in *Aspartame: Physiology and Biochemistry* (L. D. Stegink and L. J. Filer, Jr., eds.), Marcel Dekker, Inc., New York, 1984.
55. H. Ishii, Incidence of brain tumors in rats fed aspartame, *Toxicol. Lett., 7:* 433–437 (1981).

56. H. Ishii, T. Kochimizu, S. Usami, and T. Fujimoto, Toxicity of aspartame and its diketopiperazine for Wistar rats by dietary administration for 194 weeks, *Toxicology, 21:* 91–94 (1981).
57. A. Kedo, Prediction of aspartame stability under real life temperature conditions, *Industrial Data Acquisition and Control 1*(1): 53 (August, 1989).
58. S. E. Keller, T. C. Nash, S. S. Newberg, and W. H. Shazer, Degradation of aspartame in chocolate milk related to microbial contamination, Annual Meeting ADSA, Raleigh, NC (June 24–27, 1990).
59. S. E. Keller, S. S. Newberg, T. M. Krieger, and W. H. Shazer, Degradation of aspartame in yogurt related to microbial growth, *J. Food Sci.* (1991).
60. R. H. Knopp, K. Brandt, and R. A. Arkey, Effect of aspartame in young persons during weight reduction, *J. Toxicol. Envir. Health, 292:* 596 (1976).
61. P. J. Kobe, J. K. Bauer-Nehrling, M. McLane, and D. L. Horowitz, Exogenous insulin treatment of noninsulin dependent diabetes mellitus (NIDDM) is associated with increased insulin resistance, presented at American Diabetes Association Annual Meeting, June 14, 1983.
62. R. Koch, Safety of aspartame as a sweetener, *N. Engl. J. Med., 292:* 596 (March 13, 1975).
63. R. Koch, G. Schaeffler, and K. H. F. Shaw, Results of loading doses of aspartame by two phenylketonuric (PKU) children compared with two normal children, *J. Toxicol. Envir. Health, 2:* 453–457 (1976).
64. R. Koch, K. H. F. Shaw, M. Williamson, and M. Haber, Use of aspartame in phenylketonuric heterozygous adults, *J. Toxicol. Envir. Health, 2:* 453–457 (1976).
65. N. Larson-Powers, and R. M. Pangborn, Paired comparisons and time-intensity measurements of the sensory properties of beverages and gelatins containing sucrose or synthetic sweeteners, *J. Food Sci., 43:* 41–46 (1978).
66. N. Larson-Powers, and R. M. Pangborn, Descriptive analysis of the sensory properties of beverages and gelatins containing sucrose or synthetic sweeteners, *J. Food Sci., 43:* 47–51 (1978).
67. H. D. Lennon, L. E. Metcalf, S. E. Mares, J. H. Smith, E. F. Nutting, and F. J. Saunders, The biological properties of aspartame, IV. Effects of reproduction and lactation, *J. Envir. Pathol. Toxicol., 3:* 375–386 (1980).
68. Levels of amino acids in lactating women following ingestion of the sweetener aspartame, *Nutr. Rev., 38:* 183–184 (May 1980).
69. R. H. Mazur, Aspartame—a sweet surprise, *J. Toxicol. Envir. Health, 2:* 243 (1976).
70. R. H. Mazur, Aspartic acid-based sweeteners, in *Symposium Sweeteners* (G. E. Inglett, ed.), AVI Publishing, Westport, CT, 1974, Chap. 14.
71. R. H. Mazur, and A. Ripper, Peptide-based sweeteners, in *Developments in Sweeteners* (C. A. M. Hough, K. J. Parker, and A. J. Vlitos, eds.), Applied Science Publishers, London, 1979, pp. 125–134.
72. R. H. Mazur, J. M. Schlatter, and A. H. Goldkamp, Structure-taste relationships of some dipeptides, *J. Am. Chem. Soc., 91:* 2684–2691 (1969).
73. R. D. McCormick, Extensive test program confirms safety for aspartame, new dipeptide sweetener, *Food Prod. Devel., 8:* 30, 45 (Dec. 1974).

74. B. A. McPherson, L. A. McGill, and F. W. Bodyfelt, Effect of stabilizing agents and aspartame on the sensory properties of orange sherbert, *J. Food Sci., 43:* 935–939 (1978).
75. H. R. Moskowitz, and C. Bubose, Taste intensity, pleasantness and quality of aspartame, sugars, and their mixtures, *Can. Inst. Food Sci. Technol. J., 10:* 126–131 (1977).
76. H. R. Moskowitz, K. Wolfe, and C. I. Beck, Sweetness and acceptance optimization in cola flavored beverages using combinations of artificial sweeteners—psychophysical approach, *J. Food Quality, 2:* 17–26 (1978).
77. J. W. Olney, L-Glutamic and L-aspartic acids—a question of hazard, *Food Cosmet. Toxicol., 13:* 595–596 (1975).
78. J. W. Olney, Another view of aspartame, in *Sweeteners, Issues and Uncertainties,* National Academy of Sciences, Washington, DC, 1975, pp. 189–195.
79. J. A. Oppermann, E. Muldoon, and R. E. Ranney, Metabolism of aspartame in monkeys, *J. Nutr., 103:* 1454–1459 (1973).
80. A. Oppermann, E. Muldoon, and R. E. Ranney, Effect of aspartame on phenylalanine metabolism in the monkey, *J. Nutr., 103:* 1460–1466 (1973).
81. Peryam & Kroll Research Corporation, Chicago, IL, Effect of varying concentrations of aspartame on consumer acceptance of soft drinks, study commissioned by Searle Biochemics, Skokie, IL, 1975.
82. K. P. Porikos, G. Booth, and T. B. Van Itallie, Effect of covert nutritive dilution on the spontaneous food intake of obese individuals: A pilot study, *Am. J. Clin. Nutr., 30* (Oct.): 1638–1644 (1977).
83. A. B. Porter, Effectiveness of multiple sweeteners and other ingredients in food formulation, *Chem. Ind.:* 696–699 (Sept. 19, 1983).
84. W. J. Potts, J. L. Bloss, and E. F. Nutting, Biological properties of aspartame, I. Evaluation of central nervous system effects, *J. Envir. Pathol. Toxicol., 3:* 341–353 (1980).
85. R. E. Ranney, S. E. Mares, R. E. Schroeder, T. D. Hutsell, and F. M. Radzialowski, The phenylalanine and tyrosine content of maternal and fetal body fluids from rabbits fed aspartame, *Toxicol. Appl. Pharmacol., 32:* 339–346 (1975).
86. R. E. Ranney, J. A. Oppermann, E. Muldoon, and F. G. McMahon, Comparative metabolism of aspartame in experimental animals and humans, *J. Toxicol. Envir. Health, 2:* 441–451 (1976).
87. K. S. Rao and R. G. McConnell, Evaluation of the toxicologic potential of a nutritive sweetener (aspartame) in the dog, *Toxicol. Appl. Pharmacol., 29:* 155 (1974).
88. F. E. Reno, R. G. McConnell, J. F. Ferrell, J. A. Trutler, and K. S. Rao, A tumorigenic evaluation of aspartame, a new sweetener, in the mouse, *Toxicol. Appl. Pharmacol., 33:* 182 (1975).
89. W. A. Reynolds, A. F. Bauman, E. Renn, S. Naidu, L. D. Stegink, and L. J. Filer, Jr., Effect of excessive aspartame ingestion by the infant nonhuman primate: Behavior studies, in *Aspartame:* Physiology and Biochemistry (L. D. Stegink and L. J. Filer, Jr., eds.), Marcel Dekker, New York, 1984.

90. W. A. Reynolds, V. Butler, and N. Lemkey-Johnston, Hypothalamic morphology following ingestion of aspartame or MSG in the neonatal rodent and primate: A preliminary report, *J. Toxicol. Envir. Health, 2:* 471–480 (1976).
91. W. A. Reynolds, L. D. Stegink, L. J. Filer, Jr., and E. Renn, Aspartame administration to the infant monkey: Hypothalamic morphology and plasma amino acid levels, *Anat. Rec., 198:* 73–85 (1980).
92. F. J. Saunders, W. F. Pautsch, and E. F. Nutting, The biological properties of aspartame, III. Examination for endocrine-like activities, *J. Envir. Pathol. Toxicol., 3:* 363–373 (1981).
93. H. R. Schade, Taste modifier for artificial sweeteners, U.S. Patent No. 3,934,047, to General Foods Corp., White Plains, NY (Jan. 20, 1976).
94. S. S. Schiffman, Comparison of the taste properties of aspartame with other sweeteners, in *Aspartame: Physiology and Biochemistry* (L. D. Stegink and L. F. Filer, Jr., eds.), Marcel Dekker, New York, 1984.
95. S. S. Schiffman, H. Cahn, and M. G. Lindley, Multiple receptor sites mediate sweetness: Evidence from cross adaptation, *Pharmacol. Biochem. Behavior., 15:* 377–388 (1981).
96. S. S. Schiffman, D. A. Reilly, and T. B. Clark, III, Qualitative differences among sweeteners, *Physiol. Behav., 23:* 1–9 (1979).
97. J. M. Schlatter, Peptide sweetening agents, U.S. Patent 3,492,131, to G. D. Searle & Co., Chicago, IL (Jan. 27, 1970).
98. D. Scott, Sweetening composition and method for use thereof, U.S. Patent 3,780,189, to G. D. Searle & Co., Skokie, IL (Dec. 18, 1973).
99. G.D. Searle & Co., The NutraSweet Breakthrough: A Technical View, Skokie, IL, 1983.
100. G.D. Searle & Co., Skokie, IL, Aspartame in carbonated beverages, Vol. 2, Food Additive Petition No. 2A3661 filed with the U.S. Food & Drug Administration (Aug. 13, 1982).
101. G.D. Searle & Co., Skokie, IL, Master file #134 for Food Additive Petition 3A 2885.
102. W. H. Shazer, A. Kedo, L. Hill, and L. Metcalf, Stability of aspartame dairy aseptic beverage processing, Society of Soft Drink Technologists Processing, Tampa, FL (March, 1988).
103. M. D. Shoaf and L. D. Pischke, Stabilized methyl-L-aspartyl-L-phenylalanine preparation by heating with a matrix forming material and surfactant, then grinding, U.S. Patent 3,956,507, to General Foods Corp., White Plains, NY (May 11, 1976).
104. M. D. Shoaf and L. D. Pischke, Spray-dried aspartic acid derivatives, U.S. Patent 4,059,706, to General Foods Corp., White Plains, NY (Nov. 22, 1977).
105. L. D. Stegink, M. C. Brummel, L. J. Filer, Jr., and G. L. Baker, Blood methanol concentrations in one-year-old infants administered graded doses of aspartame, *J. Nutr., 113:* 1600–1606 (1983).
106. L. D. Stegink, M. C. Brummel, K. McMartin, G. Martin-Amat, L. J. Filer,

Jr., G. L. Baker, and T. R. Tephly, Blood methanol concentrations in normal adult subjects administered abuse doses of aspartame, *J. Toxicol. Envir. Health, 7:* 281–290 (1981).

107. L. D. Stegink, and L. J. Filer, Jr., eds., *Aspartame: Physiology and Biochemistry,* Marcel Dekker, New York, 1984.
108. L. D. Stegink, L. J. Filer, Jr., and G. L. Baker, Plasma, amino acid concentrations in normal adults fed meals with added monosodium L-glutamate and aspartame, *J. Nutr., 113:*1851–1860 (1983).
109. L. D. Stegink, L. J. Filer, Jr., and G. L. Baker, Effect of aspartame plus monosodium L-glutamate ingestion on plasma and erythrocyte amino acid levels in normal adult subjects fed a high protein meal, *Am. J. Clin. Nutr., 36:* 1145–1152 (1982).
110. L. D. Stegink, L. J. Filer, Jr., and G. L. Baker, Effect of aspartame and sucrose loading in glutamate-susceptible subjects, *Am. J. Clin. Nutr., 34:* 1899–1905 (1981).
111. L. D. Stegink, L. J. Filer, Jr., and G. L. Baker, Plasma and erythrocyte concentrations of free amino acids in adult humans administered abuse doses of aspartame, *J. Toxicol. Envir. Health, 7:* 291–305 (1981).
112. L. D. Stegink, L. J. Filer, Jr., and G. L. Baker, Effect of plasma amino acid levels of monosodium L-glutamate (MSG) plus aspartame added to a high protein meal, *Am. J. Clin. Nutr., 33* (Suppl.): 947 (1980).
113. L. D. Stegink, L. J. Filer, Jr., and G. L. Baker, Plasma, erythrocyte and human milk levels of free amino acids in lactating women administered aspartame or lactose, *J. Nutr., 109:* 2173–2181 (1979).
114. L. D. Stegink, L. J. Filer, Jr., and G. L. Baker, Effect of aspartame and aspartate loading upon plasma and erythrocyte free amino acid levels in normal adult volunteers, *J. Nutr., 107:* 1837–1845 (1977).
115. L. D. Stegink, L. J. Filer, Jr., and G. L. Baker, Effect of aspartame and monosodium glutamate (MSG) upon plasma amino acid levels when ingested as part of a soup-beverage meal, *Am. J. Clin. Nutr., 34:* 649 (1981).
116. L. D. Stegink, L. J. Filer, Jr., G. L. Baker, and M. C. Brummel., Plasma and erythrocyte amino acid levels of adult humans given 100 mg/kg body weight aspartame, *Toxicology, 14:* 131–140 (1979).
117. L. D. Stegink, L. J. Filer, Jr., G. L. Baker, M. C. Brummel, and T. R. Tephly, Aspartame metabolism in human subjects, in *Health and Sugar Substitutes* (B. Guggenheim, ed.), Proceedings of ERGOB Conference, Geneva, S. Karger, New York, 1979, pp. 160–165.
118. L. D. Stegink, L. J. Filer, Jr., G. L. Baker, and J. E. McDonnell, Effect of an abuse dose of aspartame upon plasma and erythrocyte levels of amino acids in phenylketonuric heterozygous and normal adults, *J. Nutr., 110:* 2216–2224 (1980).
119. L. D. Stegink, L. J. Filer, Jr., G. L. Baker, and J. E. McDonnell, Effect of aspartame loading upon plasma and erythrocyte amino acid levels in phenylketonuric heterozygotes and normal subjects, *J. Nutr., 109:* 709–717 (1979).

120. L. D. Stegink, R. T. Koch, M. E. Blaskovics, L. J. Filer, Jr., G. L. Baker, and J. E. McDonnell, Plasma phenylalanine levels in phenylketonuric heterozygous and normal adults administered aspartame at 34 mg/kg body weight, *Toxicology, 20:* 81–90 (1981).
121. S. B. Stern, S. J. Bleicher, A. Flores, G. Gombos, D. Recitas, and J. Shu, Administration of aspartame in non-insulin dependent diabetes, *J. Toxicol. Envir. Health, 2:* 429–439 (1976).
122. H. Stone and S. M. Oliver, Measurement of relative sweetness of selected sweeteners and sweetener mixtures, *J. Food Sci., 34:* 215 (1969).
123. F. M. Sturtevant, Aspartame, a new sweetening agent: Toxicological questions discussed at a scientific hearing, *Trends Pharmacol. Sci., 1:* 1–2 (Apr. 1980).
124. S. J. Suomi, Effects of aspartame on the learning test performance of young stumptail macaques, in *Aspartame: Physiology and Biochemistry* (L. D. Stegink and L. J. Filer, Jr., eds.), Marcel Dekker, New York, 1984.
125. J. G. Surka and K. A. Robertson, Aspartame effects on *Tetrahymena pyriformis, J. Food Safety, 1:* 165–171 (1978).
126. J. H. Tsau and J. G. Young, Depeptide sweetener sulfate and sulfonate salts, U. S. Patent 4,439,460 to G.D. Searle & Co., Skokie, Il (March 27, 1984).
127. S. Yamaguchi, A. Shimizu, J. Kirimura, and T. Ninomiya, Method of improving the taste of artificial sweeteners, Japanese Patent 7,223,389, to Ajinomoto Co., Inc., Tokyo, Japan (June 26, 1972).

# 5

# Cyclamate

**Barbara A. Bopp and Paul Price** *Abbott Laboratories, Abbott Park, Illinois*

## INTRODUCTION

Cyclamate (Fig. 1) was synthesized in 1937 by a University of Illinois graduate student, Michael Sveda, who accidentally discovered its sweet taste (1,2). The patent for cyclamate eventually became the property of Abbott Laboratories, which performed the necessary studies and submitted a New Drug Application for the sodium salt in 1950. Cyclamate was initially marketed as tablets that were recommended for use as a table-top sweetener for diabetics and others who had to restrict their use of sugar. In 1958, after enactment of the Food Additive Amendment to the Food, Drug and Cosmetic Act, the Food and Drug Administration (FDA) of the United States classified cyclamate as a GRAS, or Generally Recognized as Safe, sweetener. A mixture of cyclamate and saccharin, which had been found to have synergistic sweetening properties and an improved taste (3,4), was subsequently marketed for use in special dietary foods. When soft drinks sweetened with the cyclamate-saccharin mixture became popular in the United States during the 1960s, the consumption of cyclamate increased dramatically. Prompted by the growing use of cyclamate, additional studies were initiated. It was found that cyclamate could be metabolized to cyclohexylamine (Fig. 1) by bacteria in the intestine (5). Then in 1969, the results of a chronic toxicity study with a 10:1 cyclamate–saccharin mixture were interpreted as implicating cyclamate as a bladder carcinogen in rats (6). Cyclamate was removed from GRAS status (7), its use was initially

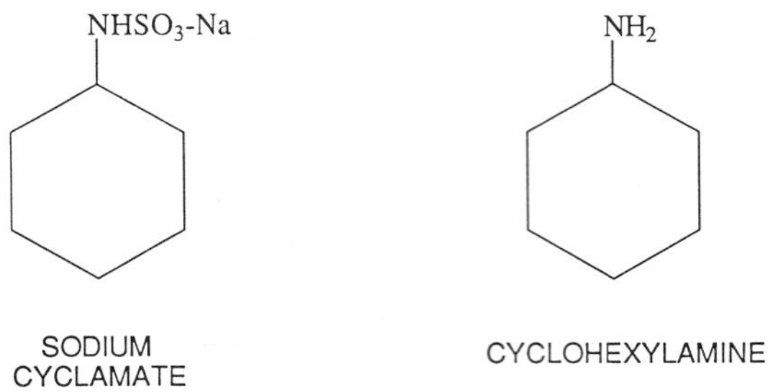


**Figure 1** Structures of cyclamate and cyclohexylamine.

restricted, and then in 1970, it was banned in the United States from use in all foods, beverages, and drugs (8). However, many foreign governments did not act as precipitously, and cyclamate continued to be used as a sweetener in these countries.

In the next few years, many additional toxicity and carcinogenicity studies were conducted with cyclamate, the 10:1 cyclamate–saccharin mixture, and cyclohexylamine, but they failed to confirm the original findings. Based on the results of these studies, Abbott Laboratories filed a Food Additive Petition for cyclamate in 1973 (9). After a lengthy review and an administrative hearing, cyclamate was denied food additive status in 1980 (10). Subsequently, in September 1982, the Calorie Control Council and Abbott Laboratories filed a second Food Additive Petition, containing new studies that confirmed the safety of cyclamate and assisted in determining the safe level of intake (11). Action on this petition is still pending.

## PRODUCTION

The production of cyclamate, which is accomplished by the sulfonation of cyclohexylamine, was originally limited to Abbott Laboratories (12). However, as cyclamate became more widely used in foods and beverages, other companies, including The Pillsbury Company, Pfizer Inc., Cyclamate Corporation of America, and Miles Laboratories, entered the market. By 1970, several foreign countries, including Japan, Taiwan, and Korea, also had production capabilities. When cyclamate was banned in the United States, all domestic producers, except Abbott Laboratories, ceased production. Production was also banned in Japan, and thus the Taiwanese and Koreans became the major producers. Subsequently, production facilities were established in Brazil, Germany, Spain, South Africa, and Indonesia, in addition to Taiwan.

## PHYSICAL CHARACTERISTICS (13–16)

Cyclamic acid, or cyclohexylsulfamic acid ($C_6H_{13}NO_3S$; MW = 179.24), is a white crystalline powder, with a melting point of 169–170°C, good aqueous solubility (1 g/7.5 ml) and a lemon-sour sweetness. It is a strong acid, and the pH of a 10% aqueous solution is about 0.8–1.6. Sodium and calcium cyclamate are strong electrolytes, which are highly ionized in solution, fairly neutral in character, and have little buffering capacity. Both salts exist as white crystals or white crystalline powders. The molecular weight of the sodium salt ($C_6H_{12}NO_3S \cdot Na$) is 201.22 and that of the calcium salt ($C_{12}H_{24}N_2O_6S_2 \cdot Ca$) is 396.54 (432.58 as the dihydrate). They are freely soluble in water (1 g/4–5 ml) at concentrations far in excess of those required for normal use but have limited solubility in oils and nonpolar solvents. Cyclamate solutions are stable to heat, light, and air throughout a wide pH range.

Cyclohexylamine, which is the starting material in the synthesis of cyclamate and is also a metabolite of cyclamate, has distinctly different properties. Cyclohexylamine (MW = 99.17) is a base, not an acid; it has a fishy odor and a bitter taste, not a sweet taste. It is a clear colorless liquid, which is miscible with water, alcohol, and nonpolar solvents and has a boiling point of 134.5°C. The pH of a 0.01% aqueous solution is 10.5. Cyclohexylamine has many industrial applications, including use in water treatment and rubber acceleration.

## RELATIVE SWEETNESS AND UTILITY

In contrast to the sweet taste from sucrose, which appears quickly and has a sharp, clean cut-off, the sweetness from cyclamate builds to its maximal level more slowly and persists for a longer time (13). Cyclamate is generally accepted as being 30 times sweeter than sucrose, but the relative sweetness of cyclamate tends to decrease at higher sweetness intensities. For example, cyclamate is about 40 times as sweet as a 2% sucrose solution but only 24 times as sweet as a 20% sucrose solution (13). This trend may be at least partially due to the increasing levels of bitterness and aftertaste that characteristically appear at very high cyclamate concentrations (3). This off-taste is, however, not a problem at normal use concentrations.

Calcium cyclamate is somewhat less sweet than sodium cyclamate, and the off-taste response starts at a lower concentration of the calcium salt than of the sodium salt. Vincent et al. (3) suggested that the differences between the two salts might be related to differences in ionization, since the sweet taste is due to the cyclamate ion while the off-taste may be associated with the undissociated salt. The sweetening power of cyclamate also varies with the medium, and its actual relative sweetness should be determined in each different product (13). Most notably, the sweetening power of cyclamate is considerably enhanced in fruits. More detailed information on the relative sweetness and other properties of cyclamate can be found in papers by Beck (13,14).

The primary use of cyclamate is as a noncaloric sweetener, generally in combination with other sweeteners, but cyclamate can also be used as a flavoring agent (i.e., to mask the unpalatable taste of drugs) (12,17). Prior to its being banned, cyclamate was used in the United States in table-top sweeteners, in a variety of foods, in beverages, and in both liquid and tablet pharmaceuticals, and it is found in numerous products in over 50 countries today (see Table 1). Many of these same applications are proposed in the current food additive petition (11) as the potential uses of cyclamate:

table-top sweeteners, in tablet, powder or liquid form
beverages, fruit juice drinks, beverage bases or mixes
processed fruits
chewing gum and confections
salad dressings
gelatin desserts, jellies, jams, and toppings

**Table 1** Regulatory Status of Cyclamate

| Country | Food | Beverage | Table-top |
|---|---|---|---|
| Angola | − | − | + |
| Antigua | + | + | |
| Argentina | + | + | + |
| Australia | + | + | + |
| Austria | − | − | + |
| Bahamas | + | | |
| Bahrain | − | − | − |
| Belgium | − | − | + |
| Bermuda | − | − | − |
| Bolivia | − | − | − |
| Brazil | + | − | + |
| Bulgaria | − | − | − |
| Burundi | − | − | − |
| Canada | − | − | + |
| Caribbean (Ind.) | + | + | + |
| Chile | − | + | − |

**Table 1** Continued

| Country | Food | Beverage | Table-top |
|---|---|---|---|
| Colombia | – | – | – |
| Costa Rica | – | – | – |
| Cyprus | – | – | – |
| Czechoslovakia | | – | |
| Denmark | – | + | + |
| Dominica | + | + | + |
| Ecuador | – | – | + |
| El Salvador | – | – | – |
| FRG | + | + | + |
| Fiji | – | – | – |
| Finland | + | + | + |
| France | – | – | – |
| French Guiana | | – | |
| Greece | – | – | + |
| Guadaloupe | | + | |
| Guam | – | – | – |
| Guatemala | – | – | + |
| Guiana | | – | |
| Haiti | + | + | |
| Hong Kong | – | – | – |
| Hungary | + | + | + |
| Iceland | – | + | + |
| India | – | – | – |
| Indonesia | + | – | + |
| Iran | | – | |
| Ireland | + | + | + |
| Israel | + | + | + |
| Italy | + | + | + |
| Jamaica | – | – | – |
| Japan | – | – | – |
| Korea | – | – | – |
| Kuwait | | | + |
| Luxembourg | + | + | + |
| Malaysia | | – | |
| Martinique | | + | |
| Mexico | – | – | – |
| Monserrat | | + | |
| Netherlands | – | + | + |
| New Zealand | + | + | + |
| Nicaragua | + | – | + |
| Norway | + | + | + |
| Oman | – | – | + |

**Table 1** Continued

| Country | Food | Beverage | Table-top |
|---|---|---|---|
| Qatar | – | – | – |
| Pakistan | – | – | |
| Panama | – | – | – |
| Papua New Guinea | + | + | + |
| Paraguay | + | + | + |
| Peru | + | + | – |
| Philippines | – | – | – |
| Poland | | – | |
| Portugal | – | – | + |
| Puerto Rico | – | – | – |
| Romania | | – | |
| Rwanda | – | – | – |
| Samoa | – | – | – |
| Saudi Arabia | – | – | – |
| Sierra Leone | + | + | + |
| Sinapore | – | – | – |
| South Africa | + | + | + |
| Spain | + | + | + |
| Srilanka | | – | + |
| Sudan | – | – | – |
| Sweden | – | – | + |
| Switzerland | + | + | + |
| Taiwan | – | – | – |
| Thailand | | – | + |
| Trinidad | + | + | |
| Turkey | – | – | + |
| Uganda | – | – | – |
| United Arab Emirates | | | + |
| United Kingdom | – | – | – |
| USSR | – | – | – |
| USA | – | – | – |
| Uruguay | + | + | + |
| Venezuela | – | – | + |
| Yugoslavia | | + | + |
| Zaire | – | – | – |
| Zimbabwe | + | + | + |

+ = Permitted; – = Prohibited; Blank = no information.
Information current as of 1989 but may be subject to change. Information provided by the Calorie Control Council.

## ADMIXTURE POTENTIAL

Cyclamate has most frequently been used in combination with saccharin. In the 1950s, it was shown that cyclamate–saccharin mixtures are sweeter than would be expected from the known sweetness of either component alone, and that any off-taste is minimized with the mixtures (3,4). About a 10–20% synergism is observed when cyclamate and saccharin are used together. For example, a combination of 5 mg saccharin + 50 mg cyclamate in a table-top sweetener is as sweet as 125 mg cyclamate alone or 12.5 mg saccharin alone (13). Although the ratio of cyclamate to saccharin can vary considerably from product to product, the 10:1 mixture is used most frequently. With this combination, each component contributes about equally to the sweetness of the mixture since saccharin is about 10 times sweeter than cyclamate.

More recently, numerous patents have described the use of cyclamate in combination with aspartame or aspartame and saccharin (e.g., 18,19). Applications have also been reported for cyclamate in combination with acesulfame-K and other sweeteners (20). If and when cyclamate is again approved for use in the United States, it would undoubtedly be used primarily in combination with other more potent sweeteners since no one sweetener appears to meet all the technological requirements and mixtures of sweeteners generally appear to offer improved taste.

## TECHNICAL QUALITIES

Cyclamate has a number of technical qualities that make it a good alternative sweetener (13,17,21). It is noncaloric and noncariogenic. Although its relative sweetness is less than that of saccharin or aspartame, its sweetening power is adequate, especially when used in combination with other more intense sweeteners. Cyclamate has a favorable taste profile and does not leave an unpleasant aftertaste at normal use concentrations. It is better than sugar in masking bitterness, and it enhances fruit flavors. Cyclamate is compatible with most foods and food ingredients, as well as with natural and artificial flavoring agents, chemical preservatives, and other sweeteners. Its solubility is more than adequate for most uses, and at normal concentrations it does not change the viscosity or density of solutions. The stability of cyclamate is excellent, at both high and low temperatures, over a wide pH range, and in the presence of light, oxygen, and other food chemicals. Cyclamate is nonhygroscopic and will not support mold or bacterial growth. However, like most noncaloric sweeteners, cyclamate does not provide the bulk, texture, or body associated with sugar.

## UTILITY

Although most of the product development work with cyclamate was done in the 1950s and 1960s, Beck has recently reviewed some of the applications and presented typical formulations sweetened with cyclamate or a cyclamate–saccharin mixture (13). Perhaps the primary rule for the development of low-calorie foods and beverages is that cyclamate (or any other nonnutritive sweetener) cannot simply be substituted for the sugar; instead, the product must be reformulated (13). The two most critical aspects for a successful product are its flavor and texture. Since a flavor may not taste the same in systems sweetened with sucrose and cyclamate (or mixtures of cyclamate and other nonnutritive sweeteners), the flavoring of a product frequently has to be modified. A proper balance between the taste effects of the acid and sweetener components of a product must also be achieved. For example, if the lingering sweetness from noncaloric sweeteners results in a higher sweetness intensity than desired, some compensation can be achieved by increasing the level of acidity. On the other hand, if an aftertaste develops at the level of sweetness desired, decreasing the acidity may permit a reduction in the sweetness level, hence minimizing the aftertaste. The other major problem in the development of low-calorie products with cyclamate is that of texture or body. This largely results from the elimination of sugar solids and can frequently be solved by the addition of a suitable hydrocolloid or bulking agent. An alternative approach that has been used in soft drinks involves achieving the proper balance between sweetness and tartness, the use of special flavors, and the adjustment of the carbonation level.

Cyclamate has always been particularly useful in fruit products, since it enhances fruit flavors and, even at low concentrations, can mask the natural tartness of some citrus fruits (13,17,21–23). The cyclamate solutions used for canned fruits have a lower specific gravity and osmotic pressure than sucrose syrups and hence do not draw water out of the fruit. Thus, fruits packed in cyclamate solutions tend to have a greater drained weight than those packed in sucrose. Cyclamate-sweetened gelatins are reasonably easy to formulate, requiring the use of high-bloom gelatins and crystalline sorbitol or mannitol as a bodying agent, dispersant, and filler (13,22). Thickening and consistency represent the major problems with jams, jellies, and puddings sweetened with cyclamate. Low-methoxy pectin is usually employed as a gelling agent in jams and jellies, since it does not require sugar for gel formation (13,21–23). However, low-methoxy pectin needs more calcium than is normally present, and hence calcium cyclamate may be preferable to the sodium salt for this application. Due to the lower concentrations of osmotically active compounds, jams and jellies containing cyclamate may require a preservative to extend their shelf life. Body and thickening of puddings can be achieved with starches or a combination of nonnutritive gums and thickeners (13,22). Low calorie salad dressings require the substitution of two basic ingredients: cyclamate (or a mixture of cyclamate and another sweetener) for the sugar and a hydrocolloid or thickener for the oil (13,22).

Baked goods are probably the most difficult foods to reformulate with noncaloric sweeteners (13,22,24). In addition to sweetness, sugar provides bulk and texture, has a tenderizing effect on gluten, and is important in the browning reaction. Cyclamate cannot furnish these properties, and hence the formulations must be modified to include bulking agents (e.g., modified starch or dextrins, carboxymethylcellulose) and a tenderizing agent (e.g., lecithin). Although proper browning is difficult to achieve, some success can be obtained by application of a caramel solution onto the surface. In yeast doughs, sugar also acts an energy source for the fermentation reaction. However, satisfactory products containing cyclamate and only a small amount of sugar ($\leq 1\%$) can be prepared if the salt content and the fermentation time are reduced (24). As the amount of sugar in chemically aerated products is decreased, more liquid must be added to retain the proper consistency of the batter and the eggs must be used to their best advantage for structure and aeration. Good results were obtained with a slurry technique, in which the flour and other dry ingredients were mixed with water prior to being added to the whipped eggs (24). However, preparation of high quality, low calorie baked goods still represents a major technical challenge.

In contrast to baked goods, the lack of a browning reaction with cyclamate can be an advantage in cured meats (17,22,25). When sugar-cured bacon or ham is fried, the sugar tends to caramelize, losing its sweet taste and giving the meats a darkened appearance. Because cyclamate has a higher melting point than sucrose, cyclamate-cured meats taste better, have an improved color, and do not scorch or stick in the frying pan.

Cyclamate has also found applications in pharmaceutical and oral hygiene products. It is particularly good at masking the bitterness and unpalatable taste of many drugs and hence is especially useful in syrups, other liquid formulations, and chewable tablets (16). Cyclamate imparts a high level of sweetness with a low solid content, thus providing suspensions that are more fluid and have fewer problems with caking (16) or tablets that disintegrate rapidly and have less bulk (26). Cyclamate is also useful as a sweetener in both film coating and compression coating of tablets, and the acid form can be used as an effervescent agent (26). Since cyclamate is noncariogenic, it is suitable for use in toothpastes and mouthwashes.

## AVAILABILITY

Cyclamate is still available in the United States, as well as from foreign suppliers. The use of cyclamate as a food additive is currently prohibited in the United States, and therefore it may not be used in foods or beverages in this country at the present time. It is, however, legal to manufacture in the United States cyclamate-containing foods for export to foreign countries where the use of cyclamate as a sweetener is permitted. This must, however, be done in compliance with Section 801(d) of the Federal Food, Drug, and Cosmetic Act 21 USC 381(d).

## SHELF LIFE

Samples of tablets containing a cyclamate and saccharin mixture, which were manufactured in 1969 or before, did not show any diminution in sweetening ability or any physical deterioration after at least 7 years (12). It would seem, therefore, that cyclamate in tablet form has an extremely long shelf life. Information about the possible shelf life of cyclamate in other applications, such as soft drinks or canned foods, was not available to the authors. However, when cyclamate was widely used in the United States, the shelf life was more than adequate to allow the products to be sold in the ordinary course of business. There is no known instance of a recall of products due to the degradation of the sweetening content from cyclamate. The cyclamate stability data indicate that an expiration date is not needed to ensure the identity, strength, quality, and purity of either the bulk food additive or foods and beverages containing cyclamate.

## TRANSPORT

There is no known problem with the transport of the bulk material, and in the United States cyclamate is nonregulated with respect to transport.

## GENERAL COST/ECONOMICS

It is anticipated that should the Food and Drug Administration once again allow the use of cyclamate as a food additive in the United States, the price would be approximately $2.00 per pound, estimated at today's production costs. The price from foreign producers has varied considerably from year to year, depending largely on product availability. As the price of sugar also fluctuates considerably, the only means of determining the economics of cyclamate usage is to compare its cost at the time of production to the cost of equivalent sweetening from sugar at its prevailing price. Such comparisons generally indicate that cyclamate is one of the most economical noncaloric sweeteners.

## METABOLISM

Cyclamate is slowly and incompletely absorbed from the gastrointestinal tract (15). In one study involving almost 200 subjects, the absorption of cyclamate averaged only 37% (27). The volume of distribution for cyclamate in rats is approximately equal to the total body water content, and hence cyclamate does not concentrate in most tissues (15). Once absorbed, cyclamate is excreted unchanged in the urine by both glomerular filtration and active tubular secretion (15).

Although early studies had indicated that cyclamate was not metabolized to any appreciable extent, in 1966 Kojima and Ichibagase (5) detected cyclohexylamine in urine samples from humans and dogs receiving cyclamate. Subsequently, conversion was also demonstrated in mouse, rat, guinea pig, rabbit, monkey, and pig, as well as in dog and human (15). Cyclamate is, however, not metabolized to cyclohexylamine by mammalian tissues, but rather the cyclohexylamine is formed by the action of the microflora on the nonabsorbed cyclamate remaining in the intestinal tract (15).

Probably the most important feature of cyclamate metabolism is the extreme variability in cyclohexylamine formation. Not all individuals are able to convert cyclamate to cyclohexylamine, and even among converters, the extent of conversion varies greatly and often changes in the same individual over time. Retrospective analyses of studies which attempted to define the incidence of converters indicated that only about 25% of the subjects were able to metabolize cyclamate to cyclohexylamine (15,27). The incidence of converters was slightly lower (~20%) among Europeans and North Americans who were given at least three daily doses of cyclamate, but appeared to be higher among the Japanese (≥80% in studies involving about 60 subjects).

As noted above, even among those individuals who can convert cyclamate to cyclohexylamine, there is substantial intersubject and day-to-day variability in the extent of conversion, which can range from <0.1% to >60%. The frequency distribution curve for cyclamate conversion is, however, strongly skewed, and only a few individuals are able to form large amounts of cyclohexylamine. It has been estimated that only about 3% of a population converts over 20% of a cyclamate dose to cyclohexylamine and ≤1% converts 60% or more (11,28,29). The 60% level approaches, on the average, the maximal conversion possible, since only the nonabsorbed cyclamate can be metabolized, and the absorption of cyclamate averages about 40%.

The conversion of cyclamate to cyclohexylamine also appears to be dependent on continuous exposure to the sweetener (15). A single dose of cyclamate will frequently not be metabolized, and daily ingestion of cyclamate is usually necessary to induce and maintain the converting ability at a high but still variable level. Furthermore, if cyclamate is withdrawn from the diet for even a few days, the ability to metabolize cyclamate is diminished and gradually lost. Hence, intermittent cyclamate use (which might be a typical pattern of use by some people) would tend to limit the converting ability of an individual.

Since the no-observed effect level (NOEL) in animal toxicity studies is based on cyclohexylamine, not cyclamate, determination of the extent of conversion is critical for establishing the safe level of use for the sweetener. However, the skewed nature of the cyclamate conversion curve makes this more difficult. It is not appropriate to use the mean conversion by an entire group of converters, since most of these subjects would be forming only small amounts of cyclohexylamine. Instead, the best estimate can probably be derived from the average conversion by a subgroup of high converters. The available data suggest that among those subjects who converted at least 1% of the dose to cyclohexylamine, conversion averaged slightly under 20% (11,15). If the group was further restricted to only those subjects converting at least 5% of the dose, the average level increased only slightly (~25%) (15,30). A few individuals would, of course, be converting at higher levels at least some of the time. However, it is probably not necessary to use the maximal conversion rate in establishing the acceptable daily intake (ADI) since the large safety factor applied to food additives compensates for considerable intersubject variability.

In contrast to cyclamate, cyclohexylamine is rapidly and completely absorbed from the gastrointestinal tract, even from the large intestine where it is formed (15). The plasma half-life of cyclohexylamine in humans is dose-dependent, increasing from 3.5 hours with a 2.5 mg/kg oral dose to 4.8 hours with a 10 mg/kg dose (31). The apparent volume of distribution of cyclohexylamine is 2–3 liters/kg, and tissue concentrations typically exceed those in plasma (15,31).

Cyclohexylamine is primarily excreted unchanged in the urine by both glomerular filtration and a saturable transport mechanism (31,32). Although cyclohexylamine is not extensively metabolized, its biotransformation shows some species differences (15). The principal metabolic pathway in rats is ring hydroxylation, leading to the formation of the isomeric 3- or 4-aminocyclohexanols which account for 5–20% of the dose (33–35). Mice and humans, however, form negligible quantities of these metabolites (33–35). Only the deamination products, cyclohexanol and *trans*-cyclohexane-1,2-diol, are found in humans given cyclohexylamine, and these metabolites represent only 1–2% of the dose (33). Both ring hydroxylation and deamination occur in guinea pigs and rabbits (33).

## CARCINOGENICITY

In 1969, a 2-year chronic toxicity study with a cyclamate–saccharin mixture was nearing completion at the Food and Drug Research Laboratories (FDRL) in Maspeth, New York. In that study, rats were fed diets containing a 10:1 mixture of sodium cyclamate and sodium saccharin to provide daily doses of 500, 1120, or 2500 mg/kg/day (6,36). The ability of some of the rats from this study to convert cyclamate to cyclohexylamine had been documented (37), and after 78 weeks, the diets of half of the animals were supplemented with cyclohexylamine at levels corresponding to about 10% conversion of the cyclamate dose (i.e., 25, 56, and 125 mg/kg/day). At the conclusion of the study, tumors were found in the urinary bladder of eight high dose rats, with four to eight of the tumors being classified as carcinomas by the different pathologists who reviewed the slides (6). In subsequent analyses, the number of tumors increased to 12 and all were considered carcinomas (36). Calcification of the urinary tract and bladder parasites were observed in some of these rats and possibly could have affected the results of this study since both of those factors are known to contribute to the development of bladder tumors in rats. Furthermore, six of the tumors occurred in consecutively numbered and presumably consecutively housed male rats, suggesting that some extraneous environmental factor could also have been involved in the development of the tumors. Nevertheless, the results of this study were interpreted as implicating cyclamate as a bladder carcinogen in rats (6) and led to its removal from GRAS status and finally to its being banned for use in foods and beverages in the United States (7,8).

In the years following this observation, the carcinogenic potential of cyclamate and cyclohexylamine was reevaluated in a group of well-designed and well-controlled bioassays, which were performed by independent investigators throughout the world. Cyclamate, including both the sodium and calcium salts, was tested in at least four separate studies with rats given doses up to 2.5 g/kg/day (38–43), in three separate studies with mice given doses up to 7–9 g/kg/day (43–45), and in one study with hamsters given doses up to 3 g/kg/day (46). The 10:1 cyclamate–saccharin mixture that had been used in the original FDRL study has been tested twice in rats (38,39,41,42) and once in mice (45). Two studies, one in rats (40) and one in mice (45), even included in utero exposure of the animals to cyclamate. In addition to these conventional rodent bioassays, two studies have been conducted in monkeys given cyclamate, with treatment lasting 8 years in one (200 mg/kg/day) (47) and 20 years in the other (100 or 500 mg/kg/day), which is still ongoing (48,49). Cyclohexylamine, as the hydrochloride or sulfate salt, was tested in three studies with rats given maximal doses of 150–300 mg/kg/day (38,50,51) and in two studies with mice given maximal doses of 400–600 mg/kg/day (45,52). A 9-year study was also performed in dogs (53). None of these studies confirmed the original findings that implicated cyclamate as a bladder carcinogen. Instead, when the results of these studies are evaluated in accordance with recognized toxicological and statistical procedures, they provide strong evidence that neither cyclamate nor cyclohexylamine is carcinogenic in animals.

In 1976, the Temporary Committee of the National Cancer Institute that was charged with reviewing the cyclamate data concluded "the present evidence does not establish the carcinogenicity of cyclamate or its principal metabolite, cyclohexylamine, in experimental animals" (54). Upon receipt of this report, the Bureau of Foods of the U.S. FDA apparently concurred and reported to the Commissioner that "we think the issue of the carcinogenicity of cyclamate is settled" (55). In 1977, the World Health Organization's Joint Expert Committee on Food Additives (JECFA) reached a similar conclusion, stating that "it is now possible to conclude that cyclamate has been demonstrated to be non-carcinogenic in a variety of species" (56). Nevertheless, carcinogenicity was a major issue in the 1980 decision of the FDA commissioner, who stated that "cyclamate has not been shown not to cause cancer" and thus denied cyclamate food additive status (10). However, some toxicological and statistical principles used in this decision were subsequently challenged by the American Statistical Association (57) and the Task Force of the Past Presidents of the Society of Toxicology (58).

After submission of the second Food Additive Petition for cyclamate in 1982, the Cancer Assessment Committee (CAC) of the U.S. FDA's Center for Food Safety and Applied Nutrition reviewed all of the cyclamate data and concluded that "there is very little credible data to implicate cyclamate as a carcinogen at any organ/tissue site to either sex of any animal species tested," that "the collective weight of the many experiments . . . indicates that cyclamate is not carcinogenic," and further that "no newly discovered toxic effects of cyclamate are likely to be revealed if additional standardized studies were performed" (59). In 1985, the National Academy of Sciences reaffirmed the CAC conclusion stating that ". . . the totality of the evidence from studies in animals does not indicate that cyclamate or its major metabolite cyclohexylamine is carcinogenic by itself" (60). The NAS report, however, raised some concern about a possible role of cyclamate as a promoter. This was largely predicated on two studies involving direct exposure of the urinary bladder to cyclamate. One involved implantation of a pellet containing cyclamate in the bladder of mice (61) and the other involved intravesicular instillation of *N*-methyl-*N*-nitrosourea into the urinary bladder of female rats which were then given cyclamate in either their food or water (62). At the request of the FDA, the Mitre Corporation evaluated these models and concluded that "both types of direct bladder exposure studies, pellet implantation and intravesicular catherization, are considered unsuitable for predicting human carcinogenic risk" (63). Thus, once again, it would appear that the carcinogenicity issue has been settled, at least from a scientific point of view, and that it can be concluded that cyclamate and cyclohexylamine are not carcinogenic in animals.

The possible association between cancer, particularly bladder cancer, and the consumption of noncaloric sweeteners by humans has been extensively studied during the past 10–20 years. Since the widespread use of cyclamate in foods and beverages was restricted to a relatively short time span in many countries, these studies are more applicable to the assessment of the possible carcinogenicity of saccharin than cyclamate. However, the cyclamate issue was specifically addressed in some of the studies. The epidemiology studies with the noncaloric sweeteners have been reviewed by others (64–66), and it has generally been agreed that there is no conclusive evidence of an increased risk of bladder cancer associated with the use of these sweeteners.

## MUTAGENICITY

In the 1960s and 1970s, the mutagenic potential of cyclamate and cyclohexylamine was evaluated in many different test systems, including the Ames test and other microbial gene mutation assays, studies in Drosophila, in vitro cytogenetic studies, in vivo cytogenetic studies with both somatic and germ cells, dominant lethal tests, and others. Although there were instances of conflicting or discordant results, the preponderance of evidence provided by this battery of tests suggested that neither cyclamate nor cyclohexylamine represented a significant mutagenic hazard (15,67,68). Nevertheless, in the 1980 decision the Commissioner of the FDA still concluded that "cyclamate has not been shown not to cause heritable genetic damage" (10).

In 1985 the NAS Committee evaluated the mutagenicity tests with cyclamate and cyclohexylamine and found little evidence to conclude that either compound was a DNA-reactive carcinogen (60). They, however, recommended that additional tests for mammalian cell DNA damage, mammalian cell gene mutation tests, and more definitive cytogenetic studies be conducted to complete and strengthen the database for the two compounds. After consultation with the FDA, both calcium cyclamate and cyclohexylamine were tested in the Chinese hamster ovary HGPRT forward mutation assay, for unscheduled DNA synthesis in rat hepatocytes, and in the Drosophila sex-linked recessive lethal assay (69). The results of these three tests gave no evidence of any intrinsic genotoxicity from either compound. In addition, dominant lethal and heritable translocation tests were conducted in male mice given the maximal tolerated dose of calcium cyclamate for 6 weeks, since the sodium salt had been used in almost all of the previous studies with mammalian germ cells (70). The results of these two tests were also negative, indicating that the calcium salt did not induce any transmissible chromosomal aberrations in the germ cells of mice. Thus, these recent studies, performed according to the currently accepted standards, confirm the previous conclusions that cyclamate and cyclohexylamine are not mutagenic and do not cause heritable genetic damage.

## OTHER TOXICITY

### Cyclamate

Although a great many toxicity studies have been conducted with cyclamate and the cyclamate–saccharin mixture, they have revealed very few pathophysiological effects attributable to the sweetener (15). Perhaps the most frequently reported effects in animals and humans given cyclamate are softening of the stools and diarrhea. These effects, however, only occur at excessively high doses and are due to the osmotic activity of the nonabsorbed cyclamate remaining in the gastrointestinal tract (71).

### Cyclohexylamine

Cyclohexylamine is considerably more toxic than cyclamate, and its toxicity limits the use of the sweetener. Since the toxicity studies with cyclohexylamine have recently been reviewed (15), only the two areas of major concern—testicular atrophy and cardiovascular effects—will be discussed herein.

*Testicular Effects of Cyclohexylamine*

Numerous toxicological studies have shown that the testis of rats is the organ that is most sensitive to any adverse effects from cyclohexylamine, and this effect has been used by JECFA and others as the basis for establishing the acceptable daily intake for cyclamate. The results of three 90-day studies in rats given cyclohexylamine in the diet indicated that the testes were not adversely affected at concentrations of 0.2% (approximately 100 mg/kg/day) and below, but at concentrations of 0.6% (approximately 300 mg/kg/day) and above testicular atrophy, characterized by decreased organ weight, decreased spermatogenesis, and/or degeneration of the tubular epithelium, was observed (72–75). Since the concentration of cyclohexylamine in the food of these animals was held constant, the mg/kg/day dose ingested by the animals declined over the 3-month study period as the food consumption of the rats relative to their body weight progressively decreased. To circumvent this problem and to more precisely define the no-adverse effect dose, an additional study was conducted by Dr. Horst Brune (76) in Germany. Groups of 100 young male rats were fed diets containing cyclohexylamine to provide constant daily doses of 50, 100, 200, or 300 mg/kg/day for 3 months. Ad libitum and pair-fed control groups were also included in the experimental design of this study. The results clearly demonstrated that 100 mg/kg/day was a no-observed effect level. Slight, but statistically significant, histopathological changes were noted in the testes of the rats receiving the 200 mg/kg/day dose, and marked effects were seen at 300 mg/kg/day. Analysis of all the studies involving an effect of cyclohexylamine on the rat testes showed that the data were quite consistent and confirmed the steep nature of the dose-response curve, as demonstrated by the Brune study (15). The lack of significant histopathological changes in the testes of rats given a 150 mg/kg/day dose of cyclohexylamine for 2 years (51) and the presence of only mild, sporadic effects in rats given ~175 mg/kg/day dose for 90 days (72) are consistent with the steep dose-response curve and suggest that the no-effect level might be as high as 150 mg/kg/day (11). However, 100 mg/kg, which was clearly shown to be a no-observed effect level in the study of Brune, was adopted as the NOEL by JECFA (77).

Early studies had indicated that mice were less sensitive to the effects of cyclohexylamine than rats, since no testicular changes had been seen in mice given doses of ~300 mg/kg/day (0.3% in the diet) (52). Renwick and Roberts suggested that this species difference in testicular toxicity might be related to differences in metabolism since the ring-hydroxylated metabolites of cyclohexylamine (i.e., 3- or 4-aminocyclohexanols) account for about 15–20% of the dose in male rats but are formed in negligible amounts in mice (34–35). To test this hypothesis, the metabolism and testicular effects of a 400 mg/kg/day dose of cyclohexylamine were compared in mice, Wistar rats, and DA rats, a strain deficit in hydroxylating ability (35). Consistent with the previous findings, the mice formed only small amounts of the ring-hydroxylated metabolites and did not develop testicular toxicity. Although the Wistar rats had higher concentrations of the aminocyclohexanol metabolites in their plasma and testes than the DA rats, the DA strain appeared to be more sensitive to the cyclohexylamine-induced testicular effects than the Wistar rats. Thus, the development of testicular atrophy could not be attributed to the ring-hydroxylated metabolites. Instead, the species differences in toxicity were related to differences in the pharmacokinetics of cyclohexylamine (32). The concentrations of cyclohexylamine in the plasma and testes of rats showed a nonlinear relationship to dietary intake, with substantially elevated levels occurring at intakes greater than 200 mg/kg/day. This nonlinear accumulation of cyclohexylamine in the rat testes correlated well with the previously noted steep dose-response curve for the testicular atrophy. In contrast, both the plasma and testicular cyclohexylamine concentrations in mice increased linearly with intake, even at doses far in excess of 200 mg/kg/day. Furthermore, the dose-normalized cyclohexylamine concentrations at levels greater than 200 mg/kg/day were considerably higher in rats than in mice, suggesting that the greater exposure of the rat testes to cyclohexylamine probably contributed to the greater toxicity in that species. Since these studies clearly demonstrated that the testicular toxicity of cyclohexylamine was attributable to the parent compound per se and that the rat was more sensitive to this effect than the mouse, it is appropriate to use the no-observed effect level in rats for determining the acceptable daily intake of cyclamate in human.

### *Cardiovascular Effects of Cyclohexylamine*

The other major question about the safety of cyclohexylamine involves its possible effects on blood pressure. Cyclohexylamine is an indirectly acting sympathomimetic agent, similar to tyramine but over 100 times less potent than tyramine (15). Intravenous administration of cyclohexylamine to anesthetized animals causes vasoconstriction and increases both the blood pressure and heart rate. In spite of these acute effects, a rise in blood pressure was not seen in subchronic or chronic toxicity studies with orally administered cyclohexylamine in rats, even at high doses (0.4–1% in the diet or approximately 200–400 mg/kg/day) (38,72).

The cardiovascular effects of single oral doses of cyclohexylamine in humans have been thoroughly characterized by Eichelbaum et al (31).

They found that a 10 mg/kg bolus oral dose of cyclohexylamine caused a 30 mm rise in the mean arterial blood pressure of healthy human volunteers. A smaller increase was seen following a 5 mg/kg dose, but no significant change in blood pressure occurred with a 2.5 mg/kg dose. A reflex-mediated, slight decrease in heart rate accompanied the vasopressor effects of the two higher doses. The cyclohexylamine levels in the plasma of these subjects were closely correlated with the increases in blood pressure, and it was estimated that the lowest cyclohexylamine concentration to cause a significant hypertensive effect was about 0.7–0.8 $\mu$g/ml.

Despite the potential of cyclohexylamine to increase blood pressure, there is no evidence to suggest that such effects actually occur after the administration of cyclamate. Periodic cardiovascular monitoring in some subjects participating in cyclamate metabolism studies revealed no changes in blood pressure or heart rate, even in those individuals who received high doses of cyclamate and were forming large amounts of cyclohexylamine (27,29,78,79). Moreover, the cyclohexylamine blood levels in some of these subjects approached the concentrations associated with a rise in blood pressure following a bolus oral dose of cyclohexylamine, yet their blood pressure was apparently not affected (29).

Two human studies conducted in Germany during the 1970s also addressed the questions of cyclamate conversion, cyclohexylamine blood levels, and possible cardiovascular effects. In one study the high converters among a group of 44 regular users of cyclamate were initially identified based on their urinary cyclohexylamine levels. After ingestion of 1.2–1.7 g of sodium cyclamate, the cyclohexylamine blood levels in three of the high converters ranged from 0.03 to 0.42 $\mu$g/ml and were correlated with the urinary cyclohexylamine levels (80). In the other study, the blood pressure and heart rate of 20 subjects were monitored three times a day during a 10-day dosing period with sodium cyclamate (81). Half of these subjects did not convert cyclamate to cyclohexylamine, five were low converters, and the other five were high converters. However, no significant changes in blood pressure or heart rate were seen, even in the high converters. Furthermore, noninvasive hemodynamic studies, which were conducted in six subjects known to be good converters, failed to reveal any changes that were attributable to cyclamate or cyclohexylamine, even though the cyclohexylamine levels in serum ranged from about 0.2–0.6 $\mu$g/ml in most subjects and reached 1 $\mu$g/ml in one subject (81).

Although the available data involve only a few subjects and were not always derived from well-controlled clinical trials, they give no indication that the cyclohexylamine formed from cyclamate causes any significant cardiovascular changes. The cyclohexylamine blood levels generally appear to be low, but occasionally concentrations that approximate or even exceed those associated with the hypertensive effects from an acute bolus dose of cyclohexylamine were apparently achieved, seemingly without eliciting a change in blood pressure. One possible explanation for this apparent discrepancy could involve the differences in the cyclohexylamine blood level profiles from the two different routes of delivery. A bolus oral dose leads to a rapid rise and subsequent fall in the circulating levels, while the cyclohexylamine concentrations following cyclamate administration would rise slowly and fluctuate less. It has recently been reported that slow attainment and rapid attainment of steady state levels of nifedipine, a calcium channel blocker, led to very different hemodynamic effects (82). Hence, there is some precedent for the concept that the rate of change of blood levels can be one determinant of the cardiovascular effects elicited by a compound.

A controlled clinical study investigating the conversion of cyclamate to cyclohexylamine and any possible cardiovascular effects is currently in progress under the direction of Dr. A. G. Renwick at the University of Southampton (83 and personal communication). Such a study requires a large number of subjects since the majority of the population will not even convert cyclamate to cyclohexylamine and there will be only a few high converters, in whom a rise in blood pressure would be most likely to occur. Preliminary results indicate that the distribution and extent of cyclamate metabolism in the group of diabetic subjects from this study are similar to those reported during the late 1960s, and that the cyclohexylamine levels in the plasma of the subjects are correlated with the extent of cyclohexylamine formation but are not correlated with any changes in blood pressure or heart rate. The final results from this study and further evaluations of some high converters should conclusively demonstrate whether there is any risk of hypertensive effects associated with the ingestion of cyclamate and its conversion to cyclohexylamine, even in those individuals who are among the highest converters.

## REGULATORY STATUS

The regulatory status of cyclamate in the United States is quite simply expressed. It is still banned from usage. Whether the FDA will alter that status will not be known until their evaluation of the cyclamate Food Additive Petition is completed. Cyclamate is, however, currently available for use in over 50 countries (84). As summarized in Table 1, some countries allow the use of cyclamate in foods, beverages, or both, while others only permit table-top use. Pharmaceutical use is also permitted in some countries. In 1982, JECFA (77) increased the acceptable daily intake for cyclamate almost threefold and established an ADI of 0–11 mg/kg, expressed as cyclamic acid. However, the ADI for cyclamate may vary from country to country.

## REFERENCES


1. L. F. Audrieth, and M. Sveda, Preparation and properties of some N-substituted sulfamic acids, *J. Org. Chem.*, *9:* 89–101 (1944).
2. L. F. Audrieth and M. Sveda, U.S. Patent No. 2,275,125 (March 3, 1942).
3. H. C. Vincent, M. J. Lynch, F. M. Pohley, F. J. Helgren, and F. J. Kirchmeyer, A taste panel study of cyclamate–saccharin mixture and of its components, *J. Amer. Pharm. Assoc.*, *44:* 442–446 (1955).
4. U.S. Patent No. 2,803,551 (August 20, 1957).
5. S. Kojima and H. Ichibagase, Studies on synthetic sweetening agents. VIII. Cyclohexylamine, a metabolite of sodium cyclamate, *Chem. Pharm. Bull.*, *14:* 971–974 (1966).
6. J. M. Price, C. G. Biava, B. L. Oser, E. E. Vogin, J. Steinfeld, and H. C. Ley, Bladder tumors in rats fed cyclohexylamine or high doses of a mixture of cyclamate and saccharin, *Science*, *167:* 1131–1132 (1970).
7. *Fed. Reg.*, *34*(202): 17063 (October 21, 1969).
8. *Fed. Reg.*, *35*(167): 13644 (August 27, 1970).
9. Abbott Laboratories, Cyclamic acid and its salts, Food Additive Petition 4A2975 (November 15, 1973).
10. *Fed. Reg.*, *45:*61474 (September 16, 1980).
11. The Calorie Control Council and Abbott Laboratories, Food Additive Petition for Cyclamate 2A3672 (September 22, 1982).
12. R. W. Kasperson, and N. Primack, Cyclamate, In *Alternative Sweeteners*, L. O'Brien-Nabors and R. C. Gelardi (Eds.), Marcel Dekker, New York, 1986, pp. 71–87.
13. K. M. Beck, Nonnutritive sweeteners: saccharin and cyclamate, In *CRC Handbook of Food Additives*, Vol. II, 2nd ed. T. E. Furia, (Ed.) CRC Press, Boca Raton, FL 1980.
14. K. M. Beck, Properties of the synthetic sweetening agent, cyclamate, *Food Technol.*, *11:* 156–158 (1957).
15. B. A. Bopp, R. C. Sonders, and J. W. Kesterson, Toxicological aspects of cyclamate and cyclohexylamine, *CRC Critical Reviews in Toxicology*, *16:* 213–306 (1986).
16. M. J. Lynch, and H. M. Gross, Artificial sweetening of liquid pharmaceuticals, *Drug and Cosmetic Industry*, *87:* 324–326, 412–413 (1960).
17. K. M. Beck, and A. S. Nelson, Latest uses of synthetic sweeteners, *Food Engineering*, *35:* 96–97 (1963).
18. D. Scott (G. D. Searle), Saccharin-dipeptide sweetening compositions, British Patent 1,256,995 (1971).

19. T. P. Finucane (General Foods), Sweetening compositions containing APM and other adjuncts, Canadian Patent 1,043,158 (November 28, 1978).
20. Hoechst AG, U.S. Patent 4,158,068 (June 12, 1979).
21. J. A. Stein, Technical aspects of non-nutritive sweeteners in dietary products, *Food Technol.* (March, 1966).
22. K. M. Beck and J. V. Ziemba, Are you using sweeteners correctly? *Food Engineering, 38:* 71–73 (1966).
23. D. K. Salunkhe, R. L. McLaughlin, S. L. Day, and M. B. Merkley, Preparation and quality evaluation of processed fruits and fruit products with sucrose and synthetic sweeteners, *Food Technol., 17:* 85–91 (1963).
24. C. D. Stone, Sucaryl. Calorie reduction in baked products, *Bakers Digest, 36:* 53–56, 59–64 (1962).
25. K. M. Beck, R. Jones, and L. W. Murphy, New sweetener for cured meats, *Food Engineering, 30:* 114 (1958).
26. C. J. Endicott and H. M. Gross, Artificial sweetening of tablets, *Drug and Cosmetic Industry, 85:* 176–177, 254–256 (1959).
27. R. C. Sonders, unpublished studies, Abbott Laboratories (1967–1968).
28. A. G. Renwick, The fate of cyclamate in man and rat, Third European Toxicology Forum, Geneva, Switzerland, October, 1983.
29. A. J. Collings, Metabolism of cyclamate and its conversion to cyclohexylamine, *Diabetes Care, 12* (Suppl. 1): 50–55 (1989).
30. A. G. Renwick, The metabolism, distribution and elimination of non-nutritive sweeteners, In *Health and Sugar Substitutes* B. Guggenheim, (ed), S. Karger, Basel, 1979, pp. 41–47.
31. M. Eichelbaum, J. H. Hengstmann, H. D. Rost, T. Brecht, and H. J. Dengler, Pharmacokinetics, cardiovascular and metabolic actions of cyclohexylamine in man, *Arch. Toxikol, 31:* 243–263 (1974).
32. A. Roberts and A. G. Renwick, The pharmacokinetics and tissue concentrations of cyclohexylamine in rats and mice, *Toxicol. Appl. Pharmacol., 98:* 230–242 (1989).
33. A. G. Renwick and R. T. Williams, The metabolites of cyclohexylamine in man and certain animals. *Biochem. J., 129:* 857–867 (1972).
34. A. Roberts and A. G. Renwick, The metabolism of $^{14}$C-cyclohexylamine in mice and two strains of rat, *Xenobiotica, 15:* 477–483 (1985).
35. A. Roberts, A. G. Renwick, G. Ford, D. M. Creasy, and I. Gaunt, The metabolism and testicular toxicity of cyclohexylamine in rats and mice during chronic dietary administration, *Toxicol. Appl. Pharmacol., 98:* 216–229 (1989).
36. B. L. Oser, S. Carson, G. E. Cox, E. E. Vogin, and S. S. Sternberg, Chronic toxicity study of cyclamate:saccharin (10:1) in rats, *Toxicology, 4:* 315–330 (1975).
37. B. L. Oser, S. Carson, E. E. Vogin, and R. C. Sonders, Conversion of cyclamate to cyclohexylamine in rats, *Nature, 220:* 178–179 (1968).
38. D. Schmähl, Absence of carcinogenic activity of cyclamate, cyclohexylamine, and saccharin in rats, *Arzneim-Forsch/Drug Research, 23:* 1466–1470 (1973).

39. D. Schmähl and M. Habs, Investigations on the carcinogenicity of the artificial sweeteners sodium cyclamate and sodium saccharin in rats in a two-generation experiment, *Arzneim-Forsch/Drug Research, 34:* 604–606 (1984).
40. J. M. Taylor, M. A. Weinbuger, and L. Friedman, Chronic toxicity and carcinogenicity to the urinary bladder of sodium saccharin in the in utero exposed rat, *Toxicol. Appl. Pharmacol., 54:* 57–75 (1980).
41. T. Furuya, K. Kawamata, T. Kaneko, O. Uchida, S. Horiuchi, and Y. Ikeda, Long-term toxicity study of sodium cyclamate and sodium saccharin in rats, Japan. *J. Pharmacol., 25* (Suppl. 1): 55P (1975).
42. Y. Ikeda, S. Horiuchi, T. Furuya, K. Kawamata, T. Kaneko, and O. Uchida, Long-term toxicity study of sodium cyclamate and sodium saccharin in rats, National Institute of Hygienic Sciences, Tokyo, Japan, unpublished report, 1973.
43. F. Homburger, Studies on saccharin and cyclamate, Bio-Research Consultants, Inc., unpublished report, 1973.
44. P. G. Brantom, I. F. Gaunt, and P. Grasso, Long-term toxicity of sodium cyclamate in mice, *Food Cosmet. Toxicol., 11:* 735–746 (1973).
45. R. Kroes, P. W. J. Peters, J. M. Berkvens, H. G. Verschuuren, T. deVries, and G. J. Van Esch, Long-term toxicity and reproduction study (including a teratogenicity study) with cyclamate, saccharin, and cyclohexylamine, *Toxicology, 8:* 285–300 (1977).
46. J. Althoff, A. Cardesa, P. Pour, and P. A. Shubik, A chronic study of artificial sweeteners in Syrian golden hamsters, *Cancer Lett., 1:* 21–24 (1975).
47. F. Coulston, E. W. McChesney and K.-F. Benitz, Eight-year study of cyclamate in Rhesus monkeys, *Toxicol. Appl. Pharmacol., 41:* 164–165 (1977).
48. S. M. Sieber and R. H. Adamson, Long-term studies on the potential carcinogenicity of artificial sweeteners in non-human primates, In *Health and Sugar Substitutes* (B. Guggenheim, ed)., S. Karger, Basel, 1979, pp. 266–271.
49. R. Adamson, Quoted in *Food Chemical News,* February 27, 1989.
50. I. F. Gaunt, J. Hardy, P. Grasso, S. D. Gangolli, and K. R. Butterworth, Long-term toxicity of cyclohexylamine hydrochloride in the rat, *Food Cosmet. Toxicol., 14:* 255–267 (1976).
51. B. L. Oser, S. Carson, G. E. Cox, E. E. Vogin, and S. S. Sternberg, Long-term and multigeneration toxicity studies with cyclohexylamine hydrochloride, *Toxicology, 6:* 47–65 (1976).
52. J. Hardy, I. F. Gaunt, J. Hooson, R. J. Hendy, and K. R. Butterworth, Long-term toxicity of cyclohexylamine hydrochloride in mice, *Food Cosmet. Toxicol., 14:* 269–276 (1976).
53. Industrial Bio-Test Laboratories, Chronic oral toxicity study with cyclohexylamine sulfate in beagle dogs, unpublished report, April 21, 1981.
54. Report of the Temporary Committee for the Review of Data on Carcinogenicity of Cyclamate, DHEW Publication No. (NIH)77-1437, Department of Health, Education and Welfare, February, 1976.
55. H. Roberts, Actions Memorandum, Food Additive Petition for Cyclamate (FAP 4A2975), undated and unsigned to Commissioner Schmidt.
56. World Health Organization, 21st Report of the Joint FAO/WHO Expert Com-

mittee on Food Additives, Evaluation of Certain Food Additives, April 18–27, 1977, Technical Report Series, 617, 1978.
57. American Statistical Association, Letter to the Commissioner of the Food and Drug Administration, April 7, 1981.
58. Task Force of Past Presidents of the Society of Toxicology, Animal data in hazard evaluation: Paths and pitfalls, *Fundam. Appl. Toxicol., 2:* 101–107 (1982).
59. Cancer Assessment Committee (CAC), Center for Food Safety and Applied Nutrition, Food and Drug Administration, Scientific review of the long-term carcinogen bioassays performed on the artificial sweetener, cyclamate, April, 1984.
60. National Academy of Sciences, National Research Council, Committee on the Evaluation of Cyclamate for Carcinogenicity, Evaluation of cyclamate for carcinogenicity, National Academy Press, Washington, D.C., 1985.
61. G. T. Byran and E. Erturk, Production of mouse urinary bladder carcinomas by sodium cyclamate, *Science, 167:* 996–998 (1970).
62. R. M. Hicks, J. St. J. Wakefield, and J. Chowaniec, Evaluation of a new model to detect bladder carcinogens or co-carcinogens: Results obtained with saccharin, cyclamate and cyclophosphamide, *Chem.-Biol. Interactions, 11:* 225–233 (1975).
63. J. M. DeSesso, J. M. Kelley, and B. B. Fuller, Relevance of direct bladder exposure studies to human health concerns, The MITRE Corporation, unpublished report, May, 1987.
64. R. W. Morgan and O. Wang, A review of epidemiological studies of artificial sweeteners and bladder cancer, *Food Chem. Toxicol., 23:* 529–533 (1985).
65. O. M. Jensen, Artificial sweeteners and bladder cancer: Epidemiological evidence, Third European Toxicology Forum, Geneva, Switzerland, October, 1983.
66. D. L. Arnold, D. Krewski, and I. C. Munro, Saccharin: A toxicological and historical perspective, *Toxicology, 27:* 179–256 (1983).
67. B. M. Cattanach, The mutagenicity of cyclamates and their metabolites, *Mutat. Res., 39:* 1–28 (1976).
68. P. Cooper, Resolving the cyclamate question, *Food Cosmet. Toxicol., 15:* 69–70 (1977).
69. D. Brusick, M. Cifone, R. Young, and S. Benson, Assessment of the genotoxicity of calcium cyclamate and cyclohexylamine, *Environ. Mol. Mutagen., 14:* 188–199 (1989).
70. K. T. Cain, C. V. Cornett, N. L. A. Cacheiro, L. A. Hughes, J. G. Owens, and W. M. Generoso, No evidence found for induction of dominant lethal mutations and heritable translocations in male mice by calcium cyclamate, *Environ. Mol. Mutagen., 11:* 207–213 (1988).
71. K. Hwang, Mechanism of the laxative effect of sodium sulfate, sodium cyclamate and calcium cyclamate, *Arch. Int. Pharmacodyn. Ther., 163:* 302–340 (1966).
72. A. J. Collings and W. W. Kirkby, The toxicity of cyclohexylamine hydrochlo-

ride in the rat, 90-day feeding study, Unilever Research Laboratory, unpublished report, 1974.
73. I. F. Gaunt, M. Sharratt, P. Grasso, A. B. G. Lansdown, and S. D. Gangolli, Short-term toxicity of cyclohexylamine hydrochloride in the rat, *Food Cosmet. Toxicol., 12:* 609–624 (1974).
74. R. F. Crampton, British Industrial Biological Research Association, correspondence to H. Blumenthal, Food and Drug Administration, December 19, 1975.
75. P. L. Mason and G. R. Thompson, Testicular effects of cyclohexylamine hydrochloride in the rat, *Toxicology, 8:* 143–156 (1977).
76. H. Brune, U. Mohr, and R. P. Deutsch-Wenzel, Establishment of the no-effect dosage of cyclohexylamine hydrochloride in male Sprague-Dawley rats with respect to growth and testicular atrophy, unpublished report, 1978.
77. World Health Organization, 26th Report of the Joint FAO/WHO Expert Committee on Food Additives, Evaluation of certain food additives, April 19–28, 1982, Technical Report Series, 1982, p. 683.
78. M. H. Litchfield and A. A. B. Swan, Cyclohexylamine production and physiological measurements in subjects ingesting sodium cyclamate, *Toxicol. Appl. Pharmacol., 18:* 535–541 (1971).
79. A. J. Collings, Effect of dietary level and the role of intestinal flora on the conversion of cyclamate to cyclohexylamine, Unilever Research Laboratory, unpublished report, 1969.
80. U. Schmidt, Cyclamate conversion study in humans, Supplement to Food Additive Petition 2A3672, December 14, 1987.
81. H. J. Dengler, J. H. Hengstmann and H. Lydtin, Extent and frequency of conversion of cyclamate into cyclohexylamine in a German patient group, Supplement to Food Additive Petition 2A3672, December 14, 1987.
82. C. H. Kleinbloesem, P. van Brummelen, M. Dunhof, H. Faber, J. Urquhart, and D. D. Breimer, Rate of increase in the plasma concentration of nifedipine as a major determinant of its hemodynamic effects in humans, *Clin. Pharmacol. Ther., 41:* 26–30 (1987).
83. A. G. Renwick, N. E. Buss, and K. Donaldson, The formation and cardiovascular actions of cyclohexylamine in diabetic patients given cyclamate orally, Abstracts from IV World Conference on Clinical Pharmacology and Therapeutics, Mannheim-Heidelberg, July 1989.
84. Calorie Control Council, Worldwide Status of Cyclamate, unpublished material, 1989.

# 6

# Dihydrochalcone Sweeteners from Citrus Flavanones

**Robert M. Horowitz and Bruno Gentili*** *Agricultural Research Service, U.S. Department of Agriculture, Pasadena, California*

The first dihydrochalcones recognized as having useful sweetening properties were prepared 30 years ago from the bitter citrus flavanones naringin and neohesperidin (1–3). Since then, a very large number of variants of the original sweeteners have been synthesized. Some of these represent only small variations on the theme; others are more radical departures. All were made to gain a better understanding of taste–structure relations, or to improve taste quality, raise solubility, or ensure minimal toxicity. From a practical standpoint, it appears that even the best of the new derivatives are not significantly better than the original compounds. For this reason we shall focus primarily on three of the original compounds, all of which are derived from citrus flavanones: naringin dihydrochalcone (I), neohesperidin dihydrochalcone (II), and hesperetin dihydrochalcone 4′-β-D-glucoside (III). A further reason to concentrate on these compounds is that virtually all the toxicology studies of this group of sweeteners have been done on a single compound, neohesperidin dihydrochalcone (II), which is the only dihydrochalcone currently in use.

## ORIGIN AND PREPARATION

Dihydrochalcones are almost always prepared by catalytic reduction of the corresponding chalcones. These, in turn, are derived from flava-

*Retired.

nones by a ring-opening reaction that occurs in alkali. The process is shown in Figure 1.

In most cases the reactions proceed almost quantitatively, so that the overall yield from flavanone to chalcone to dihydrochalcone is high. As seen in Table 1, the flavanones used for preparing sweeteners I, II, and III (Fig. 2) are found in citrus fruit, where they are often major constituents of the peel. Two of the flavanones (IV and V) are glycosides that contain β-neohesperidose (2-*O*-α-L-rhamnopyranosyl-β-D-glucopyranose), whereas the third (VI) contains β-rutinose (6-*O*-α-L-rhamnopyranosyl-β-D-glucopyranose) (see Fig. 3). It has been shown (2,4,5) that many phenolic or flavonoid glycosides are taste-eliciting (bitter, sweet, or bittersweet) if their sugar component is β-neohesperidose or β-D-glucose, but tasteless if the sugar component is β-rutinose. Because the β-rutinosyl group abolishes taste, it is essential to alter the group in some way if a taste-eliciting substance is to be obtained. One way to do this is by converting the β-rutinosyl group to the β-glucoside through hydrolytic removal of rhamnose. Thus, hesperidin (VI), which is tasteless and yields a tasteless dihydrochalcone, can be converted to the sweet dihydrochalcone III by hydrolytic cleavage of rhamnose from hesperidin dihydrochalcone. More detailed descriptions of the laboratory preparation of sweeteners I–III are given below.

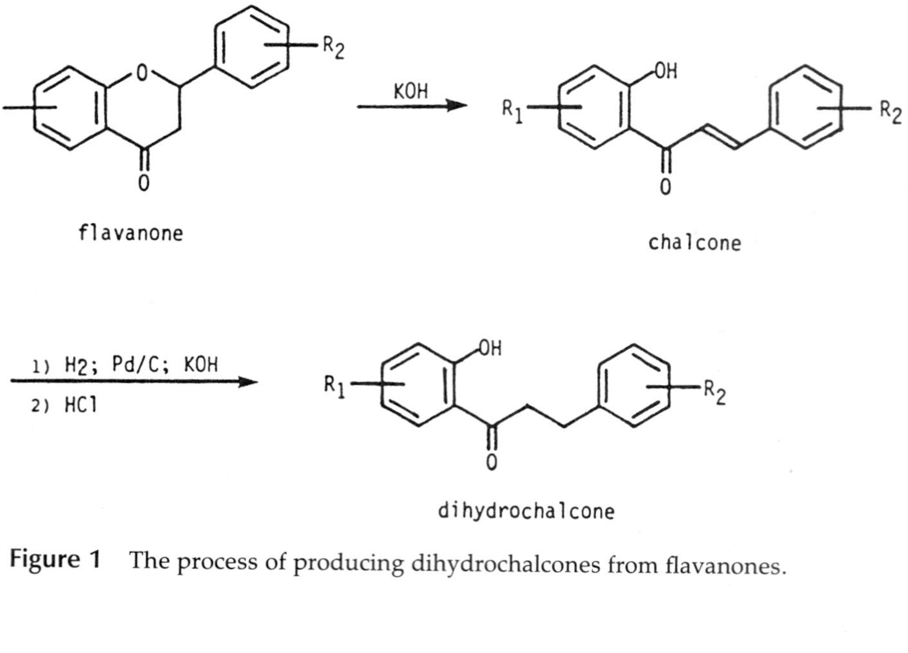


**Figure 1** The process of producing dihydrochalcones from flavanones.

**Table 1** Dihydrochalcone Sweeteners and Their Sources

| Sweetener | Starting flavanone | Flavanone source |
|---|---|---|
| Naringin dihydro-chalcone (I) | Naringin (IV) (bitter) | *Citrus paradisi* (grapefruit) |
| Neohesperidin dihydrochalcone (II) | Neohesperidin (V) (bitter) | *Citrus aurantium* (bitter orange) |
| Hesperetin dihydrochalcone 4′-β-D-glucoside (III) | Hesperidin (VI) (tasteless) | *Citrus sinensis* (sweet orange) |
| | | *Citrus limon* (lemon) |

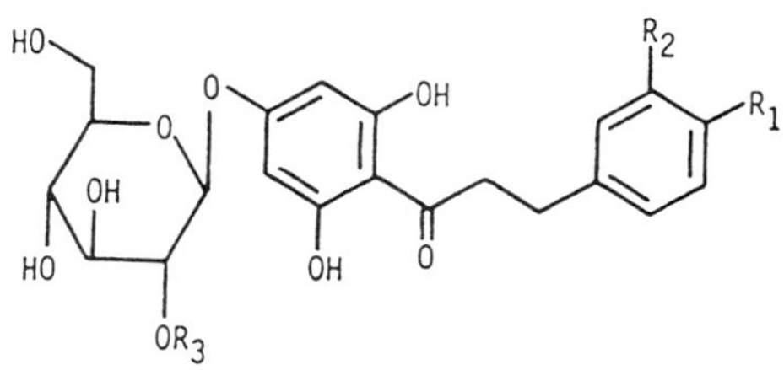

I: $R_1$ = OH; $R_2$ = H; $R_3$ = α-L-Rhamnosyl

II: $R_1$ = $OCH_3$; $R_2$ = OH; $R_3$ = α-L-Rhamnosyl

III: $R_1$ = $OCH_3$; $R_2$ = OH; $R_3$ = H

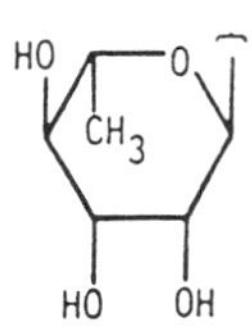

α-L-Rhamnosyl

**Figure 2** The chemical structures of naringin dihydrochalcone (I), neohesperidin dihydrochalcone (II), and hesperetin dihydrochalcone 4′-β-D-glucoside (III).

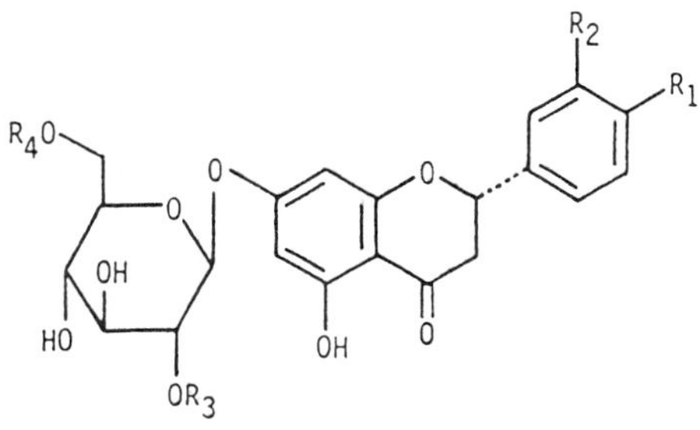


IV: $R_1$ = OH; $R_2$ = $R_4$ = H; $R_3$ = α-L-Rhamnosyl

V: $R_1$ = $OCH_3$; $R_2$ = OH; $R_3$ = α-L-Rhamnosyl; $R_4$ = H

VI: $R_1$ = $OCH_3$; $R_2$ = OH; $R_3$ = H; $R_4$ = α-L-Rhamnosyl

**Figure 3** The chemical structures of naringin (IV), neohesperidin (V), and hesperidin (VI).

### Naringin Dihydrochalcone (I) (1)

The starting material, *naringin,* is available in quantity from commercial sources. Gram amounts can be obtained by extracting grapefruit peel albedo in a blender for a few seconds with a small volume of cold aqueous 50% methanol. The filtered extract is kept in a refrigerator for several days until crystallization is complete. Recrystallized from water, naringin melts at about 80°C, changing to 171°C after drying in a vacuum oven.

To prepare *naringin dihydrochalcone,* recrystallized naringin (110 g) is dissolved in cold 10% aqueous potassium hydroxide (61 g KOH in 549 ml water) to which is added 10% palladium-carbon hydrogenation catalyst (10 g). The mixture is hydrogenated at room temperature at about 30 lb/in$^2$ hydrogen in a Parr shaker or similar apparatus until the uptake of hydrogen stops (usually < 2 hr). The catalyst is removed by filtration, and the filtrate is placed in an ice bath, acidified with concentrated hydrochloric acid, and kept in a refrigerator until crystallization is complete. The crystals of naringin dihydrochalcone are collected, water-washed, and recrystallized from water, thereby obtaining 104 g (94%) after vacuum drying (melting point, 170°C). Further recrystallization from acetone gives a different crystalline form (melting point, 212°C; yield, 98 g).

### Neohesperidin Dihydrochalcone (II) (1,6,7)

Neohesperidin can either be extracted from citrus peel or synthesized from naringin. The best natural source is the bitter (or Seville) orange

(*Citrus aurantium*). Taxonomists have lumped together a number of diverse, chemically disparate bitter oranges under the banner of *C. aurantium*, and some of these oranges are of little value as a source of neohesperidin. The more useful varieties contain either neohesperidin and naringin, or neohesperidin and hesperidin as the major flavanones. An alcohol extract of small, immature fruit is likely to give the best yield of neohesperidin, which can be separated from naringin or hesperidin by taking advantage of the fact that neohesperidin is less water-soluble than naringin but more water-soluble than hesperidin.

In Spain neohesperidin is available in quantity from Seville oranges, but these are a relatively minor crop in most citrus-growing regions of the world. In the absence of natural neohesperidin, large quantities can be prepared by the conversion of naringin (IV). The reactions are shown in Figure 4. Phloracetophenone 4′-β-neohesperidoside (VII) (5) is a useful intermediate for the preparation of substituted flavanones and their derivatives. It is made by heating a solution of naringin (17 g), water (175 ml), and potassium hydroxide (50 g) under reflux for 3.5 hr in a nitrogen atmosphere. The chilled solution, when neutralized, yields VII as a pale yellow precipitate, which is then recrystallized twice from water to give colorless needles (melting point, 164–165°C; yield, 6.6 g). Brought out of acetone, it yields a different crystalline form (melting point, 256–257°C).

Neohesperidin can be obtained from the condensation of VII with isovanillin (3-hydroxy-4-methoxybenzaldehyde). In a typical experiment, VII (5.64 g) is added to a hot solution of potassium hydroxide (12 g) in water (12 ml), and the whole is placed in a bath of boiling water. Isovanillin (2.82 g) is added in portions and heating continued for about 7 min. The solution is poured onto ice (70 g), the pH adjusted to 5.7, and the mixture kept overnight in an ice bath. The precipitate that forms is mainly *neohesperidin chalcone*, which is collected, water-washed and air-dried. In order to cyclize it to the flavanone, it is redissolved in water (70 ml) and kept first at 80–90°C (30 min), then at 40°C (3 hr). Pure neohesperidin (V)

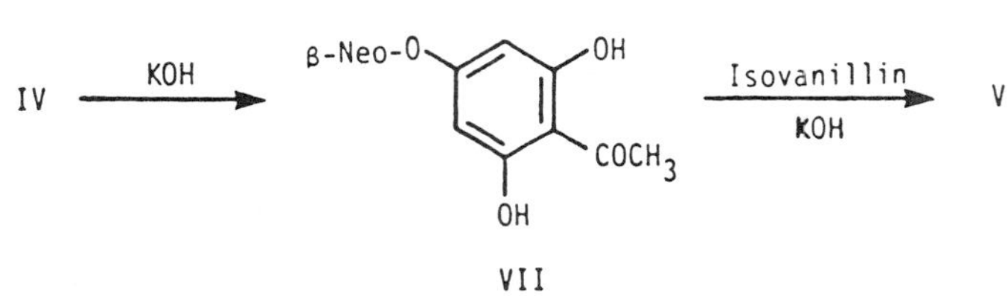


**Figure 4** The conversion of naringin to neohesperidin.

crystallizes from the 40° solution (melting point, 241–242°C; yield, 3.17 g (44%) after vacuum drying). The filtrate affords a small quantity of unreacted VII when cooled in an ice bath.

The conversion of neohesperidin to neohesperidin dihydrochalcone (II) is carried out in the same way as the conversion of naringin to its dihydrochalcone. Thus, neohesperidin (250 g), 10% aqueous potassium hydroxide (1250 ml), and 10% palladium-carbon (23 g) is hydrogenated, filtered, acidified, and diluted with water to a volume of 2900 ml. After keeping the mixture for several days in a refrigerator, the product is collected, water-washed, and recrystallized from water (1000 ml) (yield, 206 g (82%) after drying; melting point, 150°C). Recrystallized from methanol, the sweetener has a melting point of 172–174°C; the yield is 199 g. The final recrystallization from methanol is recommended for obtaining maximum purity.

A detailed and useful description has been published by Robertson and coworkers (8) of the large-scale preparation of II (100 kg quantities, including nuclear magnetic resonance (NMR) analysis of the products). The patent literature (9,10) claims the conversion of naringin to VII and II in much higher yields than reported here; we have been unable to duplicate these claims.

A novel biotechnical approach to the synthesis of neohesperidin dihydrochalcone is currently being explored by Jonathan Gressel and coworkers at the Weizmann Institute in Israel (11–13). They have discovered that cell-free extracts of young pummelo leaves (*C. maxima*) contain enzymes that are able to catalyze (1) the conversion of UDP-glucose to UDP-rhamnose and (2) the transfer of rhamnose from UDP-rhamnose to the 2-position of the glucose in either naringenin 7-β-D-glucoside (prunin) or hesperetin 7-β-D-glucoside. In other words, the flavanone 7-β-D-glucosides can be converted in one step to the corresponding flavanone 7-β-neohesperidosides, naringin (IV), and neohesperidin (V), respectively. It is interesting that the rhamnose-transferring enzyme cannot use hesperetin dihydrochalcone 4′-β-D-glucoside (III) as a substrate. The properties of this rather unstable enzyme have been studied and efforts to purify it further are underway. Gressel et al. have pointed out that if the gene for the enzyme could be found, it might, at least in principle, be possible to reintroduce it into cultured citrus cells, which could then be used to produce neohesperidin (V) (and, hence, neohesperidin dihydrochalcone (II)) from hesperetin 7-β-D-glucoside. The latter compound would be obtained by partially hydrolyzing hesperidin (VI) to remove the rhamnose attached at position 6 of the glucose. Hesperidin, a readily available byproduct of the orange processing industry, has had few significant uses.

### Hesperetin Dihydrochalcone 4′-β-D-Glucoside (HDG) (III) (14)

*Hesperidin* (50.0 g) in 10% aqueous potassium hydroxide (250 ml) is allowed to stand for 30 min at room temperature and then is hydrogenated (1.5 hr) using 10% palladium-carbon catalyst (4.0 g). (Crude commercial hesperidin can be used, but the recrystallized compound is preferable.) The reaction mixture is filtered and the pH adjusted to 7 by addition of hydrochloric acid. The neutralized solution of *hesperidin dihydrochalcone* is diluted with water to 600 ml. To this is added concentrated (39%) hydrochloric acid (5 ml) and the solution brought rapidly to a boil and kept under reflux for 2.5–2.75 hr. Paper chromatography (Whatman No. 1, 10% aqueous acetic acid) now shows the virtual absence of hesperidin dihydrochalcone ($R_f$ = 0.59) and the presence of an approximately 1:1 mixture of HDG (III) ($R_f$ = 0.44) and its aglycone, hesperetin dihydrochalcone ($R_f$ = 0.18). The cooled oily reaction mixture is extracted with ether (4 × 100 ml), evaporation of which yields hesperetin dihydrochalcone (16.0 g; melting point, 195–198°C from ethanol-water). The raffinate (still somewhat oily) crystallizes easily in the refrigerator to give pure hesperetin dihydrochalcone 4′-β-D-glucoside (III) (10.3 g); an additional quantity (1.5 g) can be extracted from the aqueous filtrate with ethyl acetate (total yield 31%). HDG forms colorless needles from ethanol (melting point, 119–121°C). The hydrolytic removal of rhamnose can be achieved in higher yield, but less conveniently, by using the enzyme α-L-rhamnosidase (15,16).

## STRUCTURE

If we are ever to understand why sweet compounds are sweet, it will first be necessary to have a detailed, intimate knowledge of the structures of the compounds in question. With this in mind, Wong and Horowitz (17) studied the structure of neohesperidin dihydrochalcone (II) by means of single crystal X-ray analysis. Suitable crystals of II were obtained by slow evaporation of a solution made up in methanol–water. The main features of the crystal structure were found to be: (1) the rhamnosyl and glucosyl moieties exist in chair conformations; (2) both aromatic rings are essentially planar; (3) the overall shape of the aglycone moiety, though somewhat zigzag, is essentially linear; and (4) the compound occurs as two independent conformers, A and B, in one of which the isovanillyl ring is rotated approximately 90° with respect to the isovanillyl ring of the other conformer. The molecular structures and absolute configurations of conformers A and B are shown in the stereoscopic drawings in Figure 5 and the superimposed conformers in Figure 6.

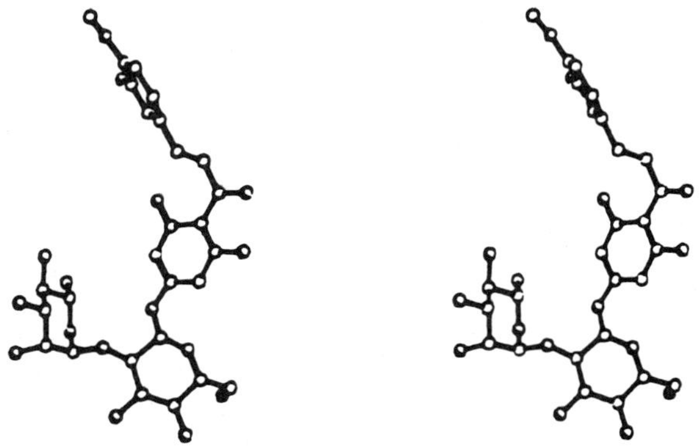

(A)

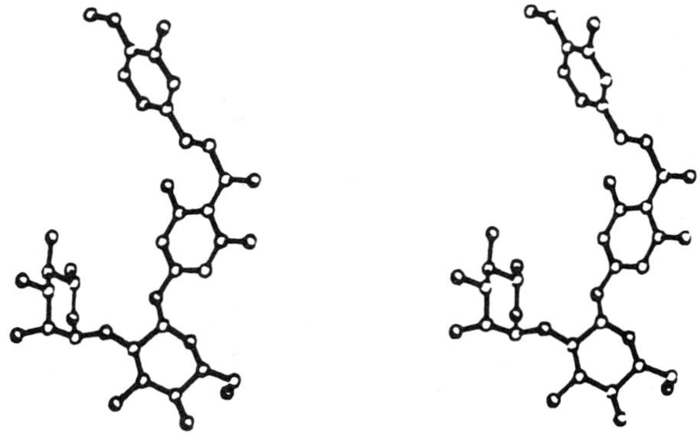

(B)

**Figure 5** Stereoscopic views of neohesperidin dihydrochalcone conformers, A and B.

The gross structural features of the two molecules, A and B, are closely similar. The entire neohesperidin dihydrochalcone molecule is more or less 'J'-shaped with the $\beta$-neohesperidosyl residue forming the curved part of the 'J' and the hesperetin dihydrochalcone aglycone forming the linear segment. Assuming that the solid-state conformers are

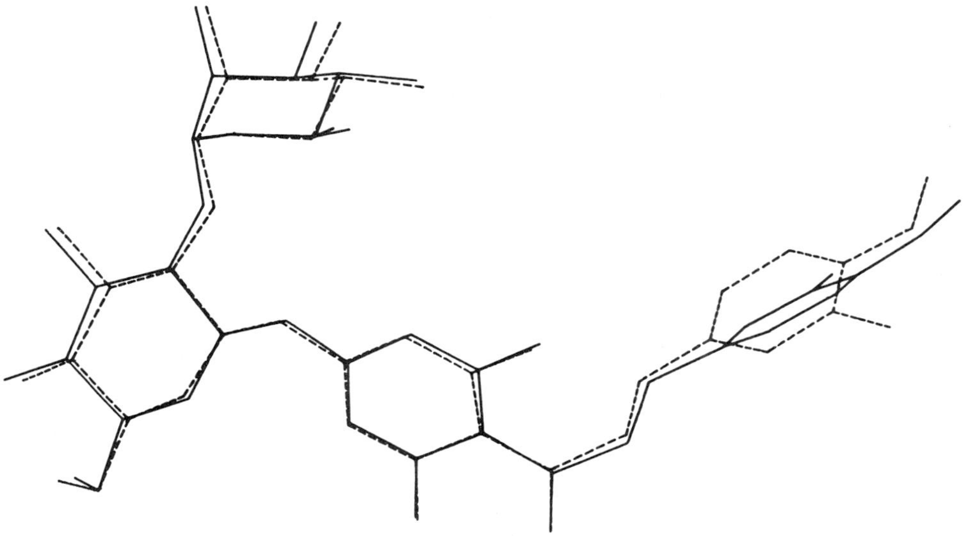

**Figure 6** Neohesperidin dihydrochalcone conformers A and B superimposed. (The sugars are at the left.)

similar to those that exist in solution, the X-ray studies contradict earlier suggestions (18–20) that the "active" (i.e., sweet) conformation of dihydrochalcones is one in which the aglycone moiety is bent or folded in such a way that the planes of the two aromatic rings are approximately at right angles to each other.

## PROPERTIES

### Taste

The dihydrochalcones described here, and the many others that have been synthesized, have most of the characteristics common to high-potency sweeteners: slow onset of sweetness, a lingering aftertaste (which may or may not be a fault, depending on the application), and an apparent decrease in sweetening power relative to that of sucrose as concentration increases. Among the special virtues of dihydrochalcones are their pleasant, agreeable sweetness, lack of bitter aftertaste, and ability (in varying degrees) to decrease the perception of bitterness in beverages and pharmaceuticals.

Several studies of the sweetness intensity of dihydrochalcones have been published. The results of Guadagni and co-workers (21) are shown in Table 2. At or near threshold, neohesperidin dihydrochalcone (II) is

**Table 2** Sucrose Sweetness and Dihydrochalcone Sweetness Compared[a]

| Sweeteners | Threshold concentrations (mg/kg) | Isosweet concentrations (mg/kg) | | | | | |
|---|---|---|---|---|---|---|---|
| Sucrose | 1300 | 5,000 | 10,000 | 20,000 | 25,000 | 50,000 | 100,000 |
| Neohesperidin dihydrochalcone (II) | 0.7 (1857) | 5.4 (927) | 15.0 (667) | 50 (400) | 67.9 (368) | 200 (250) | 600 (167) |
| HDG (III) | 4.5 (289) | 25.9 (193) | 74.1 (135) | 180 (111) | 250 (100) | 641 (78) | — |
| Naringin dihydrochalcone (I) | 4.5 (289) | 32.1 (156) | 98.0 (102) | 289 (69) | 397 (63) | 1000 (50) | — |

[a]In water solution. Numbers in parentheses are (sucrose)/(DHC) ratios.
*Source:* Adapted from Ref. 21.

about 1800 times sweeter than sucrose on a weight basis. As concentration increases, the sweetness of II decreases relative to that of sucrose, so that at the 5% sucrose level (50,000 ppm), II is about 250 times sweeter. The same downward trend is observed for naringin dihydrochalcone (I) and HDG (III), though the relative sweetness of III declines less rapidly. The relation between sweetness and concentration (in milligrams per kilogram) can be approximated (21) by the function $S = Kc^a$, where $S$ is the concentration of sucrose, $c$ is the concentration of dihydrochalcone that has the same level of sweetness, and $K$ and $a$ are constants. For I, II, and III, $K$ has the values 532, 1703, and 456, respectively; the values for $a$ are 0.639, 0.644, and 0.734.

In other studies, DuBois et al. reported the sweetness of neohesperidin dihydrochalcone (II) to be 663 (22) and 612 (18) times greater than that of sucrose at the 8.5% sucrose level, but later revised the figure to 340 (23). The last value can be compared with the results of Guadagni (21), from whose data we calculate that the sweetness of II would be about 196 times that of sucrose at the 8.5% level. These values agree reasonably well, considering the differences in taste panels and methods of testing.

On a mole for mole basis neohesperidin dihydrochalcone (II) is about 330 times sweeter than sucrose, naringin dihydrochalcone (I) about 490 times sweeter, and HDG (III) about 390 times sweeter. For comparison, saccharin near threshold is less than 300 times sweeter, and cyclamate only about 20 times sweeter.

The sweetness of dihydrochalcones is perceived more slowly and decays more slowly than that of sucrose or saccharin. According to the patent literature (24), the initial lag in perception of the sweetness can be overcome by admixture with β-gluconolactone. The lingering aftertaste is sometimes described as having a menthol or licorice-like quality. Even distilled water may seem sweet after tasting a dihydrochalcone.

DuBois and associates (23,25) attempted to estimate the length of time required for the sweetness to reach a maximum and then decay to a level of "faint." According to their data (Table 3), neohesperidin dihydrochalcone (II) is somewhat slower than sucrose or saccharin in reaching maximum intensity, and it persists a good deal longer. On the other hand, II is much closer to sucrose in temporal qualities than is, say, ammonium glycyrrhizinate, the sweet principle derived from licorice root. The data in Table 3 are approximations, and it is worth noting that the decay times are heavily dependent on the concentrations of the substances used. Qualitative observations suggest that the sweetness of HDG (III) is less persistent than that of neohesperidin dihydrochalcone (II).

Guadagni et al. (21) have studied the effect of dihydrochalcones in reducing the bitterness of the citrus bitter principles, limonin and naringin. The thresholds for detecting the bitterness of these substances are about 1 mg/kg and 20 mg/kg, respectively, in aqueous solution. At 1% sucrose equivalence, neohesperidin dihydrochalcone (II) increased the threshold for limonin to 1.4 mg/kg; HDG (III) increased it to 3.2 mg/kg. At 5% sucrose equivalence, the threshold for naringin was increased to 49 mg/kg and 56 mg/kg by II and III, respectively. Sucrose was much less effective than either of the dihydrochalcones, whereas naringin dihydrochalcone (I) not only failed to suppress limonin bitterness but

**Table 3** Temporal Characteristics of Sweeteners

| | Time (sec) for taste to | | | |
|---|---|---|---|---|
| | Reach a maximum | | Decay | |
| Compound | (23)[a] | (25)[a] | (23) | (25) |
| Sucrose | 5 | 4 | 24 | 13 |
| Saccharin | 5 | 4 | 42 | 14 |
| Neohesperidin dihydrochalcone (II) | 7 | 9 | 57 | 41 |
| Ammonium glycyrrhizinate | 23 | 16 | 133 | 69 |

[a]Data adapted from Refs. 23 and 25.

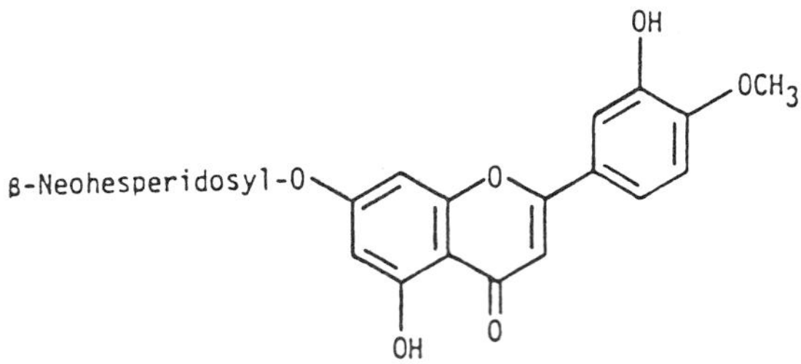


VIII

**Figure 7** Structure of neodiosmin.

may even have enhanced it. More effective than dihydrochalcones II or III is neodiosmin (VIII) (26–28) (Fig. 7), the tasteless flavone analog of neohesperidin (V). As little as 10 ppm of neodiosmin raised the threshold of limonin 3.5- to 4-fold. (This effect can be rationalized by assuming that neodiosmin competes for sites on the bitterness receptors, although it does not produce a taste response of its own.)

## Solubility and Stability

The solubility of dihydrochalcones I, II, and III in distilled water at room temperature is as follows:

Naringin dihydrochalcone (I): $1.6 \times 10^{-3}$ M; 0.94 g/liter
Neohesperidin dihydrochalcone (II): $8.1 \times 10^{-4}$ M; 0.50 g/liter, increasing to ~ 653 g/liter at 80°C (29)
HDG (III): $7.2 \times 10^{-4}$ M; 0.30 g/liter

The sodium and calcium salts of II have been described in the patent literature (30); their solubility is said to be > 1 g/ml.

The stability of neohesperidin dihydrochalcone at various pHs and temperatures has been studied (29,31). Aqueous solutions at room temperature are quite resistant to hydrolysis to the free sugars and the aglycone as long as the pH does not fall below 2. (Even if hydrolysis were to occur, there would not be complete loss of sweetness since the aglycone, hesperetin dihydrochalcone, is itself sweet, though not very soluble.) Buffered solutions between pHs 2 and 10 are stable for at least 8 hr at 100°C (29). In our laboratory, aqueous solutions have been kept for years exposed to light at room temperature; some yellowing occurs, but there is little perceptible change in sweetness.

### Caloric Value

We estimate the caloric value of neohesperidin dihydrochalcone (II) to be not more than ~2 cal/gram. This estimate is based on the assumption (thus far unproven) that the sugar residues are hydrolytically split and metabolized and that the aglycone is not extensively metabolized. Because of its high potency, II would probably afford not more than ~ 1/1000 as many calories as an equivalent amount of sucrose.

## APPLICATIONS

Dihydrochalcone sweeteners have been found to be useful in various foods, beverages, confections, and pharmaceuticals. Their long-lasting sweetness and stability make them especially well suited for chewing gum, candies, toothpastes, mouthwashes, and similar products. Compound II has been proposed as a noncariogenic, nonfermentable sweetening agent based on the finding that it is relatively inert to the action of cariogenic bacteria (32). The patent literature contains many examples of applications. Pratter (33) has pointed out that because of its high potency neohesperidin dihydrochalcone must be used judiciously in flavor formulations and that suitable bulking agents, such as sugar alcohols, are needed. Mixtures of dihydrochalcones with sucrose, nonsugar sweeteners, gluconolactone, and vanillin are also described in the patent literature; various advantages are claimed.

The ability of dihydrochalcones II and III to reduce bitterness as well as provide sweetness suggests that they might be used in the formulation of bitter drugs. A related application, which has been studied in detail by Guadagni (34), is the effect of adding small amounts of neohesperidin dihydrochalcone to grapefruit juice. For example, a batch of grapefruit juice containing 380 ppm naringin and 5.1 ppm limonin was treated with just enough neohesperidin dihydrochalcone (II) to give a concentration of 12 ppm. Before the addition of II, it was determined that 38% of the members of a taste panel disliked the juice, 33% liked it, and the remaining 29% neither liked nor disliked it. After the addition, only 5% disliked the juice, 76% liked it, and 19% neither liked nor disliked it. Thus, neohesperidin dihydrochalcone was effective in increasing the average hedonic rating, whereas adding either sucrose or aspartame caused no significant improvement. Further work showed that as little as 6–12 ppm of II significantly improved the acceptability of this lot of grapefruit juice, and that the improvement was due to reduction of bitterness as well as to increased sweetness. The addition of 18 ppm neohesperidin dihydrochalcone caused the juice to be less acceptable, probably because of excessive sweetness or aftertaste. In similar work carried out by Fellers (35), it was found that 8–12 ppm II was useful in improving the flavor of early-season Florida grapefruit juice.

## TOXICOLOGY AND REGULATORY STATUS

The toxicology of neohesperidin dihydrochalcone (II) has been investigated in detail. The most extensive studies are those of Gumbmann et al. (36) of the Western Regional Research Laboratory (WRRL) of the U.S. Department of Agriculture. Feeding studies requiring more than 400 pounds of the sweetener were carried out in laboratory animals for periods of up to 2 years.

In one series of experiments, II was fed to groups of rats for either 22–170 days at 5% of the diet, 1 year at 5%, or 11 months at 10%. Some growth inhibition and increased testes weight relative to body weight were observed in the animals kept on the 10% diet. In another trial, a 2-year carcinogenicity study in rats fed 0, 0.5, 2.5, or 5.0% II, no evidence was obtained of any increased incidence of tumors that could be associated with the ingestion of II. One effect uncovered in this work was a decrease in growth rate of the rats receiving 5% neohesperidin dihydrochalcone which, at the outset of the experiment, had been added to a basal diet containing minimal quantities of vitamins and minerals. When brewer's yeast and mineral supplements were added to this diet, a dramatic resumption in growth rate occurred, so that at 100 weeks there was no significant difference in body weight between the rats fed the supplemented diet and those in the control group. In a three-generation reproduction study, no detrimental effects due to the ingestion of neohesperidin dihydrochalcone were noted, nor were there any teratological effects.

In a second series of experiments at WRRL, neohesperidin dihydrochalcone was fed to young beagle dogs for a period of 2 years. The dogs—three of each sex per group—were given 0, 0.2, 1.0, or 2.0 g of sweetener per kilogram of body weight. (For the high-dose group, the concentration of II was estimated to be ~6% of the diet.) At the conclusion of the study, liver and thyroid weights were found to be elevated in the 2.0 g/kg group and mild thyroid hypertrophy and hyperplasia were detected in some of the dogs in this group. It was not clear, however, whether these changes were due to toxic effects or were merely reversible adaptive responses associated with the metabolism of large quantities of the dihydrochalcone. Testicular atrophy and degeneration occurred in one of three dogs in each of the two highest-dose groups; the incidence of testicular atrophy may have exceeded the normal expectation rate in beagles, although it is not possible from the available information to estimate this rate accurately.

Neohesperidin dihydrochalcone has been checked for mutagenicity in the Ames test; it is nonmutagenic, regardless of which of the various *Salmonella* tester strains is used (37–40). In mice, the compound causes no increase in the normal frequency of micronucleated polychromatic erythrocytes in bone marrow (41).

The metabolism of the unlabeled dihydrochalcones I and II has been examined in several species (42). As with other flavonoids, it appears that metabolism is brought about largely by the action of intestinal microflora. A major pathway is cleavage between the substituted phloroglucinal A-ring and the adjacent carbonyl group to yield a series of $C_3$-$C_6$ acids derived from the B ring. These include (1) *p*-hydroxyphenylpropionic acid and *p*-coumaric acid from naringin dihydrochalcone, and (2) isoferulic acid, *m*-hydroxyphenylpropionic acid, and *m*-hydroxycinnamic acid from neohesperidin dihydrochalcone. Not much is known about the fate of the A ring, though it is thought to be oxidized, possibly to carbon dioxide. Using [$^{14}$C]-neohesperidin dihydrochalcone labeled on the methylene group beta to the carbonyl, it has been shown (43) that, besides the acids already mentioned, rats produce several other major urinary metabolites, as yet unidentified. In rats, the radioactive metabolites from either $^{14}$C-labeled neohesperidin (V) or its $^{14}$C-labeled dihydrochalcone (II) appear to be remarkably similar on thin-layer chromatograms.

Excretion studies of the [$^{14}$C]-neohesperidin dihydrochalcone have also been carried out (36). When oral doses of up to 100 mg/kg body weight were administered to rats, more than 90% of the radioactivity was excreted in the first 24 hr, primarily in the urine. After 24 hr, only traces of radioactivity could be detected in various tissues.

In addition to long-term studies on II, several short-term feeding trials of naringin dihydrochalcone (I) and HDG (III) were undertaken by the group at the Western Regional Research Laboratory (44). Compound I was fed to rats at a level of 5% of the diet for 170 days, and compound III at 1% for 91 days. No abnormalities were detected, as judged by growth, feed intake, hematology, reproductive performance, organ weights, gross pathology, histopathology, photosensitivity, urine analyses, and water balance.

Most of the work outlined above was completed in the 1960s and 1970s. In 1985–86 a new, detailed study on the "subchronic (13 wk) oral toxicity of neohesperidin dihydrochalcone in rats" was carried out at the TNO-CIVO Toxicology and Nutrition Institute in the Netherlands (45). Compound II was fed to groups of 20 male and 20 female Wistar rats at dietary levels of 0, 0.2, 1.0, and 5.0% for 91 days. Only at the 5% level were there any treatment-related effects, i.e., marginal changes in body weight and food consumption, cecal enlargement, and slight changes in some of the clinical chemistry variables. These phenomena were judged to be of little, if any, toxicological significance. Neither the low- nor intermediate-dose groups showed any compound-related effects, and none of the groups showed any ophthalmoscopical, haematological, or histopathological changes. It was concluded that the intermediate dose, which translates to about 750 mg of neohesperidin dihydrochalcone per kg body weight per day, was the no-effect level. Based on these and the earlier results, the Scientific Committee for Food of the European Community recognized neohesperidin dihydrochalcone as "toxicologically acceptable" in 1987 and assigned it an ADI (acceptable daily intake) of 5 mg per kg body weight (46,47). This ADI is adequate for a wide range of uses.

Neohesperidin dihydrochalcone is currently allowed for use as a sweetener in several countries, including Belgium, where it may be added to chewing gum and some beverages. The currently pending European Community Council "directive on sweeteners for use in foodstuffs" would provide for neohesperidin dihydrochalcone's use throughout the European Community. The compound is listed in the National Food Codex of Argentina. It has not been cleared by the U.S. Food and Drug Administration, which, in 1979, took the position that the following additional toxicological tests would be needed: (1) a new rat study in which more animals would be used than in the original work; (2) a lifetime carcinogenicity study in mice; and (3) a complete metabolic study (48).

It is worth emphasizing that dihydrochalcones are typical examples of that large group of plant products known as flavonoids and defined by the presence of two $C_6$ aromatic rings joined by a $C_3$ bridge. Although the dihydrochalcones discussed in this chapter are not known to occur naturally, they are simply the reduced, open-chain analogs of the citrus flavanones naringin, neohesperidin, and hesperidin (the latter compound minus the rhamnose group). Dihydrochalcones have rarely been reported to occur in plant foods, possibly because until recently they were relatively difficult to detect; some *C*-glycosyldihydrochalcones do occur in kumquats (43).

Because flavonoids are produced by all higher plants, they are, and always have been, a common constituent of the diet. According to present knowledge, the citrus-based dihydrochalcone sweeteners are nonmutagenic, noncarcinogenic, and noncariogenic; no clearly defined toxic effects have been attributed to them; they are excreted rapidly from the body; their metabolism seems to be closely parallel to that of the naturally occurring parent flavanones; and, lastly, their high potency ensures that only minor quantities would be needed in any foreseeable applications. Thus, on toxicological grounds, there should be no serious impediment to their use as sweeteners.

## ACKNOWLEDGMENT

We thank Dr. M. R. Gumbmann for critically reviewing the section on toxicology and Drs. M. Berhow (USDA) and F. Borrego (Zoster, S.A., Zeneta-Murcia, Spain) for providing us with several references. We thank Ruth Olson for invaluable help in preparing the manuscript.

### REFERENCES

1. R. M. Horowitz and B. Gentili, Dihydrochalcone derivatives and their use as sweetening agents, U.S. Patent 3,087,821 (April 30, 1963).
2. R. M. Horowitz, Relations between the taste and structure of some phenolic glycosides, in *Biochemistry of Phenolic Compounds,* (J.B. Harborne, ed.) Academic Press, New York, 1964, pp. 545–571.
3. R. M. Horowitz and B. Gentili, Taste and structure in phenolic glycosides, *J. Agric. Food Chem., 17:* 696–700 (1969).
4. R. M. Horowitz and B. Gentili, Phenolic glycosides of grapefruit: A relation between bitterness and structure, *Arch. Biochem. Biophys., 92:* 191–192 (1961).
5. R. M. Horowitz and B. Gentili, Flavonoids of citrus. VI. The structure of neohesperidose, *Tetrahedron, 19:* 773–782 (1963).
6. R. M. Horowitz and B. Gentili, Conversion of naringin to neohesperidin and neohesperidin dihydrochalcone, U.S. Patent 3,375,242 (March 26, 1968).
7. L. Krbechek, G. Inglett, M. Holik, B. Dowling, R. Wagner, and R. Riter. Dihydrochalcones: Synthesis of potential sweetening agents, *J. Agric. Food Chem., 16:* 108–112 (1968).
8. G. H. Robertson, J. P. Clark, and R. Lundin, Dihydrochalcone sweeteners: Preparation of neohesperidin dihydrochalcone, *Ind. Eng. Chem., Prod. Res. Devel., 13:* 125–129 (1974).
9. K. Veda and H. Odawara, Neohesperidin from naringin, Jpn. Kokai 75, 154,261 (December 12, 1975) [*Chem. Abstr., 84:* 180574 (1976)].
10. K. Veda and H. Odawara, Hydrolysis of naringin to phloracetophenone 4′-$\beta$-neohesperidoside, Jpn. Kokai 75,149,635 (Nov. 29, 1975) [*Chem. Abstr., 84:* 180577 (1976)].
11. E. Lewinsohn, H. Gavish, E. Berman, R. Fluhr, Y. Mazur, and J. Gressel, Towards biotechnologically synthesized neohesperidin for the production of its dihydrochalcone sweetener, International Conference on Sweeteners. American Chemical Society, Los Angeles, CA, September 22–25, 1988, Abstract No. 39.
12. E. Lewinsohn, L. Britsch, Y. Mazur, and J. Gressel, Flavanone glycoside biosynthesis in Citrus, *Plant Physiol., 91:* 1323–1328 (1989).

13. H. Gavish, E. Lewinsohn, A. Vardi, and R. Fluhr, Production of flavanone neohesperidosides in citrus embryos, *Plant Cell Reports, 8:* 391–394.
14. R. M. Horowitz and B. Gentili, Preparation of hesperetin dihydrochalcone glucoside, U.S. Patent 3,429,873 (Feb. 25, 1969).
15. R. M. Horowitz and B. Gentili, Enzyme preparation of hesperetin dihydrochalcone glucoside, U.S. Patent 3,583,894 (June 8, 1971).
16. V. Krasnobaev, Immobilized hesperidinase in hesperetin dihydrochalcone glucoside manufacture, Ger. Offen. 2,402,221 (August 29, 1974) [*Chem. Abstr., 82:* 15272 (1975)].
17. R. Y. Wong and R. M. Horowitz, The X-ray crystal and molecular structure of neohesperidin dihydrochalcone sweetener, *J. Chem. Soc. Perkin Trans. I:* 843–848 (1986).
18. G. E. DuBois, G. A. Crosby, R. A. Stephenson, and R. E. Wingard, Jr., Dihydrochalcone sweeteners: Synthesis and sensory evaluation of sulfonate derivatives, *J. Agric. Food Chem., 25:* 763–771 (1977).
19. G. A. Crosby, G. E. DuBois, and R. E. Wingard, The design of synthetic sweeteners, in *Drug Design,* Vol. 8 (E. J. Ariëns, ed.), Academic Press, New York, 1979, pp. 215–310.
20. J. E. Hodge and G. E. Inglett, Structural aspects of glycosidic sweeteners containing (1′-2)-linked disaccharides, in *Symposium: Sweeteners* (G. E. Inglett, ed.), Avi Publishing Co., Inc., Westport, CT, 1974, pp. 216–234.
21. D. G. Guadagni, V. P. Maier, and J. H. Turnbaugh, Some factors affecting sensory thresholds and relative bitterness of limonin and naringin, *J. Sci. Food Agric., 25:* 1199–1205 (1974).
22. G. E. DuBois, G. A. Crosby, and P. Saffron, Nonnutritive sweeteners: Taste-structure relationships for some new simple dihydrochalcones, *Science, 195:* 397–399 (1977).
23. G. E. DuBois, G. A. Crosby and R. A. Stephenson, Dihydrochalcone sweeteners. A study of the atypical temporal phenomena, *J. Med. Chem., 24:* 408–428 (1981).
24. L. Givaudan et Cie S.A., Sweetener composition. Netherlands Patent Appl. 74 12,072 (April 8, 1975) [*Chem. Abstr.,* 83: 204966 (1975)].
25. G. E. DuBois and J. F. Lee, A simple technique for the evaluation of temporal taste properties, *Chem. Senses, 7:* 237–247 (1983).
26. D. G. Guadagni, V. P. Maier, and J. G. Turnbaugh, Effect of neodiosmin on threshold and bitterness of limonin in water and orange juice, *J. Food Sci., 41:* 681–684 (1976).
27. D. Guadagni, R. M. Horowitz, B. Gentili, and V. P. Maier, Method of reducing bitterness in citrus juices, U.S. Patent 4,031,265 (July 21, 1977).
28. D. Guadagni, R. M. Horowitz, B. Gentili, and V. P. Maier, Method of reducing bitterness and off-aftertaste, U.S. Patent 4,154,862 (May 15, 1979).
29. G. A. Crosby and T. E. Furia, New sweeteners, in *CRC Handbook of Food Additives,* 2nd ed., Vol. II (T. E. Furia, ed.), CRC Press, Inc., Boca Raton, FL, 1980, p. 204.
30. E. B. Westall and A. W. Messing, Salts of dihydrochalcone derivatives and their use as sweeteners, U.S. Patent 4,031,260 (June 21, 1977).

31. G. E. Inglett, L. Krbechek, B. Dowling and R. Wagner, Dihydrochalcone sweeteners—sensory and stability evaluation, *J. Food Sci., 34:* 101–103 (1969).
32. C. W. Berry and C. A. Henry, Baylor College of Dentistry, Meeting of the American Association for Dental Research, as reported in *Food Chemical News*, p. 25 (March 21, 1983).
33. P. J. Pratter, Neohesperidin dihydrochalcone, *Perfum. Flavor, 5:* 12–18 (1981).
34. D. G. Guadagni, unpublished data.
35. P. J. Fellers, The possible use of neohesperidin dihydrochalcone as sweetener in grapefruit juice, *Abstracts of the 27th Annual Citrus Processors' Meeting*, Lakeland, FL, September 30, 1976, p. 13.
36. M. R. Gumbmann, D. H. Gould, D. J. Robbins, and A. N. Booth, Toxicity studies of neohesperidin dihydrochalcone, *Proceedings, Sweeteners and Dental Caries* (J. H. Shaw and G. G. Roussos, eds.), Information Retrieval, Inc., Washington, D.C., 1978, pp. 301–310.
37. R. P. Batzinger, S.-Y. L. Ou, and E. Bueding, Saccharin and other sweeteners: Mutagenic properties, *Science, 198:* 944–946 (1977).
38. J. P. Brown, P. S. Dietrich, and R. J. Brown, Frameshift mutagenicity of certain naturally occurring phenolic compounds in the *Salmonella*/microsome test: Activation of anthraquinone and flavonol glycosides by gut bacterial enzymes, *Biochem. Soc. Trans., 5:* 1489–1492 (1977).
39. J. T. MacGregor and L. Jurd, Mutagenicity of plant flavonoids: Structural requirements for mutagenic activity in *Salmonella typhimurium. Mutation Res., 54:* 297–309 (1978).
40. J. B. Brown and P. S. Dietrich, Mutagenicity of plant flavonols in the *Salmonella*/mammalian microsome test: Activation of flavonol glycosides by mixed glycosidases from rat cecal bacteria and other sources, *Mutation Res., 66:* 223–240 (1979).
41. J. T. MacGregor, Mutagenicity studies of flavonoids *in vivo* and *in vitro, Toxicol. Appl. Pharmacol., 48:* A47 (1979).
42. A. N. Booth, D. J. Robbins and W. E. Gagne, unpublished data.
43. B. Gentili, unpublished data.
44. A. N. Booth and D. J. Robbins, Dihydrochalcone subacute toxicity study in rats, U. S. Department of Agriculture, Western Regional Research Laboratory, Report CA 74-18, June, 1968.
45. B. A. R. Lina, H. C. Dreef-van der Meulen, and D. C. Leegwater, Subchronic (13 wk) oral toxicity of neohesperidin dihydrochalcone in rats, *Food Chem. Toxicol., 28:* 507–513 (1990).
46. Report of the Scientific Committee for Food on Sweeteners, Commission of the European Communities (CS/EDUL/-Final), December 11, 1987.
47. A. Bär, F. Borrego, O. Benavente, J. Castillo, and J. A. del Rio, Neohesperidin dihydrochalcone: properties and applications, *Food Science and Technology* (*Lebensm. Wiss. u Techol.*)*, 23:* 371–376 (1990).
48. Donald Kennedy, Food and Drug Commissioner, letter dated January 15, 1979, as reported in *Food Chemical News*, January 22, 1979, p. 18.

# 7

# L-Sugars: Lev-O-Cal™

**Gilbert V. Levin and Lee R. Zehner** *Biospherics Inc., Beltsville, Maryland*

## INTRODUCTION

The L-sugars comprising Lev-O-Cal™ are simple 6-carbon sugars (hexose monosaccharides). They are L-sugars by virtue of having left-handed molecular configurations at the fifth carbon atom. They offer the prospect of providing a low-calorie, true sugar flavor sweetener and bulking agent.

On April 14, 1981, U.S. Patent No. 4,262,032 (1) for use of the L-sugars L-allose, L-altrose, L-fructose, L-galactose, L-glucose, L-gulose, L-idose, L-psicose, L-tagatose, and L-talose as low-calorie sweeteners in foods, beverages, and drugs was awarded and assigned to Biospherics Inc. Corresponding patents in a number of foreign countries where obesity or sugar-implicated diseases are significant problems have also been filed.

## BACKGROUND

The natural occurrence of L-sugars is rare, although they have been variously reported as minor unquantitated constituents in natural products. Until recently, L-sugars used in research were chemically synthesized using processes that result in equal amounts of L- and D-forms (2–13). When the L/D mixtures were fed to bacteria, the D- form was consumed, but the L-form was left intact. Despite this easily demonstrated biochemical difference, with its indication that L-sugars might be noncaloric, apparently no one thought to use them as sweeteners. This was probably because of their scarcity and, also, the possibility that they would not taste sweet. However, L-sugars have long been used as chemical markers and for other research purposes.

The structural aspect of importance in L-sugars can best be described by beginning with the carbon atom, generally represented as the center of a tetrahedron with four bonding arms, each extending outward from the center to the four points (Fig. 1). Numerous other atoms or groups of atoms can bond with the carbon atom (Fig. 2). Considering two carbon atoms side by side, each bonding with an atom A at all four of its bonding sites, it can be seen that the resulting structures are identical (Fig. 3). If each carbon atom bonds with three As and a B in the manner shown in Figure 4, they may, at first glance, appear to differ. However, by rotating one of the carbon atoms, the molecules can be superimposed and become identical (Fig. 5). The same identity prevails if the carbon atoms form four bonds with four groups of three different types (e.g., ABCA, ABCB, or ABCC) regardless of which bonds are made with which types. However, if all four bonded groups are different (ABCD), then the configuration cannot be rotated into superposition (Fig. 6). If

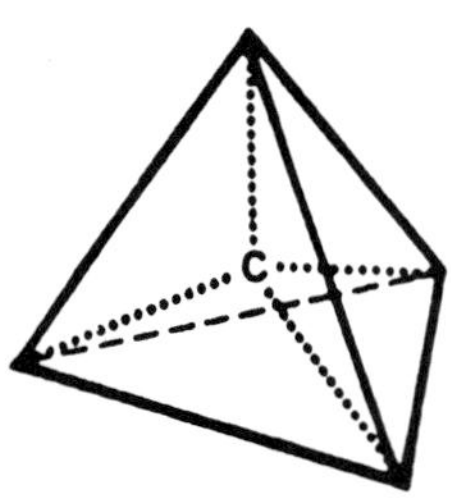


**Figure 1** Carbon atom.

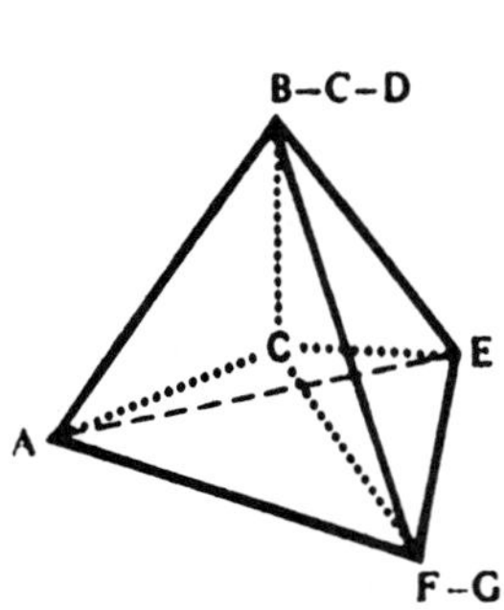


**Figure 2** Carbon molecule.

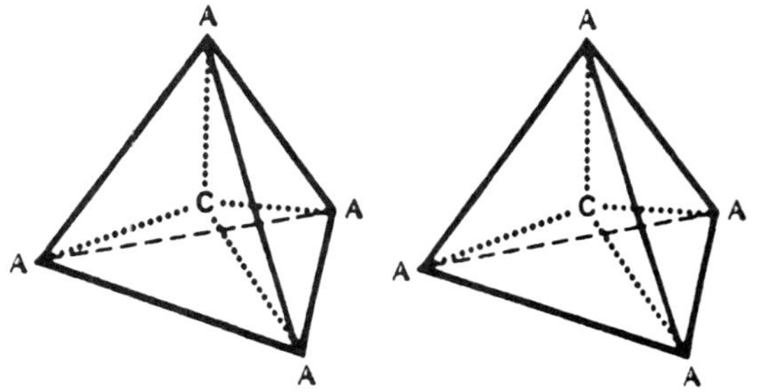


Figure 3 Identical carbon molecules.

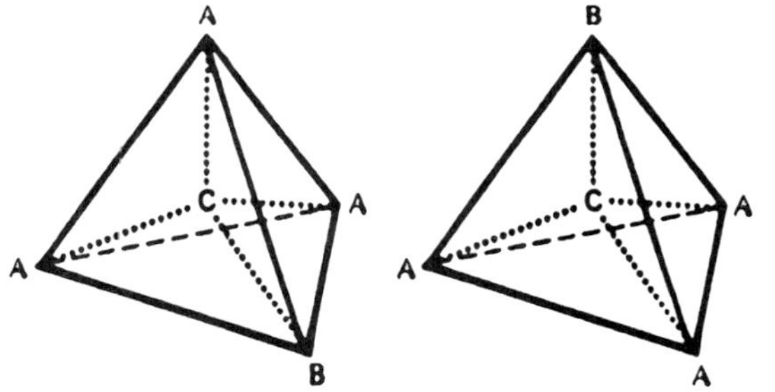


Figure 4 Two apparently different carbon molecules.

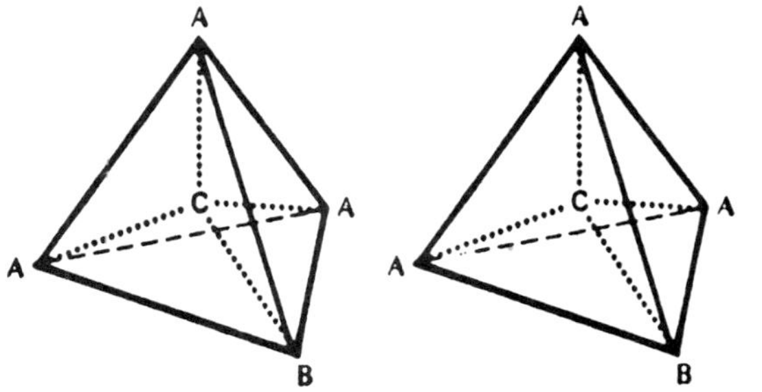


Figure 5 Superimposable mirror images.

these groups are rotated to place any two of the bonds in identical positions, the remaining two will be reversed. These configurations are thus nonsuperimposable mirror images of each other, as are our two hands, and so these compounds have come to be called "left-handed" and "right-handed" sugars. According to the universally accepted system of nomenclature, the configurations are termed "levo" and "dextro" and abbreviated L- and D- (the letters are distinct from the lowercase prefixes l- and d-, or the symbols + and −, used to indicate the direction of rotation of polarized light passing through a sugar solution).

The asymmetric carbon structures of glucose, i.e., dextro-glucose (or D-glucose) and levo-glucose (or L-glucose), are shown in Figure 7. Having the same constituents, mirror-image molecules behave almost identically in chemical reactions. However, in the realm of biochemistry, this similarity may not hold true. In biochemistry, enzymes play an essential role by promoting reactions that would otherwise not occur. This role

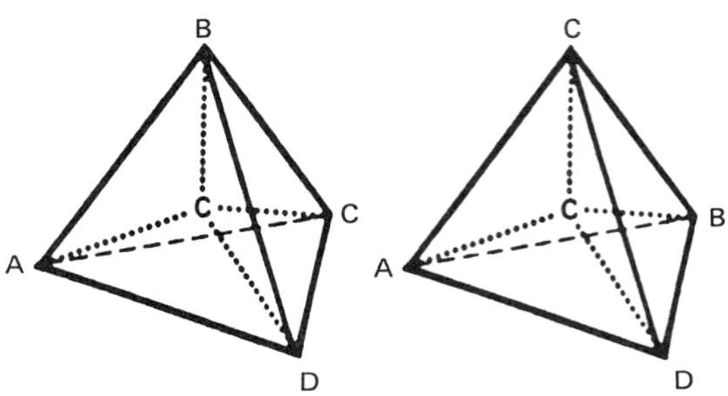


**Figure 6** Nonsuperimposable mirror images

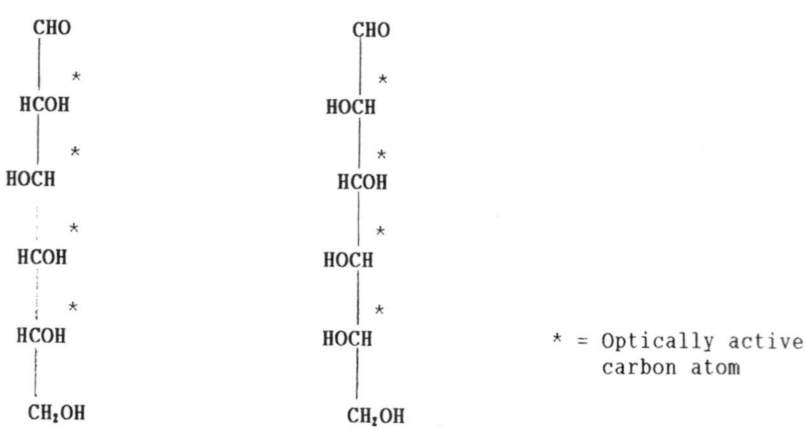


(a) (b)

**Figure 7** Asymmetric carbon structures of glucose: (a) dextro-glucose (b) levo-glucose.

requires that the enzymes physically fit the shapes of the reacting chemicals in order to bring them together for the reaction to take place. Hence, left-handed (L) sugars cannot fit the enzymes required for sugar metabolism (14) and, extrapolated to humans, should not give us their calories. This difference in caloric availability has been recognized since the time of Pasteur, and more recent findings (15) have supported the nonmetabolizable character of L-sugars.

## PHYSICAL AND CHEMICAL PROPERTIES

Differing only because of their mirror image relationship, the L- and D-forms of a particular sugar have identical physical characteristics, such as melting point, solubility, viscosity, texture, hygroscopicity, density, color, and appearance.

Chemical properties of the L- and D-forms in symmetrical (nonbiological) environments are likewise identical. For example, thermal and pH stabilities in various aqueous solutions were identical for the glucose and fructose enantiomers. Unlike all currently available low-calorie sweeteners, the L-sugars brown upon baking. Therefore, L-sugars are expected to yield food products similar to those using D-sugars, but without the calories.

## TASTE

Taste testing and sensory profiles of three L-sugars were performed by both trained and expert human panels in accordance with recognized techniques. The taste profiles were similar to those of D-hexoses. Neither cooling effect on the tongue nor any aftertaste was detected. Levels of sweetness were somewhat less than sucrose, but acceptable for product use either directly or enhanced with a small quantity of high-intensity sweetener.

## SAFETY AND EFFICACY

An essential step in bringing L-sugars to market is the approval of the U.S. Food and Drug Administration (FDA). To satisfy FDA requirements, three of the 10 L-sugars (under the Biospherics patent) were selected for extensive toxicological testing at a high percentage of diet. So far, two of the sugars have been tested in the three animal species requested by the FDA and the third in rats alone; studies ranged from single-dose acute oral administration up to 6 months feeding at high level in the diet. Organs and tissues were examined for indications of toxicity. No toxic effects were found.

The three selected L-sugars were examined for possible mutagenic activity by in vitro testing recognized and recommended by the FDA. They were all negative.

Sophisticated metabolism testing using radiolabeled L-sugar studies and retained energy balance techniques have been performed on rats fed the three L-sugars in order to gain a detailed picture of how they are dealt with by the host animal. The retained energy balance study on one of the L-sugars showed it to be completely noncaloric to the host animals.

Carefully controlled and monitored single-dose human testing has been performed on two of the three selected L-sugars in a European clinical testing facility. The findings of these studies continue to support the expected nontoxic and low-caloric nature of these L-sugars. However, human acceptability of the two L-sugars tested to date indicates that at least these two may be subject to acceptable daily intake limitations, as are currently available sugar alcohols and nonsweet bulking agents.

Two of the selected L-sugars have been tested for transplacental transfer to rat fetuses and one of these L-sugars tested for potential effect on the normal delivery of baby rats. The findings were negative.

Preliminary in vitro studies on the three selected L-sugars show them to be noncariogenic. Tests for anticariogenicity will be performed.

## PRODUCTION PROCESSES

Proprietary process technologies and engineering projections of full-scale production plants have been developed for each of the three selected L-sugars. Technoeconomic evaluations, based on these designs, have yielded detailed manufacturing cost estimates. A variety of process patents has been applied for and should add considerably to the patent protection already afforded by Biospherics' basic L-sugar "use" patent and existing process protection. Current efforts should make possible L-sugar products at reduced manufacturing costs in order to address broader markets. Starting materials are common, low-cost, agricultural materials.

Sample production runs for the three selected L-sugars have yielded quantities of up to 500 pounds used in animal testing and a variety of other research activities. Highly pure crystalline products have been obtained in all cases.

Analytical methods have been developed for a variety of L-sugars and derivatives. The samples produced have been carefully scrutinized for purity and consistency.

## MARKET PROSPECTS

Currently, there are three "intense sweeteners" that have been approved by the FDA for use in certain food products: saccharin, acesulfame-K, and aspartame. Of these, aspartame has had the greatest market success—registering close to a billion dollars in worldwide sales this year—principally in the beverage sector where product "bulk" is provided by water.

Applications for high-intensity sweeteners outside of the beverage market sector have been limited primarily because of their lack of bulk. Taste, shelf-life, and heat stability limitations also exist for many of these same uses. The all-important structural functions that natural sugars provide to products like cakes, cookies, ice creams, and candy bars cannot be provided by high-intensity sweeteners alone. While other bulking agents exist or are being developed, none is known to address fully all these needs.

In a talk given at the May 1988 meeting of the International Sweeteners Association, Landell Mills Commodities Studies, authors of *Alternative Sweeteners: Where Will Starch Sweeteners and High Intensity Sweeteners Go from Here?*, described current limitations of low-calorie sweeteners in a market place where the further development of diet sectors is "dependent upon the availability of a good tasting, heat resistant, intense sweetener and an adequate bulking agent." In looking to the future, they concluded that in nonbeverage applications "where there is a sizeable potential, the technical shortcomings of the intense sweeteners frequently restrict their use. The greatest technical difficulty is the absence of an ideal bulking agent, and if such an agent were to be developed then the market for intense sweeteners could be transformed."

L-Sugars offer the prospect of being a combined sweetening and bulking agent—thus providing a possible solution to the issues raised by Landell Mills.

L-Sugars as low-calorie "bulk sweeteners" combine several elements that may be the key to open up these currently inaccessible diet sectors in the food product market. L-Sugars provide a clean, sweet taste and lowered calories while also furnishing the bulk, texture, crystallinity, browning, and other properties so necessary for effective formulation of food products. Testing by our food science staff has produced a number of product prototypes with both low-calorie sweetness and good structural qualities.

Initial target markets include gum, confectionery, baked goods, heat-processed foods, frozen desserts, and ice cream. Considerable interest from manufacturers has been shown. Early discussions with these companies indicate a willingness to pay a premium for the very significant advantages of Lev-O-Cal.

The advantages offered to the consumer by Lev-O-Cal may be unique. Whether used alone or in combination with current or the anticipated heat-stable intense sweeteners, the L-sugars should significantly extend the range and quality of low-calorie foods. They will appeal to those constantly looking for better ways to enjoy great-tasting foods while keeping their waistlines under control and to the significant portion of the population, such as diabetics, who must deny themselves sugar for health reasons.

## REFERENCES

1. G. V. Levin, Sweetened edible formulations, U.S. Patent 4,262,032 (April 14, 1981).
2. W. C. Austin and F. L. Humoller, Preparation of two new crystalline aldohexoses, L-allose and L-altrose from L-ribose by the cyanohydrin reaction, *J. Am. Chem. Soc., 56:* 1153–1155 (1934).
3. C. S. Hudson, O. Hartley, and C. B. Purves, Convenient modification of the Kiliani synthesis of higher carbon acids (or their lactones) from reducing sugars, *J. Am. Chem. Soc., 56:* 1248–1249 (1934).
4. M. L. Wolfrom and A. Thompson, Action of diazomethane upon acyclic sugar derivatives. VIII. L-fructose, *J. Am. Chem. Soc., 68:* 791–793 (1946).
5. E. Fischer and O. Piloty, Reduction der Zuckersäure, *Berichte, 24:* 521 (1891).
6. H. H. Mottern and H. L. Cole, Enzymic preparation of D-galacturonic acid, *J. Am. Chem. Soc., 61:* 2701–2702 (1939).
7. W. W. Pigman, Improvements in the preparation of D-galacturonic acid, *J. Res. Natl. Bur. Standards, 25:* 301–303 (1940).
8. H. S. Isbell and H. L. Frush, Salts of galacturonic acid and their application to the preparation of galacturonic acid from pectic substances, *J. Res. Natl. Bur. Standards, 32:* 77 (1934).
9. W. C. Austin and F. L. Humoller, Preparation of two new crystalline aldohexoses, L-allose and L-altrose, from L-ribose by the cyanohydrin reaction, *J. Am. Chem. Soc., 56:* 1153–1155 (1934).
10. A. S. Meyer and T. Reichstein, L-Idose from D-glucose as a new method for L-idomethylose, *Helv. Chim. Acta, 29:* 152–162 (1946).
11. C. Glatthaar and T. Reichstein, L-Talose, *Helv. Chim. Acta, 21:* 3–6 (1938).
12. M. Steiger and T. Reichstein, L-Psicose (2 keto-L-ribohexose, pseudofruc-

tose), diacetone-L-psicose and diacetone-L-psicaronic acid, *Helv. Chim. Acta.*, *18:* 790–799 (1935).
13. W. Pigman, *The Carbohydrates—Chemistry, Biochemistry, Physiology,* Academic Press, Inc., New York, 1957, pp. 106–132.
14. C. R. Noller, *Chemistry of Organic Compounds,* 2nd ed., W. B. Saunders Co., Philadelphia, 1957, pp. 341–342.
15. P. Wursch, C. Wesch, and M. Arnaud, Metabolism of L-sorbose in the rat and the effect on the intestinal microflora on its utilization both in the rat and in the human, *Nutr. Metab.*, *23:* 145–155 (1979).

# 8

# Saccharin

**Mark L. Mitchell and Ronald L. Pearson** *PMC Specialties Group, A Division of PMC, Inc., Cincinnati, Ohio*

The name saccharin is aptly derived from the Latin *saccharum* for sugar. It is commercially available in three forms: acid saccharin, sodium saccharin, and calcium saccharin. These forms have been variously determined to be 200–700 times sweeter than sucrose (1). Coupled with its low cost, this intense sweetness makes saccharin, in all its forms, the most cost-effective of all presently known sweeteners. The American manufacturing process utilizes as its starting material purified, commercially manufactured methyl anthranilate, an ester which is found in a number of fruit juices, notably grape. As a pioneer alternate sweetener, saccharin has had a difficult history, requiring considerable persistence and effort to overcome its detractors. It is the only sugar substitute that has been in use for over a century. Saccharin has withstood the test of time and has earned a rightful place in our food and beverages.

## DISCOVERY

Saccharin was discovered by chemists Ira Remsen and Constantine Fahlberg in May of 1878. Fahlberg had received his doctorate at Leipzig in 1873 and was visiting Johns Hopkins University in Baltimore. There, under Professor Remsen's tutelage, he worked on the oxidation of *o*-toluenesulfonamide as part of an ongoing research project of Remsen's (2).

To the surprise of both chemists, an elemental analysis revealed that the oxidation product was the condensed heterocycle *o*-sulfobenzimide rather than the expected *o*-sulfamoylbenzoic acid. Several weeks later, Fahlberg discovered the intense sweetness of the new compound after spilling a solution on his hands and later eating bread at dinner. This serendipity was duly noted in the investigators' original report (3): "Sie schmeckt angenehm süss, sogar süsser als der Rohrzucker." (It tastes pleasantly sweet, even sweeter than cane sugar.)

Remsen had a personal disdain for commercial ventures. However, Fahlberg, an experienced sugar chemist, aggressively pursued the commercial potential of the new compound. He named it "Fahlberg's saccharin" and obtained a U.S. patent (4) without informing Remsen. This professional rudeness was a source of extreme irritation to Remsen, who summarized his feelings in 1913: "I did not want (Fahlberg's) money, but I did feel that I ought to have received a little credit for the discovery" (2).

## EARLY PRODUCTION

Fahlberg displayed the sweetening power of saccharin in a London exposition in 1885. This saccharin was manufactured in a pilot plant Fahlberg operated in New York. Two German patents were issued in 1886 to Fahlberg; these were assigned to Fahlberg, List, and Company. The plant, under the name of Saccharinfabrik A.G., was moved to Westerhusen, Germany. In 1900, Fahlberg reported an annual production of 190,000 kg. In 1902, partly at the insistence of beet sugar producers, saccharin production in Germany was brought under strict control, and saccharin was attainable only through pharmacies (5).

John F. Queeny became acquainted with saccharin sometime in the late 1890s. At that time, he was the purchase agent for Meyer Brothers Drug Company in St. Louis, Missouri, which was importing the sweetener from Germany. Queeny tried to persuade his employer to manufacture the sweetener rather than import it, but Meyer Brothers did not foresee saccharin as a profitable venture (6). In 1901, Queeny took his personal savings of $1500 and, together with $3500 obtained from Liquid Carbonic, he began to manufacturing saccharin himself. He chose his wife's maiden name in naming his new venture the Monsanto Chemical Company.

Early production of saccharin was a touch-and-go proposition with rickety, old equipment. Furthermore, saccharin was targeted by several sources, the most aggressive of which was the German cartel known as the Dye Trust. The Dye Trust waged a price war and cut the price from $4.50 a pound to $1.00 a pound (6).

Opposition to saccharin also arose from domestic sources. A prominent food editor of the time, Alfred W. McCann, wrote, "Saccharin is as false and scarlet as the glow of health transferred from the rouge pot to the cheek of a baud" (6). Saccharin was also criticized for having no food value or calories. Saccharin remained approved for use in the United States partly through the influence of President Theodore Roosevelt, who was enraged when told by a federal health official that it might be banned. The president was quoted in 1906 as saying that "anyone who says saccharin is injurious to health is an idiot!" (7). Support from the president continued, with Mr. Queeny receiving the following letter, dated July 7, 1911:

> I always completely disagreed about saccharin both as to the label and as to its being deleterious. . . . I have used it myself for many years as a substitute for sugar in tea and coffee without feeling the slightest bad effects. I am continuing to use it now: Faithfully yours, T. Roosevelt (7)

Saccharin use increased during World War I and immediately thereafter as a result of sugar rationing. This was particularly evident in Europe (8). By 1917, saccharin was a common table-top sweetener in America and in Europe. The sweetener was introduced to the Far East in 1923. The consumption of saccharin continued between the Wars, with an increase in the number of products in which it was used. The shortage of sugar during World War II again produced a significant increase in saccharin usage, firmly establishing a need for this alternative sweetener.

In the early 1950s, The Maumee Chemical Company began to manufacture saccharin by a novel process developed by O. F. Senn and G. F. Schlaudecker (9,10), who helped found Maumee Chemical Company. Maumee's introduction of calcium saccharin as a alternate soluble form soon followed. In 1966 The Sherwin-Williams Company acquired Maumee and sold this business to PMC Specialties Group in 1985.

## CURRENT PRODUCTION

A number of companies around the world manufacture saccharin. Most manufacturers use the basic synthetic route described by Remsen and Fahlberg (Fig. 1). Toluene is treated with chlorosulfonic acid to produce *ortho-* and *para-*toluenesulfonyl chloride. Subsequent treatment with ammonia forms the corresponding toluenesulfonamides. The *ortho-*toluenesulfonamide is separated from its *para* isomer (this separation is alternatively performed on the sulfonyl chlorides). The *ortho-*toluenesulfonamide is then oxidized to *ortho-*sulfamoylbenzoic acid, which on heating is cyclized to saccharin (1,5).

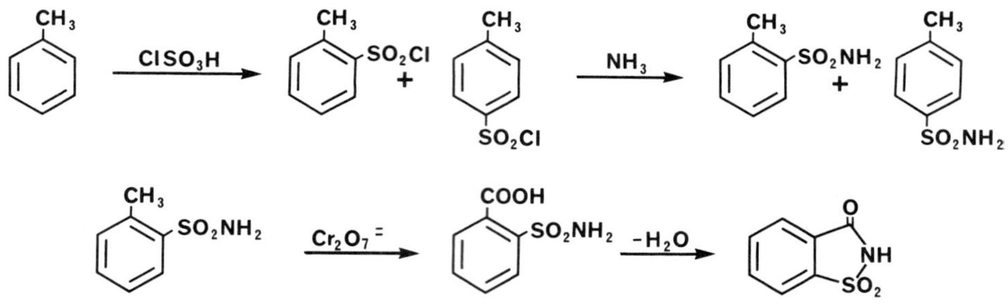


**Figure 1** Remsen-Fahlberg process.

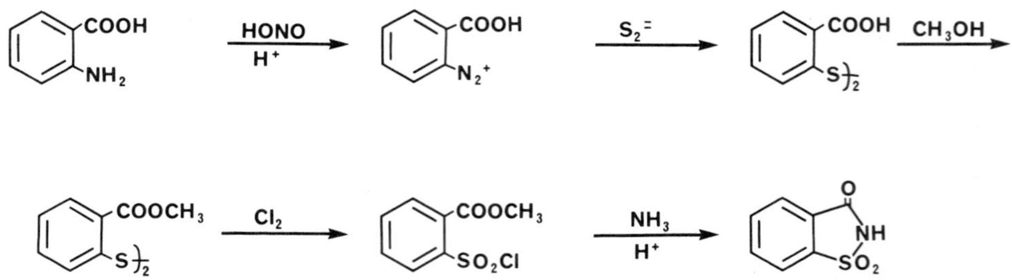


**Figure 2** Original Maumee process.

PMC Specialties Group, the sole U.S. producer, currently uses improved Maumee chemistry, producing saccharin from purified, manufactured methyl anthranilate, a substance naturally occurring in grapes (11).

The Maumee process originally diazotized anthranilic acid by adding an equivalent of sodium nitrite and sulfuric acid (Fig. 2). The resulting diazo solution was added to sodium disulfide. The mixture was then acidified to precipitate the organic disulfide, which was collected, washed, and dried. The disulfide was then esterified by treatment with methanol in sulfuric acid and oxidized with gaseous chlorine to *ortho*-carbomethoxybenzenesulfonyl chloride. Amidation with excess ammonia yielded ammonium saccharin, and neutralization with sulfuric acid produced insoluble acid saccharin (12).

Today, the Maumee process has evolved into an efficient continuous process (Fig. 3). Methyl anthranilate is diazotized to form 2-carbomethoxybenzenediazonium chloride. Sulfonation followed by oxidation yields 2-carbomethoxybenzenesulfonyl chloride. Again, amidation of this sulfonylchloride, followed by acidification, forms insoluble

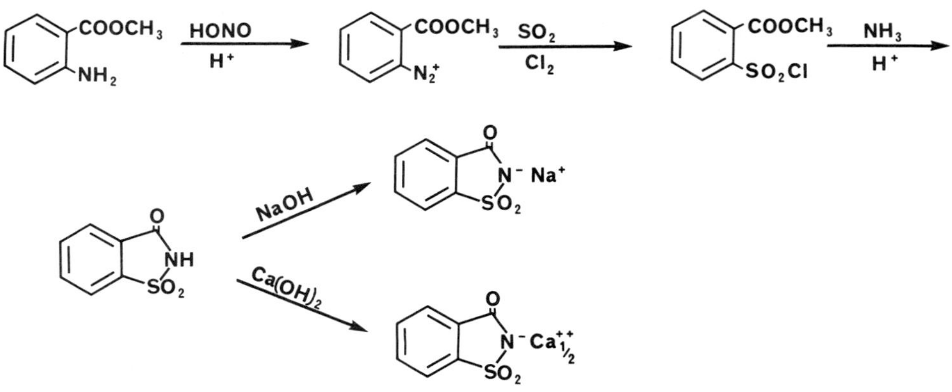


**Figure 3** Continuous Maumee process.

acid saccharin. Subsequent addition of sodium hydroxide or calcium hydroxide produces the soluble sodium and calcium forms, respectively (1,13).

## PROPERTIES

Acid saccharin occurs as a white, odorless, crystalline powder. It is a moderately strong acid, and it is slightly soluble in water. There are three commercially available saccharin forms: (1) acid, (2) sodium, and (3) calcium. All are manufactured to meet Food Chemicals Codex specifications (14). Their properties are listed in Table 1.

Several additional salts of saccharin have been reported. These include silver, ammonium, cupric, lithium, magnesium, zinc, and potassium. Although all of these are intensely sweet, none is available commercially. Substitutions on the nitrogen of saccharin eliminates its sweetness, whereas carbon substitution gives unpredictable sweetness results (17).

X-ray crystallography has shown that the acid form of saccharin exists as dimers, formed by hydrogen bonding between the imide hydrogen and the keto oxygen (18,19).

## SWEETENING POWER AND ADMIXTURE POTENTIAL

Saccharin is approximately 300 times sweeter than sugar. Coupled with saccharin's low cost (approximately \$3.00/lb), this sweetening factor means that a penny's worth of saccharin has the sweetening power of one pound of sugar. The relative economics of current commercial sweet-

**Table 1** Properties of Saccharin Forms

| | Acid saccharin | Sodium saccharin | Calcium saccharin |
|---|---|---|---|
| Molecular formula | $C_7H_5NO_3S$ | $C_7H_4NO_3SNa \cdot 2H_2O$ | $[C_7H_4NO_3S]_2Ca \cdot 2H_2O$ |
| Molecular weight | 183.18 | 241.20 | 440.48 |
| C.A.S. registry no. | 81-07-2 | 128-44-9 | 6485-34-3 |
| Melting point (°C) | 228–229 | > 300 | > 300 |
| Appearance | White solid | White solid | White solid |
| pKa[a] | 1.30 | — | — |
| Solubility (g/100 g water) at:[b] | | | |
| 20°C | 0.2 | 100 | 37 |
| 35°C | 0.4 | 143 | 82 |
| 50°C | 0.7 | 187 | 127 |
| 75°C | 1.3 | 254 | 202 |
| 90°C | — | 297 | 247 |

[a]From Ref. 15.
[b]From Refs. 16, 43.

eners are compared in Table 2, showing that both sodium and calcium saccharin are by far the least expensive per sweetness equivalent.

In 1921 Paul (20) observed that saccharin's sweetening power relative to sucrose increases with decreasing concentration. He found a similar concentration effect with dulcin, another alternative sweetener used at the time, but which is no longer approved. What is remarkable is Paul's observation that when these two sweeteners are blended, each sweetener retains its higher sweetening power in the presence of the other. For example, to achieve a sweetness equal to a 9% sucrose solution, individually 450 mg/liter of saccharin or 1250 mg/liter of dulcin are required. But this 9% sucrose solution is also equalled with a blend containing 190 mg/liter of saccharin plus 120 mg/liter of dulcin, only 310 mg/liter in total. Individually, the 190 mg/liter of saccharin is equivalent to a 6% sucrose solution while the 120 mg/liter of dulcin is equivalent to a 3% sucrose solution. Thus by blending the two sweeteners, Paul was able to "add" their individual sweetening equivalencies and achieve the

**Table 2** Economics of Commercial Sweeteners

| Sweetener | Sweetness/lb[a] (sucrose = 1) | Cost ($/lb) | Cost ($/sweetness) | Relative cost ($/sweetness) (sodium saccharin = 1) |
|---|---|---|---|---|
| Sodium saccharin·$2H_2O$ | 300 | 2.50[b] | 0.0083 | 1.0 |
| Acid saccharin | 370[c] | 4.25[b] | 0.0115 | 1.4 |
| Calcium saccharin·$2H_2O$ | 300 | 4.75[b] | 0.0158 | 1.9 |
| Aspartame (Canada) | 180 | 28.00[d] | 0.1555 | 19 |
| Sodium cyclamate | 30 | 1.45[e] | 0.0483 | 5.8 |
| Acesulfame-K | 200 | 45.00[f] | 0.2250 | 27 |
| HFCS (55%) | 1 | 0.255[g] | 0.2550 | 31 |
| Aspartame (USA) | 180 | 50.00[d] | 0.2778 | 33 |
| Sucrose | 1 | 0.325[h] | 0.3250 | 39 |

[a] L. O'Brien and R. C. Gelardi, *Chem. Tech, 11:* 274 (1981).
[b] PMC Specialties Group, fall/1990.
[c] A. Fricker, J. Gutschmidt, and E. Prochazka, *Dtsch. Lebensm-Rundsch, 71:* 138 (1975).
[d] *Beverage Digest, 17*(2) (1990).
[e] PMC International, fall/1990.
[f] Hoechst Celanese Company, fall/1990.
[g] Archer Daniels Midland, fall/1990.
[h] Various suppliers, fall/1990.

same sweetness with less total sweetener. This effect is more completely demonstrated by Paul's data presented in Table 3.

Today, this phenomenon of increasing sweetening power with decreasing concentration has been established for many sweeteners (21), and additive effects have been reported for blends of saccharin with several other sweeteners. The blend is frequently sweeter than would be predicted by the additive effect. This enhanced effect is termed "synergistic." Table 4 lists various binary blends with saccharin and their reported effect on sweetness intensity. Whether synergistic or merely additive, sweetener blends provide an obvious economic advantage.

In spite of its intense sweetness, saccharin has a bitter metallic aftertaste that becomes more noticeable with increasing concentration. At normal use concentrations this aftertaste is detectable by about 25% of the population (29). This aftertaste is apparently intrinsic to the saccharin molecule itself rather than to the presence of a potent impurity (30).

Beverage formulators can greatly reduce or eliminate this aftertaste by blending saccharin with other sweeteners. A blend permits one to

**Table 3** Additivity Effect of Saccharin–Dulcin Blends

| Sucrose | | Saccharin | | Dulcin | | Blend | | |
|---|---|---|---|---|---|---|---|---|
| Concentration (g/l) | Concentration (weight %) | Conc. (mg/l) | Sweetening Power (saccharin/sucrose) | Conc. (mg/l) | Sweetening Power (dulcin/sucrose) | Saccharin (mg/l) | Dulcin (mg/l) | Total (mg/l) |
| 10 | 1 | 20 | 500 | 30 | 333 | — | — | — |
| 20 | 2 | 30 | 667 | 55 | 364 | — | — | — |
| 30 | 3 | 55 | 545 | 120 | 250 | — | — | — |
| 40 | 4 | 100 | 400 | 290 | 138 | 55 | 30 | 85 |
| 50 | 5 | 150 | 333 | 450 | 111 | 55 | 55 | 110 |
| 60 | 6 | 190 | 316 | 665 | 90 | 100 | 55 | 155 |
| 70 | 7 | 280 | 250 | 855 | 82 | 150 | 55 | 205 |
| 80 | 8 | 370 | 216 | 1050 | 76 | 190 | 55 | 245 |
| 90 | 9 | 450 | 200 | 1250 | 72 | 190 | 120 | 310 |
| 100 | 10 | 535 | 187 | 1430 | 70 | 280 | 120 | 400 |

Solutions were of equivalent sweetness.
*Source:* Ref. 20.

**Table 4** Saccharin Admixture Potential

| Binary mixture saccharin + | Sweetness enhancement | Ref. |
|---|---|---|
| Aspartame | Synergistic | 22 |
| Acesulfame-K | Additive | 23 |
| Cyclamate | Synergistic | 24 |
| Sucralose | Synergistic | 25 |
| Alitame | Synergistic | 26 |
| Sucrose | Additive | 27 |
| Fructose | Additive | 28 |
| Dulcin | Additive | 20 |

keep saccharin's concentration below its bitterness threshold (29,31). The earliest blends were composed of saccharin and dulcin (32). Later, before its U.S. ban in 1969, sodium cyclamate was used extensively with saccharin (24) in the rapidly growing diet soft drink market. This combination is synergistic and mutually beneficial, as cyclamate alone also has objectionable taste characteristics. The taste profile of a 10:1 cyclamate:saccharin blend has been found to be similar to that of sucrose (33).

More recently blends of saccharin and aspartame have been used in low-calorie soft drinks (22). Binary saccharin blends with other dipeptides (26,34), acesulfame-K (23), sucralose (25), as well as ternary blends with cyclamate and aspartame have also been reported (35–37). Even sucrose or fructose and saccharin blends can provide reduced calorie products with good taste characteristics (38–40).

A second approach to removing saccharin's aftertaste is the use of "masking agents." A popular table-top sweetener employs cream of tartar with dextrose as masking agent (41). The patent literature abounds with reported masking agents. Many of these are described in reviews by Bakal (35) and Daniels (42).

Most food and beverage literature assessing improved taste of saccharin blends has focused on sodium saccharin. Recently the calcium salt of saccharin has been found to possess a shorter, cleaner aftertaste with less bitterness (43). In blends with other sweeteners, calcium saccharin may offer improved taste aesthetics.

No single alternative sweetener available today closely matches the full taste attributes of natural sugar. Moreover, these alternative sweeteners all suffer one or more additional disadvantages such as poor solution stability, high cost, or objectionable taste characteristics. By blending saccharin salts with other sweeteners, the food chemist can economically provide low caloric foods and beverages with optimized taste characteristics while minimizing any single sweetener's weakness (44). This multiple sweetener concept is more completely described in Chapter 20.

## STABILITY AND SHELF LIFE

In its bulk form, saccharin and its salts show no detectable degree of decomposition over periods as long as several years (43).

Another major advantage of saccharin and its salts are their high stability in aqueous solutions over a wide pH range. With saccharin the food and beverage formulator does not have to compromise taste aesthetics by altering the pH to minimize sweetener hydrolysis. Thus, taste is maintained over a longer shelf life.

In 1952, DeGarmo and co-workers studied the stability of saccharin in aqueous solutions (45). They found that saccharin solutions buffered at pHs ranging from 3.3 to 8.0 were essentially unchanged after heating for 1 hr at 150°C. More recent work employing high-performance liquid chromatographic (HPLC) techniques has confirmed their findings (43). Only under severe laboratory conditions of high temperature, high and low pH does saccharin hydrolyze to a measurable extent. The hydrolysis products are 2-sulfobenzoic acid and 2-sulfamoylbenzoic acid (Fig. 4)

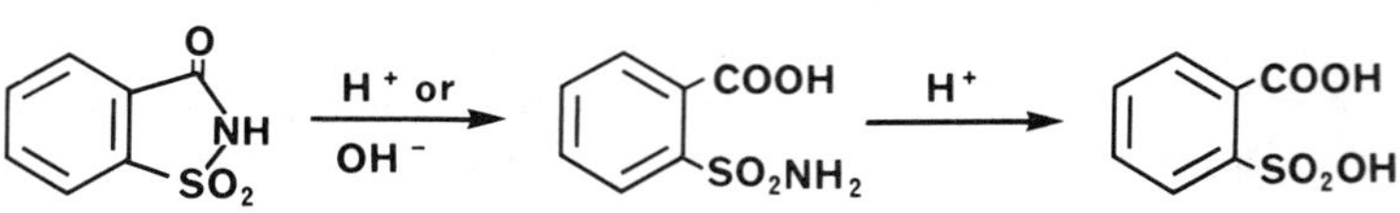


**Figure 4** Hydrolysis of saccharin.

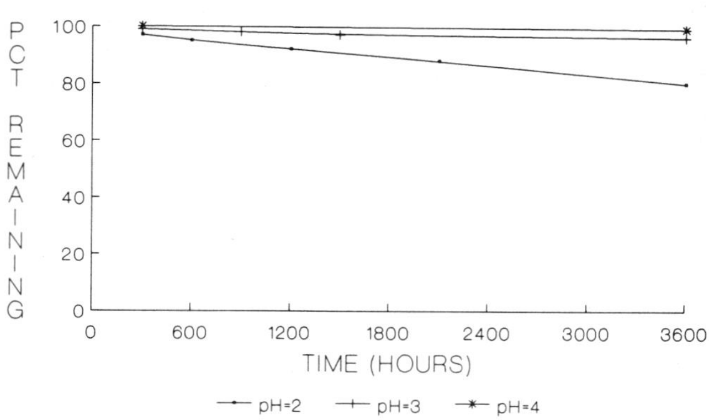


**Figure 5** Saccharin hydrolysis at 20 degrees C (68 degrees F).

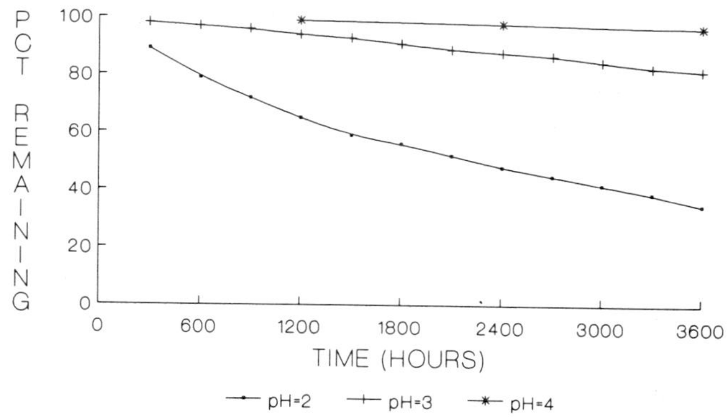


**Figure 6** Saccharin hydrolysis at 40 degrees C (104 degrees F).

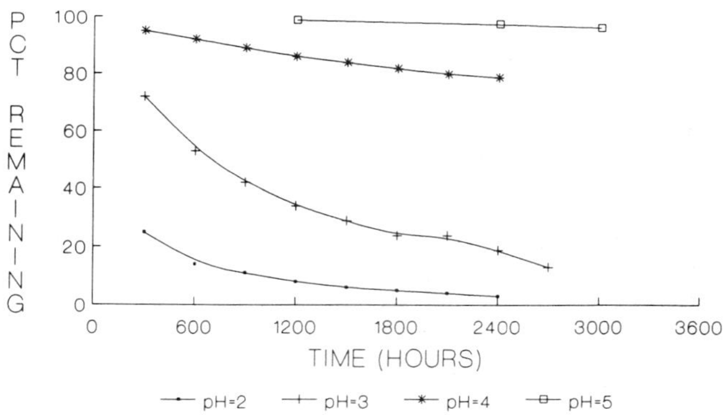


**Figure 7** Saccharin hydrolysis at 80 degrees C (176 degrees F).

(45). Hydrolysis curves for sodium saccharin at various temperatures and pHs are shown in Figures 5, 6, and 7 (43).

## FUNCTIONALITY

Over the last century saccharin and its salts have found their way into a variety of beverages, foods, cosmetics, and pharmaceuticals. Its primary function is to safely and economically provide sweetness without the attendant calories. Saccharin is used as a noncaloric sweetener in the following foods and beverages:

Soft drinks, fruit juice drinks, other beverages, and beverage bases or mixes

Table-top sweeteners in tablet, powder, or liquid form
Processed fruits
Chewing gum and confections
Gelatin desserts, juices, jams, and toppings
Sauces and dressings

A minor amount of saccharin is used in a variety of nonfood applications, including: (1) nickel electroplating brightener (46), (2) agricultural chemical intermediate (47,48), (3) animal food sweetener (49), (4) pharmaceutical chemical intermediate (50), (5) chemical intermediate for a mammalian repellent (51,52), (6) biocide intermediate (53), and (7) anaerobic adhesive accelerator (54,55). Over the years the chemistry of saccharin has been the subject of several reviews (56–58).

## METABOLISM

Chemical carcinogenesis is generally believed to be associated with a chemical interaction between the carcinogen and macromolecules (such as DNA) that initiate tumor development (59). Because the saccharin molecule is nucleophilic, negatively charged (60), and would not be expected to show affinity for binding to DNA per se, the question of the metabolism of saccharin is a major issue. In fact, [$^{35}$S] saccharin has been shown not to bind covalently to DNA in the rat liver or bladder after oral administration (61). Saccharin clearly differs from classic chemical carcinogens which act via reaction with DNA, in that the parent compound (saccharin) is chemically stable at physiological pH and temperature and undergoes little or no in vivo or in vitro metabolism.

That saccharin is not metabolized by humans is fairly certain, despite three early papers reporting the detection of metabolites. First, the studies in which metabolites were detected are open to question because of procedural flaws. Second, other workers have been unable to find detectable levels of metabolites of saccharin in the excreta of laboratory animals or humans.

Studies in which highly purified, radiolabeled saccharin was used (62–65) have all failed to detect metabolism of saccharin, despite the use of specific analytical procedures for the reported metabolites (64,65). The [3-$^{14}$C]saccharin used by Lethco et al. (66), who detected metabolism, had been purified by column chromatography under neutral conditions. However, the efficacy of this procedure in removing the impurity found by Byard and Golberg (63) or that proposed by Renwick and Williams (67), i.e., benz(*d*)isothiazoline-1, 1-dioxide (BIT), or 3-hydroxy BIT, cannot be assessed.

In those studies in which both 2-sulfamoylbenzoic acid and 2-sulfobenzoic acid were detected as metabolites (68,69), the purity of the dose was not investigated and the levels of radioactivity used were low. In both cases, metabolite identity was based on a single thin-layer chromatography (TLC) system, and the authors did not adequately rule out the possibility of artifact formation during chromatography.

In the study of Lethco and Wallace (66), the results are of greater validity, although the possibility of artifactual hydrolysis during dose preparation was not investigated. The authors concluded that the metabolites arose from slight chemical breakdown of saccharin rather than from enzymatic mechanisms. The possibility of a minor impurity in the dose cannot be excluded. The apparent uniformity of the levels of metabolites found suggests such a cause.

At any rate, the metabolites reported (2-sulfamoylbenzoic acid and 2-sulfobenzoic acid) are both more polar than saccharin and retain anionic character. Thus, both represent unlikely candidates for electrophilic carcinogenic metabolites. Similarly $CO_2$ and $CO_3^{2-}$ are not of relevance per se, although the remaining fragment (benzenesulfonamide) could be of considerable significance. However, labelling studies using [benzene ring-$^{14}$C]-, [5-$^{3}$H]-, and [$^{35}$S]saccharin have not detected any such less polar metabolites. $CO_2$ and $CO_3^{2-}$ were not detected by Ball et al. (64). Indeed, chemical decarboxylation of saccharin has not been reported, even under extreme chemical conditions. It thus appears that the decarboxylation reported is an unexplained artifact. The absence of detectable metabolism of saccharin in the rat has been demonstrated in normal males and females (70,63), pregnant females (64), females fed 1% or 5% saccharin diets for up to 12 months in a study designed to examine the possibility of subtrate induction of metabolism (64), $F_1$ males reared and maintained on 5% saccharin diet, in a study designed to determine the possibility of substrate induction of metabolism (65), and males and females treated with a 3-methylcholanthrene (65) and phenobarbitone (63) in a study designed to determine whether metabolism is induced by an increase in microsomal drug metabolism.

In conclusion, saccharin metabolism has not been detected in studies using modern analytical techniques. Treatment of rats by a variety of means has attempted to optimize the detection of minor metabolic pathways, but all have failed to reveal biotransformation. The vast majority of the data on the biotransformation of saccharin demonstrates that this compound is excreted unchanged, predominately in the urine, in both humans and laboratory animals (71). Low levels of metabolites reported in three early studies probably represent artifacts resulting from inadequate techniques and/or impurities in the saccharin used (72).

## TISSUE DISTRIBUTION AND PHARMACOKINETICS

The most thorough studies on saccharin distribution and kinetics were published by Sweatman and Renwick (71,73) in 1980 and 1982 in *Toxicology and Applied Pharmacology.*

In the 1980 study, Sweatman and Renwick examined the amount of saccharin in plasma, urine, and tissues of the rat following several dosing regimes, including diets containing saccharin at levels up to 10% (71). As might be expected for an unmetabolized chemical that is rapidly eliminated from the body within 24–48 hr after a single dose, Sweatman and Renwick reported a marked diurnal variation in the concentration of saccharin detected in plasma. As part of the same study, rats were fed diets containing 0–10% saccharin for 22 days, and the amount of saccharin in various tissues was determined. The tissue concentrations for the lung, liver, kidney, spleen, and adrenals were much higher at the 7.5–10% levels than would be predicted by linear extrapolation from lower dosages, whereas only at the 10% dietary level did the amount of saccharin in fat and muscle depart from linearity. When the tissue plasma ratio was examined, only the gut wall, kidney, and urinary bladder had a ratio greater than unity. Sweatman and Renwick suggest that the ratio in the gut wall was due to incomplete rinsing of dietary material. The concentration gradient between bladder wall and urine approximated 1:100, which the authors interpreted as suggesting that there was limited diffusion of saccharin from the urine into the bladder wall (71). This conclusion was in keeping with their previous study (74), but contrary to the results reported by Colburn (75), where significant absorption of saccharin was reported to occur from the rat bladder. Some of this apparent discrepancy may be attributable to the degree of bladder manipulation, as Renwick and Sweatman (74) noted that even manipulation of the bladder with forceps or gentle palpation in an attempt to expel urine resulted in an increase in apparent permeability. Although the permeability of human bladder epithelium to saccharin has not been specifically studied, the efficiency of normal adult bladder epithelium as a barrier between tissue fluids and urine is well known.

In 1982, Sweatman and Renwick reported the tissue levels of saccharin in the rat during two-generation feeding studies (73). A single oral dose of [$^3$H]saccharin was given to female rats in late pregnancy. The concentrations in fetal tissues, including bladder wall, decreased more slowly, so that by 48 hr, they exceeded the corresponding values obtained for maternal tissues, suggesting the possibility of accumulation during chronic intake. Despite this, the steady-state concentrations of saccharin in the liver and kidneys of fetuses from mothers fed 5% saccharin diet ad libitum were lower than the corresponding maternal values, whereas the concentrations in the fetal bladder were similar or slightly higher. The concentrations of saccharin in the tissues of rats in utero were not markedly higher than those found in adult $F_1$ animals. The turnover of saccharin in the fetuses of animals maintained on 5% saccharin diet was similar to that seen after a single dose. The results showed no evidence of excessive accumulation in the bladder wall or other tissues of male rats during in utero exposure or during lactation.

Comparison of the pharmacokinetics of saccharin in humans and rats shows that saccharin is incompletely absorbed from the gut in both species, and that plasma concentration-time curves after oral administration are determined somewhat by absorption rate (71,76). Saccharin is eliminated rapidly from the blood by the renal tubular secretion of the kidneys. Saturation of this process, which was found at plasma concentrations of 200–300 μg/ml in the rats, was not found in humans at plasma levels up to 50–60 μg/ml, which occurred after oral administration of 2g—over 12 times the mean daily consumption of saccharin users. Renwick and Sweatman concluded that their data raise doubts on the validity of toxicity and carcinogenicity studies in which the test animals are fed dietary levels of saccharin sufficient to alter normal physiological functions that would not occur at human exposure levels (71).

## CARCINOGENICITY AND TOXICITY

Saccharin has been studied extensively in rats, mice, hamsters, and monkeys in lifetime feeding experiments. An absence of evidence of carcinogenicity has characterized most of the single-generation studies. Statistically significant tumors have been found, however, in male second-generation Sprague-Dawley rats in two-generation, in utero feeding studies at the highest doses.

In single-generation feeding studies, groups of rats were fed varying concentrations of saccharin (0.005% up to 5%) over extended periods of time (24 months to lifetime). Saccharin had no effect on hematology, mortality, organ weight, life expectancy, growth rate, systolic blood pressure, or morphology of the heart (77–80).

A variety of other species (mice, hamsters, and monkeys) were also examined for bladder tumors. However, none were found, and most authors concluded that there is no relationship between saccharin and bladder cancer (78,81–90).

Four two-generation bioassays have been conducted by the Wisconsin Alumni Research Foundation (WARF), the FDA, the Canadian

Health Protection Branch (HPB), and the International Research and Development Corporation (IRDC) (91–94).

The WARF study fed groups of 20 male and 20 female Sprague-Dawley rats 0.05, 0.5, and 5.0% saccharin diets. The $F_1$ generation was weaned onto their parents' diet, and consumption continued for 100 weeks. The $F_1$ generation animals were exposed to saccharin during gestation, throughout lactation via the dam's milk, and for the remainder of their lifetime. The authors reported that neither feed consumption nor survival was affected by treatment, and that the hematological changes observed were considered to be the result of age and poor health rather than treatment. Tumors of the urinary bladder occurred at a statistically significant level only in the $F_1$ generation male rats ingesting 5% saccharin.

The FDA study was conducted during 1970–1972 as a combination three-generation reproduction study and two-generation chronic feeding study. Animals were randomly selected from the $F_1$ litter to give a total of 48 male and 48 female Sprague-Dawley rats per treatment group. The animals were fed diets containing 0, 0.01, 0.1, 1.0, 5.0, or 7.5% saccharin for a period of 2 years. The authors reported that saccharin did not have any effects on hematological values, organ weights, or survival. Body weights of rats ingesting 5.0 or 7.5% saccharin were lower than the controls. Although some of the initial weight differences were overcome, these animals had lower body weights throughout the study. The most significant pathological finding was an increased incidence of urinary bladder neoplasms in the male ($F_1$) rats receiving 7.5% saccharin. Two statistically insignificant tumors were found in the $F_1$ females, and one tumor was observed in both the $F_1$ 5.0% and $F_1$ control males. The authors pointed out that several pathologists or groups of pathologists independently confirmed this finding.

The Canadian Health Protection Branch (HPB) study employed groups of 50 male and 50 female rats, again of the Sprague-Dawley strain. The only saccharin dose level was 5.0%. Pups were weaned onto the parental diets, which both generations received for their lifetime. Rats from both generations fed 5.0% saccharin had decreased growth rates, but not lowered food consumption. There were no treatment effects on reproduction, longevity, or hematological parameters. The animals were free of bladder parasites. Although a few animals developed grossly visible bladder and kidney stones, the incidence was not treatment related. The incidence of bladder tumors in male rats fed the 5.0% saccharin diet was significantly increased in both generations, in comparison to their controls.

In 1983, a large bioassay conducted at the International Research Development Corporation (IRDC) in Mattawan, Michigan, was completed. In this bioassay, 2500 second-generation male rats were tested. Sodium saccharin dietary concentrations ranging between 1.0 and 7.5% were evaluated in 2125 second-generation male rats in the dose-response portion of the study. An unbalanced study design was used, which allowed for a larger number of animals at the lower dietary concentrations where the responses would be expected to be lower. For example, 700 animals were started in the 1.0% group, and 125 animals were fed at 7.5%. The study was designed to test the worst-case scenario by employing the use of (1) the most sensitive species (rat); (2) the most sensitive strain (Sprague-Dawley derived, Charles River CD rats); (3) the more sensitive sex (male); and (4) the most sensitive exposure period (second generation).

In addition to providing dose-response data on bladder tumor incidence, the study was designed to provide both qualitative and quantitative information with regard to the physiological changes that occur in the test animals. The role of in utero exposure, the effect of sodium, and the specificity of sodium saccharin in the occurrence of bladder tumors were also evaluated in other groups of second-generation male rats. Throughout the study, the pathology data were handled in a blind fashion, in that the pathologists were not provided information regarding treatment for the tissues they examined. One pathologist performed the gross examination of the bladder tissue and the microscopic examination of all scheduled tissues, including the bladder. A second pathologist also examined the bladder tissue.

The immense database on animals at all dose levels, including doses below 5.0%, provided a strong statistical correlation of physiological changes and tumor incidence as a function of dose. In the final study report, the researchers stated: "A clear dose response for urinary bladder tumours was observed in second-generation NaS-treated male rats. The steep slope of the dose-response curve indicated a rapid decline in tumour incidence with decreasing dose. The 1.0% dietary level (fed to 700 rats) was considered to be a no-effect level for bladder tumours" (94).

Thirty rats from each experimental group were followed in an in-life study, the objective of which was to relate the ultimate occurrence of a tumor in a rat to measurable characteristics of its urine. The osmolality of the urine and the daily urine volume appeared to be predictive of tumorigenicity (95). Mean urine osmolality for rats developing a tumor was lower than that of non–tumor-bearing rats, and rats with high urine volumes tended to develop tumors.

A panel of distinguished international scientists met at Duke University Medical Center, May 4–6, 1983, to assess the safety of saccharin. The panel reviewed relevant research, including the IRDC study. The panel concluded (96):

1. At high doses in the rat, saccharin produced profound biochemical and physiological changes that do not occur in humans under normal patterns of use.
2. The IRDC study showed that the lowest statistically significant response was at 3% of the rats' diet (equivalent in human terms to a person's drinking 750 cans of diet soft drink every day for a lifetime).
3. The appearance of tumors in rats seems to be a species- and organ-specific phenomenon for which there is at present no explanation.

Recent saccharin research has concentrated on understanding the mechanism of saccharin-related tumor formation in the male rat. The comprehensive approach has included chemical, biochemical, nutritional cell kinetic, ultrastructural, and pathology investigations. Considerable progress has been made in distinguishing between the effects of sodium saccharin and the other salt forms of saccharin and in investigations of the overall mechanism of action of saccharin in the male rat.

In an initiation promotion study by Dr. Samuel Cohen and his associates at the University of Nebraska, the promoting activity of sodium saccharin showed a clear dose response. This effect was mitigated with calcium saccharin and absent with acid saccharin. Furthermore, two other generally innocuous sodium salts, sodium chloride (table salt) and sodium ascorbate (vitamin C), had effects similar to that of sodium saccharin. Lowering the rats' elevated urinary pH by co-administering a buffering agent completely eliminated the bladder tumor activity of sodium saccharin (97).

Cohen and his colleagues have developed the following hypothesis derived from numerous studies. Sodium saccharin administered at high levels in the diet is associated with cellular proliferation and bladder tumors in the rat, especially the male. The increased proliferative response to sodium saccharin in the rat urothelium facilitates mitotic events that would be prone to spontaneous genetic errors. Tumor development could then occur under specific circumstances. These include beginning administration at the time of birth or earlier, administering saccharin after an initiating dose of a known bladder carcinogen, or co-administering it with low doses of a known carcinogen.

All of the tumorigenic protocols involving sodium saccharin in the rat can be explained in terms of altered cell proliferation (cell kinetics). The cell-proliferative effects following the administration of high doses of sodium saccharin in the diet are believed to be a consequence of the abrasive effects of a silicon-containing precipitate in the urine, the loss of urothelial cells from the mucosa, and consequent regenerative hyperplasia. The regenerative hyperplasia exceeds replacement levels because of numerous mitogenic factors in the urine, including epidermal growth factor, elevated pH, and high sodium.

Extensive research evaluating and supporting this hypothesis has been performed and is continuing (98).

## MUTAGENICITY

A substantial body of experimental data on the effects of saccharin in a variety of short-term tests has been developed. The tests include both microbial and mammalian cell genetic assays as well as mammalian cell neoplastic transformation assays. The most relevant assay, neoplastic transformation, was unequivocally negative (60,99–101). Additionally, saccharin was consistently negative in numerous Ames plate incorporation assays, and results appear to be negative for both in vivo tests and in vitro mammalian tests for gene mutation. Weakly positive responses for chromosomal aberrations in some other in vitro tests at physiologically unrealistic doses (99,102,103) provide little convincing evidence to the contrary. The validity of these assays and their ability to distinguish between possible direct effects and the adventitious physiological effects resulting from testing materials at high concentrations are subject to considerable doubt. Moreover, saccharin does not bind to DNA from the rat liver or bladder (61) and is a nucleophilic molecule that would not be expected to bind to DNA. Thus, the evidence indicates that saccharin does not act as a chemical mutagen or as an electrophilic carcinogen. Ashby (104) has suggested that the occasional alterations reported are a result of elevated dose levels where the levels of both saccharin and its accompanying cation are high, creating an osmotic effect rather than a direct chemical effect.

## EPIDEMIOLOGY

Saccharin's use in sweetening foods and beverages for over a century has allowed extensive study to determine whether it poses any real human cancer risk. Over 25 human-use studies have shown no overall association between saccharin use and human cancer.

Historically, as a group, diabetics have consumed greater amounts of saccharin than the general public and have done so for extended periods of time (25 years or longer). Three major studies (105–107) demonstrate that diabetics do not have an increased risk of cancer. As one study concluded, "saccharin is not carcinogenic to man in the amounts commonly used by diabetics" (106).

Several other studies either compared national bladder cancer rates with saccharin consumption (108–110) or examined bladder cancer patients to determine if they were heavy saccharin users (111–113). These studies show no overall association between saccharin and bladder cancer in humans.

On December 8, 1979, the National Cancer Institute released a report which specifically stated, "In the total study group (9,000 individuals), there was no evidence of increased risk with the long-term use of AS (artificial sweeteners) in any form or with use that began decades ago." The NCI study indicated that persons consuming the most saccharin and those who used it longest had no increased risk over nonusers. Also, the preliminary NCI report pointed out that increased risk may have been associated with some specific subgroups. However, as the study authors themselves point out, "the pattern of positive RR's [relative risk] described (at high doses, and particularly evident in a low-risk sub-group and among heavy smokers) could be ascribed to chance variation in subgroups of a study which, overall, shows no association between AS use and bladder cancer" (119).

One study investigated the risk of bladder cancer associated with in utero exposure to saccharin. The researchers concluded that bladder cancer rates of people born in Denmark during World War II do not differ from those of people born in the prewar decade, despite the fact that saccharin consumption was four to five times higher during the War (109).

In a large, case-control analysis of several types of cancer, Morrison showed "no support for an overall positive association of artificial sweeteners with cancer" (120).

Thus, extensive research on human populations has established no association between saccharin and cancer. Epidemiologist Dr. Robert Morgan has pointed out, "Few chemicals have been studied epidemiologically to the same extent as artificial sweeteners (saccharin and cyclamate) have been. The remarkable approximation to unity of the summary relative risk from all studies is impressive, and one wonders how many common foodstuffs would test to be that safe" (121).

Dr. O. M. Jensen of the Danish Cancer Registry, after reviewing the epidemiological data on saccharin, concluded that "for practical purposes saccharin cannot be considered a human carcinogen" (122).

A number of human studies have been reported since the reviews of Morgan and Jensen further supporting the safety of saccharin. A study of particular interest was reported by the same group of investigators who reported a controversial 1977 study (123). This subsequent study found no association between artificial sweeteners and increased risk of bladder cancer. The study included a total of 826 cancer patients and 792 controls. The researchers concluded "this study tends to confirm reports of a lack of association between the use of artificial sweeteners and subsequent risk of bladder cancer" (124).

## CARIOGENICITY

In studies with rats, findings indicate that animals fed a sucrose diet have a high incidence of caries development, as compared to animals fed diets composed of saccharin and various other carbohydrates. In rat studies by Tanzer and Slee (125), saccharin appeared to retard the growth rate of various plaque-forming microorganisms.

In a study of 24 student volunteers given diets containing low-calorie sweeteners composed of saccharin and up to 98% glucose, the amount of plaque formation and caries experience was lower than in the control group (126). Results do not clearly state that plaque reduction was due to saccharin, although saccharin has been found to suppress the growth of oral microorganisms (127,128).

In his review of the influence of sugar substitutes on oral microorganisms, Linke concludes that saccharin truly inhibits the growth and acid production of oral bacteria and therefore has potential in caries prevention (129).

## REGULATION

Saccharin remains a widely utilized noncaloric sweetener. In addition to its availability in the United States, saccharin also is used in more than 90 countries throughout the world.

In the United States, saccharin is currently approved for use under an interim food additive regulation, permitting use for special dietary purposes and in special dietary foods (130). The FDA specifically sanctioned these uses in trade correspondence and regulations issued from 1940 to 1942, reconfirming earlier U.S. approvals for saccharin (131).

In 1977, FDA proposed a ban on the use of saccharin largely because of the findings of bladder tumors in some male rats fed high doses of saccharin. The ban has been stayed five times by legislation, pending the outcome of additional scientific study.

The widespread use of saccharin has led to its review by numerous scientific groups. The U.S. National Academy of Sciences has studied saccharin on several occasions. The Academy first reviewed saccharin in 1955 and reported: "The available information concerning saccharin and long experience with its use indicates it to be relatively non-toxic" (132). In 1968, the Academy once again concluded: "There seemed to the Committee no reason to differ with the earlier implied conclusion of the (1955 NAS) committee . . . that intakes of 1 gram or less per day of saccharin . . . by the adult should present no hazard" (133).

A third review, in 1970, led to the Academy's conclusions that

> the present and projected usage of saccharin in the United States does not pose a hazard . . . considering the low level of toxicity of saccharin and the fact that there is nothing in the long history of saccharin use that indicates adverse effects other than a few reported incidences of photosensitization or allergic response. (134)

As each of these NAS reports emphasized, saccharin's safety has been assured by both its long history of human use and extensive animal testing.

By the time the NAS once again reviewed the data in 1974, bladder tumors had been found in some male second-generation rats, who, like their parents, had been fed saccharin at doses of 5.0 or 7.5% of their diet in experiments conducted by the FDA and by the Wisconsin Alumni Research Foundation (WARF) (135). Because it found a number of significant flaws in the FDA and WARF studies, the NAS Committee concluded that no regulatory actions against saccharin were warranted. The Committee stated: "The results of toxicity studies thus far reported have not established conclusively whether saccharin is or is not carcinogenic when administered orally, to test animals."

The Committee, however, recommended a variety of new studies to examine the cancer issue. Their suggestions included an investigation of the question of transplacental carcinogenesis of saccharin, the study of the changes in urine composition at high levels of saccharin intake and of the relation of such changes to the induction of bladder stones or calculi, and studies of human use of saccharin. The Committee also recommended studies on saccharin impurities.

In 1977, following FDA's proposal to ban saccharin, NAS was asked by Congress to review the saccharin data again. NAS presented a two-part report. The first part addressed saccharin per se, and the second part discussed saccharin in the broader context of a rational food policy. Although concluding that the animal results indicated that saccharin could be a potential human carcinogen, the 1978 NAS report nonetheless noted: "Research leading to a better understanding of methods for qualitative and quantitative extrapolation, in utero exposure and mechanisms of cancer promotion would be highly desirable" (136).

In transmitting the second part of the NAS report, Academy President Dr. Philip Handler agreed with Committee members that the scientific evidence on saccharin postulates only "a risk of a risk" (137). None of the Committee members favored either an immediate ban on saccharin use or marketing without some precautionary labeling.

Since the publication of the most recent NAS reports, several significant studies have been released that provide additional support for saccharin's safety. These studies, discussed earlier in this chapter, provide scientific data in precisely those areas where further study was recommended in 1978, 1979, and in earlier NAS reviews, and were evaluated by the expert panel assembled at Duke University Medical Center in 1983. In assessing saccharin's safety, the panel concluded that "the present level of exposure of humans to saccharin through its use as a food additive is unlikely to present a cancer risk" (96). Additional research conducted since 1983 further supports saccharin's safety for human use.

Saccharin has been marketed for many years as a sweetener for food, beverage, and table-top use in Japan and most European and South American nations. Continued marketing of saccharin in Great Britain was rerecommended by the government's expert Food Additives and Contaminants Committee (138).

The Joint Food and Agriculture Organization of the United Nations/World Health Organization Expert Committee on Food Additives (JECFA) has reviewed saccharin on a number of occasions, most recently in 1984. The Committee supported saccharin's continued use (139).

In 1987, the Scientific Committee for Food of the Commission of the European Economic Communities (EEC) issued a report on sweeteners. The Committee again approved the use of saccharin (140).

In summary, the case against saccharin still rests primarily on controversial high-dose rat studies in which a sensitive strain was fed the human equivalent of the sodium saccharin in hundreds of cans of diet soft drink a day for a lifetime. Even then, the tests produced bladder tumors only in some of the male rats at the highest doses. It is ironic that sodium saccharin's low acute toxicity ($LD_{50}$-oral rat = 14,200 mg/kg) is the very characteristic that permits such massive doses (141).

Saccharin is one of the most studied ingredients in the food supply. The totality of the reported research confirms saccharin is safe for human consumption. At a 1985 Senate hearing, then FDA Commissioner Frank Young, supporting an extension of the saccharin ban moratorium, noted the FDA has less concern about saccharin than it did in 1977 and added that "the actual risk, if any, of saccharin to humans still appears to be slight" (142).

## REFERENCES


1. Kirk-Othmer, *Encyclopedia of Chemical Technology*, 3rd ed., Vol. 22, Wiley-Interscience, New York, 1983, pp. 449–452.
2. G. B. Kauffman and P. M. Priebe, *Ambix, 25:* Pt III 191–207 (1978).
3. C. Fahlberg and I. Remsen, *Berichte, 12:* 469–473 (1879).
4. C. Fahlberg, U.S. Patent 319,082 (1885).
5. D. S. Tarbell and A. T. Tarbell, *J. Chem. Educ., 55:* 161–162 (1978).
6. J. Forrestal, *Faith, Hope and $5,000,* Simon and Schuster, New York, 1977, pp. 17–22.
7. E. M. Whelan, *The Conference Board Magazine, 16:* 54 (1977).
8. T. Smith, *Saccharin,* American Council on Science and Health, 1979, p. 14.
9. O. F. Senn, U.S. Patent 2,667,503 (1954).
10. G. F. Schlaudecker, U.S. Patent 2,705,242 (1955).
11. P. Bedoukian, *Perfumery and Flavoring Synthetics,* Elsevier, New York, 1967, pp. 41–47.
12. *Chemical Engineering, 61:* 128 (1954).
13. *Chemical and Engineering News, 41:* 76–78 (1963).
14. *Food Chemical Codex,* 3rd ed., National Academy Press, Washington, D.C., 1981.
15. I. H. Pitman, H.-S. Dawn, T. Higuchi, and A. A. Hussain, *J. Chem. Soc., B:* 1230 (1969).
16. A. Salant, *Handbook of Food Additives,* The Chemical Rubber Co., Cleveland, 1968, pp. 503–512.
17. O. J. Magidson and S. W. Gorbatschow, *Chem. Ber., 56B:* 1810–1817 (1923).
18. J. C. J. Bart, *J. Chem. Soc., B:* 376–382 (1968).
19. Y. Okaya, *Acta Cryst., B, 25:* 2257 (1969).
20. T. Paul, *Chem. Zeit., 45:* 38–39 (1921).
21. M. A. Amerine, R. M. Pangborn and E. B. Roessler, *Food Science and Technology,* Academic Press, New York, 1965, pp. 86–103.
22. D. Scott, U.S. Patent 3,780,189 (1973).
23. G. Von Ryman Lipinski and E. Luck, U.S. Patent 4,158,068 (1979).
24. F. J. Helgren, U.S. Patent 2,803,551 (1957).
25. P. K. Beyts and Z. Latymor, Eur. Pat. Appl. EP64361 (1982). (*Chem. Abstr., 98:* 87933b).
26. T. M. Brennan and M. E. Hendrick, U.S. Patent 4,411,925 (1983).
27. K. Hoppe, *Nahrung, 25:* 769 (1981).
28. P. Van Tornout, J. Pelgroms, and J. Vander Meeren, *J. Food Sci., 50:* 469 (1985).
29. F. J. Helgren, M. J. Lynch and F. J. Kirchmeyer, *J. Am. Pharm. Assoc., 44:* 353 (1955).

30. C. P. Rader, S. G. Tihanyi, and F. B. Zienty, *J. Food Sci., 32:* 357–360 (1967).
31. H. Schmandke, K. Hoppe, and B. Gassmann, *Nahrung, 29:* 417–420 (1985).
32. T. Paul, *Chem. Zeit., 45:* 705–706 (1921).
33. K. M. Beck, *Food Prod. Dev., 9:* 47–54 (1975).
34. T. M. Brennan and M. E. Hendrick, U.S. Patent 4,399,163 (1983).
35. A. I. Bakal, *Chem. Ind., 18:* 700 (1983).
36. H. Lim, C. S. Setser and S. S. Kim, *J. Food Sci., 54:* 625 (1989).
37. T. P. Finucane, Canadian Patent 1,043,158 (*Chem. Abstr., 90:* 102244h).
38. L. Hyvonen, P. Koivistoinen, and A. Ratilainen, *J. Food Sci., 43:* 1580 (1978).
39. L. Hyvonen, R. Kurkela, P. Koivistoinen, and A. Ratilainen, *J. Food Sci., 43:* 1577 (1978).
40. L. Hyvonen, R. Kurkela, P. Koivistoinen, and A. Ratilainan, *J. Food Sci., 43:* 251 (1978).
41. M. E. Eisenstadt, U.S. Patent 3,625,711 (1971).
42. R. Daniels, *Sugar Substitutes and Enhancers,* Noyes Data Corporation, N.J., 1973, pp. 80–103.
43. PMC Specialties Group, unpublished data.
44. A. Porter, *Chem. Ind., 18:* 696–699 (1983).
45. O. De Garmo, G. W. Ashworth, C. M. Eaker, and R. H. Munch, *J. Am. Pharm. Assoc., 41:* 17 (1952).
46. G. A. Di Bari, *Metal Finishing,* 57th Guidebook-Directory, 1989, pp. 244–265.
47. S. Seki, et. al., U.S. Patent 3,629,428 (1971).
48. G. Levitt, U.S. Patent 4,238,621 (1980).
49. M. R. Kare and G. K. Beauchamp, *Dukes' Physiology of Domestic Animals,* 9th ed., Cornell University Press, Ithaca, 1977, pp. 713–730.
50. J. G. Lombardino, U.S. Patent 4,289,879 (1981).
51. *Chem. Marketing Rep.:* 17 (November 14, 1988).
52. G. T. Hollander and M. Blum, U.S. Patent 4,661,504 (1987).
53. R. L. Wakeman and J. F. Coates, U.S. Patent 3,280,137 (1966).
54. V. K. Krieble, U.S. Patent 3,046,262 (1962).
55. V. K. Krieble, U.S. Patent 3,218,305 (1965).
56. L. L. Bambas, *The Chemistry of Heterocyclic Compounds,* Vol. 4, Interscience, New York, 1952, pp. 318–353.
57. H. Hettler, *Advances in Heterocyclic Chemistry,* Vol. 15, Academic Press, New York, 1973, 233–276.
58. J. L. Lowe, *Chem. Eng. Monogr., 15:* 285–305 (1982).
59. E. C. Miller, Some current perspectives on chemical carcinogenesis in humans and experimental animals, *Cancer Res., 38:* 1479–1496 (1978).
60. J. Ashby et al., Saccharin: An epigenetic carcinogen/mutagen?, *Food Cosmet. Toxicol., 16:* 95–103 (1978).
61. W. K. Lutz and C. Schlatter. Sacharin does not bind to DNA of liver or bladder in the rat, *Chemico-Biological Interactions, 19:* 253–257 (1977).
62. H. B. Matthews, M. Fields, and L. Fishbein, Saccharin: Distribution and excretion of a limited dose in the rat, *J. Agric. Food Chem., 21:* 916–919 (1973).
63. J. L. Byard and L. Goldberg, The metabolism of saccharin in laboratory animals, *Food Cosmet. Toxicol., 11:* 391–402 (1973).

64. L. M. Ball et al., The fate of [$^{14}$C] saccharin in man, rat and rabbit and of 2-sulphamoyl [$^{14}$C] benzoic acid in the rat, *Xenobiotica, 7:* 189–203 (1977).
65. T. W. Sweatman and A. G. Renwick, Saccharin metabolism and tumorigenicity, *Science, 205:* 1019–1020 (1979).
66. E. J. Lethco and W. C. Wallace, The metabolism of saccharin in animals, *Toxicology, 3:* 287–300 (1975).
67. A. G. Renwick and R. T. Williams, The fate of saccharin impurities, *Xenobiotica, 8*(8): 475–489 (1978).
68. R. M. Pitkin et al., Saccharin metabolism in *Macaca mulatta, Proc. Soc. Exp. Biol. Med., 137:* 803–806 (1971).
69. G. Kennedy, O. Fancher, and J. Calandra, Metabolic fate of saccharin in the albino rat, *Food Cosmet. Toxicol., 10:* 143–149 (1972).
70. K. I. Minegishi, M. Asahina, and T. Yamaha, The metabolism of saccharin and the related compounds in rats and guinea pigs, *Chem. Pharmaceut. Bull., 20:* 1351–1356 (1972).
71. T. W. Sweatman and A. G. Renwick, The tissue distribution and pharmokinetics of saccharin in the rat, *Toxicol. Appl. Pharmacol., 55:* 18–31 (1980).
72. A. G. Renwick, The metabolism of intense sweeteners, *Xenobiotica, 16*(10/11): 1057–1071 (1986).
73. T. W. Sweatman and A. G. Renwick, Tissue levels of saccharin in the rat during two-generation feeding studies, *Toxicol. Appl. Pharmacol., 62:* 465–473 (1982).
74. A. G. Renwick and T. W. Sweatman, The absorption of saccharin from the rat urinary bladder, *Comm. J. Pharm. Pharmacol., 31:* 650–652 (1979).
75. W. A. Colburn, Absorption of saccharin from rat urinary bladder, *J. Pharmaceut. Sci., 67:* 1493–1494 (1978).
76. A. G. Renwick, The fate of non-nutritive sweeteners in the body, in *Developments in Sweeteners*-2 (T. H. Grenby, K. J. Parker, and M. G. Lindley, eds.), Applied Science Publishers, London and New York, 1983, pp. 179–224.
77. O. G. Fitzhugh, A. A. Nelson, and J. P. Frawley, A comparison of the chronic toxicities of synthetic sweetening agents, *J. Am. Pharm. Assoc., 40:* 583–586 (1951).
78. B. Lessel, A two-year trial on saccharin for carcinogenic activity, Report No. 1014, Biology Division, Boots Pure Drug Co., Ltd., unpublished, 1959.
79. D. Schmahl, Absence of carcinogenic activity of cyclamate, cyclohexylamine, and saccharin in rats, *Arzneimittelforschung, 23:* 1466–1470 (1973).
80. Litton Bionetics, Inc., Carcinogenicity of chemicals present in man's environment, Final Report, unpublished, January 31, 1973.
81. F. Homburger et al., Bio-Research Consultants, Final Report. Contract No. NIH-NCI-E-68-1311, 1973, unpublished.
82. I. C. Munro et al., A carcinogenicity study of commercial saccharin in the rat, *Toxicol. Appl. Pharmacol., 32:* 513–516 (1975).
83. T. Furuya et al., Long term toxicity study of sodium cyclamate and saccharin sodium in rats, *Jpn. J. Pharmacol., 25* (Suppl.): 55–56 (1975).

84. R. Kroes et al., Long term toxicity and reproduction study (including teratogenicity study) with cyclamate, saccharin and cyclohexylamine, *Toxicology, 8:* 285 (1977).
85. J. Althoff et al., A chronic study of artificial sweeteners in Syrian Golden Hamsters, *Cancer Lett., 1:* 121–124 (1975).
86. M. F. Cranmer, *Saccharin: A Report,* American Drug Research Institute, Inc., 1980.
87. F. J. C. Roe, L. S. Levy, and R. L. Carter, Feeding studies on sodium cyclamate, saccharin and sucrose for carcinogenic and tumor-producing activity, *Food Cosmet. Toxicol., 8:* 135–145 (1970).
88. T. Miyaji et al., National Institute of Hygienic Sciences, *Chronic Toxicity Study of Sodium Saccharin: 21 Months Feeding in Mice,* Tokyo, Japan, 1973.
89. F. Coulston, E. W. McChesney, and L. Golberg, Long-term administration of artificial sweeteners to the rhesus monkey, *Food Cosmet. Toxicol., 13:* 297–300 (1975).
90. E. H. Adamson and S. M. Sieber, *Chemical Carcinogenesis Studies in Nonhuman Primates* (R. Langenbach, S. Nasnow, and J. M. Rice, eds.) Plenum Publishing Corp., New York, 1983.
91. M. O. Tisdel et al., Long-term feeding of saccharin in rats, in *Symposium: Sweeteners* (George Inglett, ed.), Avi Publishing Co., Inc., Westport, CT, 1974, pp. 145–158.
92. J. M. Taylor, M. A. Weinberger, and L. Friedman, Chronic toxicity and carcinogenicity to the urinary bladder of sodium saccharin in the *in utero*-exposed rat, *Toxicol. Appl. Pharmacol., 54:* 57–75 (1980).
93. D. L. Arnold et al., Long-term toxicity study of orthotoluenesulfanamide and sodium saccharin in the rat, *Toxicol. Appl. Pharmacol., 52:* 113–152 (1980).
94. G. P. Schoenig et al., Evaluation of the dose response and *in utero* exposure of saccharin in the rat, *Fd. Chem. Toxic., 23*(4/5): 475–490 (1985).
95. F. W. Carlborg, A statistical analysis of the IRDC experiment with saccharin, Scientific Review Conference, Duke University Medical Center, May 4, 1983.
96. L. Golberg et al., Saccharin: Current Status, Report of an Expert Panel, *Fd. Chem. Toxic., 23:* 543–546 (1985).
97. S. M. Cohen et al., Comparative bladder tumor promoting activity in rats of sodium saccharin, sodium ascorbate and related acids and calcium salts (submitted for publication).
98. L. B. Ellwein and S. M. Cohen, The health risks of saccharin revisited, *Critical Reviews in Toxicology, 20*(5): 311–326 (1990).
99. J. C. McCann, Cancer testing technology and saccharin, Appendix II, short-term tests, Office of Technology Assessment, Congress of the United States, 1977, pp. 91–143.
100. S. Mondal, D. W. Brankow, and C. Heidelberger, Enhancement of oncogenesis in C3H-10T-1/2 mouse embryo cell culture by saccharin, *Science, 201:* 1141–1142 (1978).

101. A. Sivak and A. S. Tu, Cell culture tumor promotion experiments with saccharin, phorbol myristate acetate and several common food materials, *Cancer Lett., 10:* 27–32 (1980).
102. D. Clive et al., Validation and characterization of the L5178Y/TK$^+$ / mouse lymphoma mutagen assay system, *Mutation Res., 59:* 61–108 (1979).
103. E. Evans et al., A comparative investigation of the mutagenic and clastogenic effects of saccharin, ascorbic acid, pantothenic acid and ergocalciferol in several *in vitro* mammalian and microbial assays. Unpublished Report by SRI International and Litton Bionetics, Inc., 1984.
104. J. Ashby, The genotoxicity of sodium saccharin and sodium chloride in relation to their cancer-promoting properties, *Fd. Chem. Toxic., 23*(4/5): 507–519 (1985).
105. B. Armstrong et al., Cancer mortality and saccharin consumption in diabetics, *Br. J. Prev. Soc. Med., 30:* 151–157 (1976).
106. B. Armstrong and R. Doll, Bladder cancer mortality in diabetics in relation to saccharin consumption and smoking habits, *Br. J. Prev. Soc. Med., 29:* 73–81 (1975).
107. I. I. Kessler, Cancer mortality among diabetics, *J. Nat. Cancer Inst., 44:* 673–686 (1970).
108. A. S. Morrison et al., Artificial sweeteners and bladder cancer in Manchester, U. K., and Nagoya, Japan, *Br. J. Cancer, 45:* 332–336 (1982).
109. O. M. Jensen and C. Kamby, Intra-uterine exposure to saccharin and risk of bladder cancer in man, *Int. J. Cancer, 29:* 507–509 (1982).
110. F. Burbank and J. F. Fraumen, Synthetic sweetener consumption and bladder cancer trends in the United States, *Nature, 227:* 296–297 (1970).
111. A. S. Morrison and J. E. Buring, Artificial sweeteners and cancer of the lower urinary tract, *N. Engl. J. Med., 302:* 537–541 (1980).
112. E. L. Wynder and S. D. Stellman, Artificial sweetener use and bladder cancer; a case-control study: A methodological issue, *Science, 210:* 1214–1216 (1980).
113. I. Kessler and J. P. Clark, Saccharin, cyclamate and human bladder cancer: No evidence of an association, *JAMA, 240:* 349–355 (1978).
114. J. G. Connolly et al., Relation between the use of artificial sweeteners and bladder cancer, *Can. Med. Assoc. J., 119:* 408 (1978).
115. E. L. Wynder and R. Goldsmith, The epidemiology of bladder cancer: A second look, *Cancer, 40:* 1246–1268 (1977).
116. D. Simon, S. Yen, and P. Cole, Coffee drinking and cancer of the lower urinary tract, *J. Natl. Cancer Inst., 54:* 587–591 (1975).
117. B. Armstrong and R. Doll, Bladder cancer mortality in England and Wales in relation to cigarette smoking and saccharin consumption, *Br. J. Prev. Soc. Med., 28:* 231–240 (1974).
118. R. W. Morgan and M. G. Jain, Bladder cancer: Smoking, beverages, and artificial sweeteners, *Can. Med. Assoc. J., 3:* 1067–1070 (1974).
119. R. N. Hoover and P. H. Strasser, Artificial sweeteners and human bladder cancer, *Lancet, 1*(April 19): 837(1980).

120. A. S. Morrison, Use of artificial sweeteners by cancer patients, *J. Natl. Cancer Inst., 62:* 1397–1399 (1979).
121. R. W. Morgan and O. Wong, A review of epidemiological studies of artificial sweeteners and bladder cancer, *Fd. Chem. Toxic., 23:* 529–533 (1985).
122. O. M. Jensen, Artificial sweeteners and bladder cancer: Epidemiological evidence, Third European Toxicology Forum, Geneva, Switzerland, October, 1983.
123. G. R. Howe et al., Artificial sweeteners and human bladder cancer, *Lancet, II*(September 17): 578–581(1977).
124. H. A. Risch, Dietary factors and the incidence of cancer of the urinary bladder, *Am. J. Epidem., 127:* 1179–1191 (1988).
125. J. M. Tanzer and A. M. Slee, Saccharin inhibits tooth decay in laboratory models, *J. Am. Dent. Assoc., 106*(March): 331–333(1983).
126. T. H. Grenby, Dental plaque, dental caries and sugar intake, *Br. Dent. J., 139:* 129–134 (1975).
127. J. Carlsson, Potentials of the oral microflora to utilize sugar substitutes as energy source, in *Health and Sugar Substitutes* (B. Guggenheim, ed.), 1978, pp. 205–210.
128. H. A. B. Linke, Growth-inhibition of glucose-grown cariogenic and other streptococci by saccharin *in vitro, Z. Naturforschung, 32:* 839–843 (1977).
129. H. A. B. Linke, Sweeteners and dental health: The influence of sugar substitutes on oral microorganisms, in *Developments in Sweeteners-3,* (T. H. Grenby, ed.), 1987, pp. 151–188.
130. 21 C.F.R.§ 180.37.
131. TC-311 (August 20, 1940); 6 *Fed. Reg.* 5921 (November 22, 1941); TC-381 (January 23, 1942; TC-388 (July 27, 1942).
132. Food Protection Committee of the Food and Nutrition Board, National Academy of Sciences-National Research Council, *The Safety of Artificial Sweeteners for Use in Foods,* August, 1955, p. 7.
133. Ad Hoc Committee on Non-nutritive Sweeteners, Food Protection Committee, National Academy of Sciences-National Research Council, *Non-nutritive Sweeteners: An Interim Report to the FDA,* November, 1968, p. 8.
134. Food Protection Committee of the Food Protection Board, National Academy of Sciences-National Research Council, *Safety of Saccharin for Use in Foods,* July, 1970, p. 10.
135. Sub-committee on Non-nutritive Sweeteners, Committee on Food Protection, National Academy of Sciences-National Research Council, *Safety of Saccharin and Sodium Saccharin in the Human Diet,* December, 1974, p. 63.
136. Committee for the Study of Saccharin and Food Safety Policy, Part I, National Academy of Sciences-National Research Council, *Saccharin Technical Assessment of Risks and Benefits,* November, 1978, p. ES-5.
137. Committee for a Study on Saccharin and Food Safety Policy, National Academy of Sciences, *Food Safety Policy: Scientific and Societal Considerations,* March 1, 1979, pp. iiA; 3–6; 10–6,7.

138. United Kingdom Food Additives and Contaminants Committee (FACC), *Report on the Review of Sweeteners in Food,* 1982, p. 28.
139. Joint FAO/WHO Expert Committee on Food Additives (JECFA), *Evaluation of Certain Food Additives and Contaminants,* Twenty-eighth report, Technical Report Series, World Health Organization, Geneva, 1984.
140. Scientific Committee for Food, Commission of the European Economic Communities (EEC), *Report of the Scientific Committee for Food on Sweeteners,* December, 1987.
141. N. I. Sax, *Dangerous Properties of Industrial Material,* 6th ed., Van Nostrand Reinhold Co., New York, 1984, p. 2385.
142. F. E. Young, Commissioner, Food and Drug Administration, Public Health Service, Department of Health and Human Services, Statement Before the Committee on Labor and Human Resources, United States Senate, April 2, 1985.

# 9

# Stevioside

**A. Douglas Kinghorn and D. Doel Soejarto** *University of Illinois at Chicago, Chicago, Illinois*

## INTRODUCTION

*Stevia rebaudiana*, a South American plant, is the source of the highly sweet *ent*-kaurenoid diterpene glycosides, stevioside, rebaudioside A, and several other sweet analogs. Although stevioside, the major sweet constituent of this species, was first isolated in impure form in the first years of this century, this compound was not utilized as a sweetening agent until its development for this purpose by a group of Japanese manufacturers in the early 1970s. Currently, by far the largest use of stevioside occurs in Japan. For this compilation, the present authors have had access not only to the previous version of this chapter (1), but also to several recent comprehensive reviews on stevioside and rebaudioside A (2–4). Particular emphasis will be made on the latest additions to our knowledge of stevioside.

## PRODUCTION

*Stevia rebaudiana* Bertoni is a member of a New World genus of 150–300 species that belongs to the tribe Eupatorieae of the family Asteraceae (chrysanthemum family). The plant may reach a height of 80 cm when fully grown and is native to elevated terrains in the Amambay and Iguaçu provinces on the borders of Paraguay and Brazil (2). Although high concentration levels (10% w/w or more) of stevioside and other sweet diterpene glycosides occur in dried *S. rebaudiana* leaves, this type of sweet compound appears to be rare in the genus *Stevia* as a whole. When 110 *Stevia* leaf herbarium samples were analyzed for the presence of sweet compounds, stevioside was found in only one species in addition to *S. rebaudiana*, namely, *S. phlebophylla* A. Gray, a Mexican species which may now be extinct (5).

The commercialization of *S. rebaudiana* leaves for sweetening purposes has been very rapid since its introduction to Japan. Thus, in 1988, *S. rebaudiana* extracts containing stevioside comprised 41% by value of the Japanese high-potency sweetener market (6). In 1987, a total of 1,700 metric tons of *S. rebaudiana* leaves were utilized in Japanese food products, representing an estimated 190 metric tons of pure stevioside (6). However, only a small proportion of this amount was actually cultivated in Japan, with about 75% of the *S. rebaudiana* leaves produced in the People's Republic of China, and smaller quantities being exported to Japan by Taiwan, Thailand, and Malaysia (6). In the People's Republic of China, *S. rebaudiana* is cultivated mainly in Fujian, Zhejiang, and Guangdong Provinces (7). Table 1 shows the profile of the Japanese supply of *S. rebaudiana* leaves over the 6-year period of 1982–1987.

Methods for the production of extracts of *S. rebaudiana* leaves containing stevioside in varying degrees of purity have been summarized (1–3). Much of this type of information has appeared in the patent literature, especially from Japan. Most methods for the purification of stevioside involve initial extraction into an aqueous solvent, followed by refinement involving one or more of selective extraction into a polar organic

**Table 1** Origin of *Stevia rebaudiana* Leaves Used in Japan Between 1982 and 1987

| Year | Total amount of raw leaves[a] | Sources of Raw Leaves[a] | | | | |
|---|---|---|---|---|---|---|
| | | Japan | South Korea | Taiwan | PRC[b] | Others[c] |
| 1982 | 700 | 200 | 30 | 200 | 200 | 70 |
| 1983 | 1000 | 300 | 30 | 150 | 450 | 70 |
| 1984 | 1400 | 200 | 0 | 200 | 1000 | 0 |
| 1985 | 1600 | 200 | 0 | 150 | 1200 | 50 |
| 1986 | 1500 | 200 | 0 | 150 | 1100 | 50 |
| 1987 | 1700 | 200 | 0 | 100 | 1300 | 100 |

[a]Expressed in metric tons.
[b]People's Republic of China.
[c]Constituted by Paraguay, Brazil, Thailand, and Malaysia.
*Source:* Adapted from Ref. 6.

solvent, decolorization, precipitation, coagulation, adsorption, ion exchange, and crystallization (1–3).

## PHYSICAL AND ORGANOLEPTIC PROPERTIES

While chemical work to determine the structural nature of stevioside began at the turn of the century, over 60 years elapsed before the structure of this compound was finally elucidated. A group at the National Institutes of Health, Bethesda, Maryland, was able to show that one sugar unit of stevioside occurs as a glucopyranose functionality attached $\beta$ to a carboxyl group, while the second sugar unit is a sophorose unit attached $\beta$ to an alcoholic group on the aglycone (8,9). The structure and stereochemistry of steviol (*ent*-13-hydroxykaur-16-en-19-oic acid), the aglycone obtained on the enzymatic hydrolysis of stevioside, were finally determined in 1963 (10). During the 1970s, additional sweet analogs of stevioside were identified as *Stevia rebaudiana* leaf constituents, comprising rebaudiosides A and B (11), steviolbioside (11), rebaudioside C ( = dulcoside B) (12,13), rebaudiosides D and E (14), and dulcoside A (13). The structures of the eight sweet constituents of *S. rebaudiana* leaves are shown in Figure 1. The sweet *ent*-kaurenoid glycosides occur in very high concentration levels in dried *S. rebaudiana* leaves, with approximate values for the four most abundant compounds being stevioside (5–10% w/w), rebaudioside A (2–4% w/w), rebaudioside C (1–2% w/w), and dulcoside A (0.4–0.7% w/w) (2).

In pure form, stevioside is a white, crystalline material with a melting point of 196–198° C, an optical rotation of −39.3° in water, and an elemental composition of $C_{38}H_{60}O_{18}$ (8). However, stevioside is not generally used as a sweetener in pure form, but rather as *S. rebaudiana* extracts of various degrees of purity, such as 10, 50, and 90% w/w (1). A large number of analytical methods are available for the determination of purity of stevioside (2,3).

Stevioside has been rated as possessing about 300 times the relative sweetness intensity of 0.4% w/v sucrose solution, although the compound exhibits some bitterness and an undesirable aftertaste (1). The sweetness intensities (i.e., sweetening power relative to sucrose, which is taken as 1) of the other seven *S. rebaudiana* sweet principles have been determined as follows: rebaudioside A, 250–450; rebaudioside B, 300–350; rebaudioside C, 50–120; rebaudioside D, 250–450; rebaudioside E, 150–300; dulcoside A, 50–120; and steviolbioside, 100–125 (4).

There have been many attempts to improve the sweetness hedonic parameters of stevioside by formulation with a variety of flavor-masking

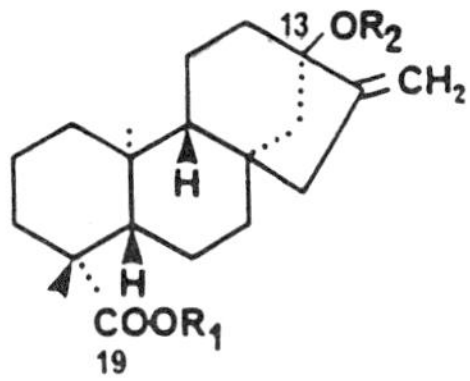


| | $R_1$ | $R_2$ |
|---|---|---|
| Stevioside | $\beta$-glc | $\beta$-glc$^{2}$-$\beta$-glc |
| Rebaudioside A | $\beta$-glc | $\beta$-glc$^{2}$-$\beta$-glc<br>$^{3}$<br>$\beta$-glc |
| Rebaudioside B | H | $\beta$-glc$^{2}$-$\beta$-glc<br>$^{3}$<br>$\beta$-glc |
| Rebaudioside C | $\beta$-glc | $\beta$-glc$^{2}$-$\alpha$-rha<br>$^{3}$<br>$\beta$-glc |
| Rebaudioside D | $\beta$-glc$^{2}$-$\beta$-glc | $\beta$-glc$^{2}$-$\beta$-glc<br>$^{3}$<br>$\beta$-glc |
| Rebaudioside E | $\beta$-glc$^{2}$-$\beta$-glc | $\beta$-glc$^{2}$-$\beta$-glc |
| Dulcoside A | $\beta$-glc | $\beta$-glc$^{2}$-$\alpha$-rha |
| Steviolbioside | H | $\beta$-glc$^{2}$-$\beta$-glc |

Glc = D-glucopyranosyl; rha = L-rhamnopyranosyl

**Figure 1** Structures of the eight sweet-tasting glycoside constituents of *Stevia rebaudiana* leaves.

and sweetness-enhancing agents (1–3). Much research has also been conducted to produce strains of *S. rebaudiana* that have a higher ratio of rebaudioside A to stevioside compared to those of the wild Paraguayan populations, because of the increased sweetness and more pleasant taste exhibited by extracts of the plant that contain proportionately more rebaudioside A than stevioside (3). In another type of strategy, enzymatic transglucosylation of steviol derivatives has led to analogs of stevioside with improved taste profiles over the parent substance (15,16). The sweetness intensity of such a *S. rebaudiana* "sugar-transferred" product of commercial use is quoted as being about 120 times that of sucrose (6).

## AVAILABILITY AND COST

It is apparent that the natural area of distribution of *Stevia rebaudiana* has been considerably reduced in recent years in its native habitat in the Sierra of Amambay in northeastern Paraguay, where the plant is found in natural grasslands on mountain slopes and valleys occurring at altitudes of 500–700 meters above sea level. The reason for this is that many of the subtropical forests surrounding the natural populations of *S. rebaudiana* have been exploited for their timber, with the newly cleared land used for other agricultural purposes. In addition, thousands of *S. rebaudiana* plants have been transplanted to other areas to start large-scale plantations. Because *S. rebaudiana* is sensitive to changes in its environment, the net effect is that there is now a danger to the germplasm of this species (17).

As noted earlier in this chapter, by far the largest cultivation areas where *S. rebaudiana* is presently produced are in the People's Republic of China. Raw *S. rebaudiana* leaves were priced at $4,500 per metric ton at the time of their first introduction from China to Japan in 1981. However, by 1985, the price of the Chinese leaves dropped to $1,400 per metric ton, though it has risen again slightly since then (6). If it is assumed that the Chinese *S. rebaudiana* leaves contain 5–10% stevioside, then, based on their 1985 value, the price of the stevioside contained within the leaves would be about $15–30 per kilogram. However, stevioside, when pure or partially purified, is somewhat more expensive than this figure. For example, 90% pure stevioside was priced at between $90 and $130 per kilogram in 1982 (1).

## USE AND ADMIXTURE POTENTIAL

Although *Stevia rebaudiana* leaves are well known to have been employed for centuries to sweeten bitter beverages such as maté in Paraguay (2–4), Phillips has suggested that this plant was not of particularly great significance to the indigenous Guaraní Indians of that country, since they placed more value on the use of honey as a sweetener (3). Indeed, currently the plant appears to be only used sporadically for sweetening beverages in Paraguay (17). However, for several years teas made from *S. rebaudiana* leaves have been prescribed by physicians in Paraguay for the treatment of diabetes (17). In neighboring Brazil, two products, namely, *S. rebaudiana* tea and *S. rebaudiana* capsules, were officially approved for sale for this purpose in 1980 (2).

*Stevia rebaudiana* extracts were introduced commercially for sweetening purposes in Japan in 1976 (6). Since then, Japanese commercialization of *S. rebaudiana* has been very rapid, which has been attributed to several facts. First, sucrose consumption has been curtailed in Japan due to health concerns related to dental caries, obesity, and diabetes. Second, certain artificial sweeteners were banned or severely restricted in Japan in the 1960s. Third, there has been a perception among the Japanese government regulatory agencies and public alike that synthetic compounds are inherently more harmful than naturally occurring substances (18). Extracts of *S. rebaudiana* containing stevioside have several particular advantages that have contributed to their development as widely used noncaloric sucrose substitutes in Japan. For example, stevioside has been found to suppress the pungency of sodium chloride, a universal preservative and flavoring agent that is commonly added to Japanese-style vegetables, dried seafoods, soy sauce, and miso (bean paste) products. Stevioside is relatively stable under normal elevated temperatures involved in food processing, and does not turn brown on heating or ferment during use. The compound does not precipitate at the acid pH levels (ca. pH 3) characteristic of many soft drinks (2). However, the undesirable hedonic effects of stevioside that have already been mentioned have limited its use for sweetening certain categories of drinks such as diet colas and ciders in Japan (6). Table 2 provides estimates of the major categories of Japanese foods in which *S. rebaudiana* extracts were used in 1988 (6).

Stevioside was recently approved as a food additive in South Korea, where it must be of at least 98% purity for use. It is not yet permitted for

**Table 2** Major Categories of Food Use of Stevioside in Japan in 1988

| Use | Estimated amount[a] |
|---|---|
| Salty foods | 66 |
| Drinks | 50 |
| Processed seafoods | 30 |
| Delicacies | 20 |
| Table-top sweeteners | 14 |
| Other uses[b] | 20 |

[a]Expressed in metric tons.
[b]Includes various sauces and pet food.
*Source:* Adapted from Ref. 6.

use in bread, baby food, white sugar, brown sugar, glucose, starch syrup, candy, honey, or milk products in Korea (19). Products manufactured from *S. rebaudiana* leaves also have some use in the People's Republic of China (7). Teas are prepared from the plant using either hot or cold water, and are recommended for increasing the appetite, as a digestant, for losing weight, for keeping young, and as a sweet-tasting low caloric tea.

Stevioside is available in Japan in combination with glycyrrhizin, with which it is synergistic, resulting in improvements of taste qualities of both natural sweeteners. In addition, stevioside has been found to be synergistic with aspartame, cyclamate, and acesulfame-K, but not with saccharin (1).

## STABILITY AND SOLUBILITY

Stevioside is a stable molecule at 100°C, when maintained in solution in the pH range 3–9, although it decomposes quite rapidly at alkaline pH levels of greater than 10 under these conditions (1). Both stevioside and its sweet analog, rebaudioside A, have been found to be stable when formulated in acidulated beverages that were stored at 60°C for 5 days (20). Detailed stability profiles have been determined for stevioside when treated with dilute mineral acids and enzymes (2).

Stevioside is only sparingly soluble in water and is slightly soluble in ethanol. Rebaudioside A, as a more polar compound, is more water soluble than stevioside (4).

## METABOLISM

There are only limited data available on the in vitro and in vivo metabolism of stevioside and the other *Stevia rebaudiana* sweet constituents. An initial investigation in which stevioside and rebaudioside A were degraded to steviol by rat intestinal flora in vitro was reviewed previously (1,2). Steviol has also been found as a major metabolite of stevioside when a tritiated form of this compound was fed to Wistar rats at an oral dose of 125 mg/kg. The biological half-life of stevioside was estimated to be 24 hours, and, 120 hours after compound administration, the highest percentages of radioactivity were found in the feces, followed by expired air, and then the urine. It was concluded that while a portion of orally administered stevioside was excreted unchanged in the feces of the rat, most of it was degraded by the intestinal bacterial flora to steviol, steviolbioside, and glucose, which were then absorbed in the cecum. Absorbed glucose was metabolized and excreted in expired air as carbon dioxide and water, while steviol was conjugated in the liver and excreted in the bile. It was also inferred from the results of biliary and fecal excretion that enterohepatic circulation of steviol occurred (21).

Despite the widespread use in Japan of *S. rebaudiana* extracts containing stevioside as sweeteners in foods, the metabolism of this substance in humans does not seem to have been investigated to date.

## SAFETY STUDIES

### Toxicity Studies in Rodents

Previous acute and subacute toxicity studies in rats on *Stevia rebaudiana* extracts with various stevioside levels have not resulted in the demonstration of any significant toxic effects (1). In addition, a chronic toxicity study of a hot water extract of *S. rebaudiana* leaves, purified to contain about 74.5% stevioside and 16.3% rebaudioside A, has been performed on male and female F344 rats (22). The animals were fed on diets incorporating up to 1.0% by weight of the purified plant extract for 22 months (males) and 24 months (females). It was found at the conclusion of this study that the mortality of the *S. rebaudiana*–treated animals did not differ significantly when compared to the control animals, although some transient growth retardation was observed for both male and female animals that were administered the highest dose (1.0% w/w) of the plant extract in their diet. Animals sacrificed after 6 months showed several changes relative to controls in terms of organ weights and urinary, hematological, and blood biochemical values. For example, animals dosed with *S. rebaudiana* extracts exhibited significant increases in the weights of the liver, kidneys, heart, prostate, and testes in males, and increases in the weights of the liver and decreases in the weight of the ovaries in females. However, no such differences were apparent in animals sacrificed after 12 months or at the conclusion of the study. A variety of nonneoplastic histopathological findings were recorded, and, in most cases, similar conditions were exhibited by both animals dosed with *S. rebaudiana* extracts and control animals. Changes more common in dosed animals than controls included reduced spermatogenesis, medullary cell proliferation in the adrenal glands, inflammatory lesions in the trachea and lungs, and pigmentation and increased hematogenesis of the spleen. A number of neoplasms were observed for all animals in the study, which were not related to dose, with the most common being interstitial cell adenoma of the testes and pheochromocytoma of the adrenal glands in the males, and interstitial polyps of the endometrium and adenoma of the anterior lobe of the pituitary in the females. It was concluded in this protracted and detailed study that no significant dose-related changes were found in the growth, general appearance, hematological and blood biochemical findings, organ weights, and macroscopic and microscopic observations, as a result of feeding male and female F344 rats with *S. rebaudiana* extracts at levels of up to 1% of their diet for about 2 years. It was also noted that almost all of the neoplasms detected could be associated with rats of the F344 strain entering old age, and it was thus concluded that such neoplasms were unrelated to the administration of *S. rebaudiana* extracts (22).

## Genetic Toxicity Studies

Partially purified extracts and pure sweet constituents (stevioside, rebaudiosides A–C, dulcoside A, steviolbioside) of *Stevia rebaudiana* leaves have been extensively tested for their mutagenic activity. None of these substances has been reported to be mutagenic, when evaluated with *Salmonella typhimurium* strains TA98, TA100, TA1538, and TM677, or *Escherichia coli* strain WP2, either in the presence or absence of metabolic activating systems (1,2).

Recent studies, however, have shown that steviol (Fig. 2), the aglycone of all of the *S. rebaudiana* sweet constituents including stevioside, is mutagenic in a forward mutation assay utilizing *S. typhimurium* strain TM677, in the presence of a metabolic activating systems derived from a 9,000 x *g* supernatant fraction from the livers of Aroclor-1254 pretreated rats. Unmetabolized steviol was inactive in this system (23).

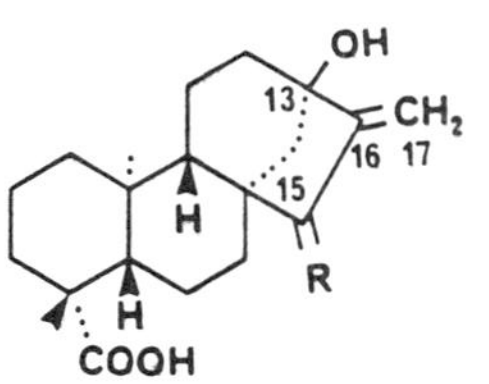


| | R |
|---|---|
| Steviol | $H_2$ |
| 15α-Hydroxysteviol | α-OH, β-H |
| 15-Oxosteviol | O |

**Figure 2** Structures of steviol, 15α-hydroxysteviol, and 15-oxosteviol.

Several steviol analogs have been assessed for mutagenicity under the same conditions used for steviol, and it is apparent that the C(13) tertiary hydroxy group and the C(16)-C(17) exomethylene unit must be present for compounds of this type to elicit a mutagenic response under the experimental protocols used (23,24). Using gas chromatography/mass spectrometry, 15α-hydroxysteviol (Fig. 2) was shown to be a major metabolite of steviol, when treated under conditions found to produce a bacterial mutagenic response by the latter compound. While 15α-hydroxysteviol was not found to be mutagenic, its closely related derivative, 15-oxosteviol (Fig. 2), was demonstrated as a direct-acting mutagen. However, 15-oxosteviol was not detected among the products of steviol metabolism, and it still remains to be shown that this compound can be enzymatically generated from either steviol or 15α-hydroxysteviol. Conversely, it is possible that other presently undefined mutagenic metabolites may be generated from steviol in these experiments (25,26). Currently, data are not available that would indicate whether stevioside is biotransformed to steviol and then to an active mutagenic substance or substances on ingestion by humans (26).

In a study in which D-6 cells of the Chinese hamster were treated with stevioside at concentrations of 0.5, 1, 2, and 4%, frequencies of chromosome aberrations and sister chromatid exchanges were determined. Cultures treated with stevioside at the 4% level produced a higher incidence of both chromosomal aberrations and sister chromatid exchanges than did controls. It was suggested by the authors of this investigation that additional screening tests on the cytotoxicity of stevioside should be carried out (27).

### Effects on Reproduction

It was documented some years ago that the Paraguayan Matto Grosso Indians employ teas produced from the leaves and stems of *Stevia rebaudiana* for their contraceptive and antifertility effects, although such effects have not been duplicated in a number of laboratories using experimental animals (1). In a more recent study, stevioside was administered to male and female rats at concentrations of 0.15, 0.75, or 3%. Male rates were given the stevioside-containing diet for 60 days both prior to mating and during mating, while females were dosed with these materials for 14 days prior to mating and for 7 days after gestation. No abnormal signs were seen in mating performance or fertility in any of the groups, and no external, internal, or skeletal anomalies attributable to stevioside were observed in the fetuses. It was therefore concluded that dietary administration of stevioside to male and female rats before mating and in the early period of gestation had no adverse effects either on fertility or on fetal development (28).

### Effects on Carbohydrate Metabolism

Earlier in the chapter, mention was made of the use of *Stevia rebaudiana* extracts as a diabetes remedy. Hypoglycemic effects in humans and rats using *S. rebaudiana* extracts have been claimed in several scientific publications from South America, thereby providing some credence for the medicinal use of this plant in Paraguay. However, a number of other laboratory investigators have found no significant effects on blood glucose levels in rats (1,2).

Recent work on the effects of *S. rebaudiana* extracts and sweet constituents are again somewhat contradictory. Stevioside has been discovered to affect the transport and metabolism of D-fructose in an isolated perfused rat liver preparation, although D-glucose was without specific effect in this regard. According to the investigators concerned, these results are compatable with previous observations of the effects of *S. rebaudiana* constituents on carbohydrate metabolism (29). In contrast, a further report described no effect on plasma glucose levels in rats by either acute or chronic administration of stevioside, and concluded that the data obtained did not support a hypoglycemic activity of stevioside (30).

### Other Biological Activities

A slight hypotensive effect has been noted in human subjects as a result of a Brazilian study wherein a *Stevia rebaudiana* tea was administered daily for 30 days (2). In a brief report, the effects of two concentrations of stevioside (8 and 16 mg/kg) were described when tested on normal Wistar rats, and it was concluded that significant alterations were obtained, in terms of both mean arterial pressure and renal function (31). A full paper describing these effects in more detail has not yet appeared in the literature.

A series of studies on the effects of stevioside and several of its glycosidic and nonglycosidic derivatives on cellular and subcellular metabolism have been performed and reviewed by Brazilian workers (32). The activities of these compounds have been probed in systems such as isolated rat liver mitochondria, rat renal cortical tubules, human and rabbit erythrocytes, rabbit reticulocytes, and in the isolated, perfused rat liver. In general, *S. rebaudiana* constituents such as stevioside and steviolbioside were either inactive in these systems, or else found to be less active than compounds such as steviol and isosteviol, which do not occur naturally (32).

A compound of apparently unknown structure that was referred to as "dihydroisosteviol" has been found to exhibit a statistically significant antiandrogenic effect in chick-comb bioassay, at a dose of 3.0 mg/comb. "Dihydroisosteviol" was found not to be effective in inhibiting the action of testosterone at total dose levels of 5.0 and 20 mg per animal when injected subcutaneously in 28-day-old Charles River male castrated rats (33). The origin of "dihydroisosteviol" was not stipulated in this study, and no compound with such a structure seems to have been determined as a metabolic or degradative product of any of the *S. rebaudiana* sweet constituents.

## CARIOGENICITY POTENTIAL

It has been suggested that stevioside could be used in oral preparations in place of common mono- and disaccharides, because it is a less favorable substrate for *Streptococcus mutans* serotypes in terms of acid production, than sucrose, glucose, or fructose (2). An in vitro study has reported that stevioside did not support the growth of cultures containing various organisms associated with dental caries, namely, *S. mutans*, *Lactobacillus plantarum*, and *L. casei* (2). These data are not conclusive evidence for a noncariogenic effect of stevioside. However, it has been put forward that sweeteners that are not carbohydrates and are used at low concentrations are unlikely to be cariogenic (34). Although stevioside has two carbohydrate moieties, this glycoside is very unlikely to be labile in the physiological conditions of the oral cavity.

## HUMAN EXPOSURE

*Stevia rebaudiana* extracts have been used for centuries in Paraguay to sweeten beverages. Also, they have been used increasingly for the last 15 years in Japan, with the human consumption of stevioside in 1987 being nearly 200 metric tons. No evidence of adverse reactions due to the ingestion of these materials has appeared in the scientific literature, and, on this basis, *S. rebaudiana* extracts and stevioside do not seem to present a potential toxicity risk to humans.

## REGULATORY STATUS

*Stevia rebaudiana* extracts containing stevioside are approved for sweetening purposes in Japan, South Korea, and Brazil, and have additional uses in Paraguay and the People's Republic of China.

## REFERENCES

1. A. I. Bakal and L. O'Brien Nabors, Stevioside, in *Alternative Sweeteners* (L. O'Brien Nabors and R. C. Gelardi, eds.), Marcel Dekker, New York, 1986, pp. 295–307.
2. A. D. Kinghorn and D. D. Soejarto, Current status of stevioside as a sweetening agent for human use, in *Economic and Medicinal Plant Research,* Vol. 1 (H. Wagner, H. Hikino, and N. R. Farnsworth, eds.), Academic Press, London, 1985, pp. 1–52.
3. K. C. Phillips, Stevia: Steps in developing a new sweetener, in *Developments in Sweeteners—3,* (T. H. Grenby, ed.), Elsevier Applied Science, London, 1987, pp. 1–43.
4. B. Crammer and R. Ikan, Progress in the chemistry and properties of the rebaudiosides, in *Developments in Sweeteners—3* (T. H. Grenby, ed.), Elsevier Applied Science, London, 1987, pp. 45–64.
5. A. D. Kinghorn, D. D. Soejarto, N. P. D. Nanayakkara, C. M. Compadre, H. C. Makapugay, J. M. Hovanec-Brown, P. J. Medon, and S. K. Kamath, A phytochemical screening procedure for sweet *ent*-kaurene glycosides in the genus *Stevia* (Compositae), *J. Nat. Prod., 47:* 439–444 (1984).
6. Anonymous, High sweeteners—Market size 7.2 billion yen: Stevia occupies 41%, but future gains will be made by aspartame, *Food Chemicals* (*Tokyo*), No. 6: 19–26 (1988).
7. J. -X. Guo, Shanghai Medical University, Shanghai, People's Republic of China, private communication.
8. H. B. Wood, Jr., R. Allerton, H. W. Diehl, and H. G. Fletcher, Jr., Stevioside. I. The structure of the glucose moieties, *J. Org. Chem., 20:* 875–883 (1955).
9. E. Vis and H. G. Fletcher, Jr., Stevioside. IV. Evidence that stevioside is a sophoroside, *J. Am. Chem. Soc., 85:* 4709–4710 (1956).
10. E. Mosettig, U. Beglinger, F. Dolder, H. Lichti, P. Quitt, and J. A. Waters, The absolute configuration of steviol and isosteviol, *J. Am. Chem. Soc., 85:* 2305–2309 (1963).
11. H. Kohda, R. Kasai, K. Yamasaki, K. Murakami, and O. Tanaka, New sweet diterpenes from *Stevia rebaudiana, Phytochemistry, 15:* 981–983 (1976).
12. I. Sakamoto, K. Yamasaki, and O. Tanaka, Application of $^{13}$C NMR spectroscopy to chemistry of natural glycosides: rebaudioside-C, a new sweet diterpene glycoside of *Stevia rebaudiana, Chem. Pharm. Bull., 24:* 844–846 (1977).
13. M. Kobayashi, S. Horikawa, I. H. Degrandi, J. Ueno, and H. Mitsuhashi, Dulcosides A and B, new diterpene glycosides from *Stevia rebaudiana, Phytochemistry, 16:* 1405–1408 (1977).
14. I. Sakamoto, K. Yamasaki, and O. Tanaka, Application of $^{13}$C NMR spectroscopy to chemistry of plant glycosides: Rebaudiosides-D and -E, new sweet diterpene-glucosides of *Stevia rebaudiana* Bertoni, *Chem. Pharm. Bull., 25:* 3437–3439 (1977).
15. M. Darise, K. Muzatani, R. Kasai, O. Tanaka, S. Kitahata, S. Okada, S. Ogawa, F. Murakami, and F.-H. Chen, Enzymic transglucosylation of ru-

busoside and the structure-sweetness relationship of steviol-bisglycosides. *Agric. Biol. Chem., 48:* 2483–2488 (1984).

16. S. Kusama, I. Kusakabe, Y. Nakamura, S. Eda, and K. Murakami, Transglucosylation into stevioside by the enzyme system from *Streptomyces* sp., *Agric. Biol. Chem., 50:* 2445–2451 (1986).
17. D. D. Soejarto, C. M. Compadre, P. J. Medon, S. K. Kamath, and A. D. Kinghorn, Potential sweetening agents of plant origin. II. Field search for sweet-tasting *Stevia* species, *Econ. Bot., 37:* 71–79 (1983).
18. H. Fujita and T. Edahiro, Safety and utilization of Stevia sweetener, *Shokuhin Kogyo, 22*(20): 65–72 (1979).
19. Anonymous, *The Korean Standards of Food Additives,* The Korea Foods Industry Association, Seoul, 1988, pp. 198–199.
20. S. S. Chang and J. M. Cook, Stability studies of stevioside and rebaudioside A in carbonated beverages, *J. Agric. Food Chem., 31:* 409–412 (1983).
21. K. Nakayama, D. Kasahara, and F. Yamamoto, Absorption, distribution, metabolism and excretion of stevioside in rats, *J. Food Hyg. Soc., Japan, 27:* 1–8 (1986).
22. A. Yamada, S. Ohgaki, T. Noda, and M. Shimizu, Chronic toxicity study of dietary Stevia extracts in F344 rats, *J. Food Hyg. Soc., Japan, 26:* 169–183 (1985).
23. J. M. Pezzuto, C. M. Compadre, S. M. Swanson, N. P. D. Nanayakkara, and A. D. Kinghorn, Metabolically activated steviol, the aglycone of stevioside, is mutagenic, *Proc. Natl. Acad. Sci., USA, 82:* 2478–2482 (1985).
24. J. M. Pezzuto, N. P. D. Nanayakkara, C. M. Compadre, S. M. Swanson, A. D. Kinghorn, T. M. Guenthner, V. L. Sparnins, and L. K. T. Lam, Characterization of bacterial mutagenicity mediated by 13-hydroxy-*ent*-kaurenoic acid (steviol) and several structurally related derivatives and evaluation of potential to induce glutathione *S*-transferase in mice, *Mutation Res., 169:* 93–103 (1986).
25. C. M. Compadre, R. A. Hussain, N. P. D. Nanayakkara, J. M. Pezzuto, and A. D. Kinghorn, Mass spectral analysis of some derivatives and in vitro metabolites of steviol, the aglycone of the natural sweeteners, stevioside, rebaudioside A, and rubusoside, *Biomed. Environ. Mass Spectrom., 15:* 211–222 (1988).
26. J. M. Pezzuto, Chemistry, metabolism, and biological activity of steviol (*ent*-13-hydroxykaur-16-en-19-oic acid), the aglycone of stevioside, in *New Trends in Natural Products Chemistry* (Atta-ur-Rahman and P. W. Le Quesne, eds.), Elsevier Applied Science, Amsterdam, 1986, pp. 371–386.
27. S. Nadamitsu, M. Segawa, Y. Sato, and K. Kondo, Effects of stevioside on the frequencies of chromosomal aberrations and sister chromatid exchanges in the D-6 cell of Chinese hamster, *Mem. Fac. Integrated Arts Sci., Hiroshima University, Ser. IV, 10:* 57–62 (1985).
28. N. Mori, S. Sakanoue, M. Takeuchi, K. Shimpo, and T. Tanabe, Effect of stevioside on fertility in rats, *Shokuhin Eiseigaku Zasshi, 22:* 409–414 (1981).
29. E. L. Ishii, A. J. Schwab, and A. Bracht, Inhibition of monosaccharide

transport in the intact liver by stevioside, *Biochemical Pharmacol., 36:* 1417–1433 (1987).

30. P. J. Medon and M. M. Zeigler, Effect of acute and chronic administration of *Stevia rebaudiana* on plasma glucose *in vivo, Pharm. Res., 3:* 162S (1986).
31. M. S. Melis, A. R. Sainati, and R. E. Marciel, Effects of two concentrations of stevioside on renal function and mean arterial pressure in rats, *IRCS Med. Sci., 14:* 973 (1986).
32. A. N. Kelmer Bracht, F. S. Kemmelmeier, E. L. Ishii, M. Alvarez, and A. Bracht, Effect of *Stevia rebaudiana* natural products on cellular and subcellular metabolism, *Arq. Biol. Tecn., 28:* 431–455 (1985).
33. R. I. Dorfmann and W. R. Nes, Anti-androgenic activity of dihydroisosteviol, *Endocrinology, 67:* 282–285 (1960).
34. W. H. Bowen, Role of sugar and other sweeteners in dental caries, in *Aspartame: Physiology and Biochemistry* (L. D. Stegink, and L. J. Filer, Jr., eds.), Marcel Dekker, New York, 1984, pp. 263–272.

# 10

# Sucralose

**Gary A. Miller** *McNeil Specialty Products Co., New Brunswick, New Jersey*

## INTRODUCTION

During the 1970s Tate & Lyle embarked upon a research program to develop new chemical entities from sugar. This research was undertaken to enhance the economic value of this abundant material. This program included the evaluation of the functionality of halogenated sucrose molecules, resulting in the discovery of the intensification of the sweetness of sucrose by selective halogenation (1–3). A long research program ensued in which a compound, sucralose (1,6-dichloro-1,6-dideoxy-*b*-D-fructofuranosyl-4-chloro-4-deoxy-*a*-D-galactopyranoside), was selected for development and commercialization as a nonnutritive sweetener (Fig. 1).

The selective chlorination of the sucrose molecule produced remarkable changes to the sweetness intensity and stability of sucrose, without compromising taste quality. The resulting sweetener has a pleasant sweet taste similar to sucrose and has no unpleasant aftertaste. Sucralose is a white, crystalline, nonhygroscopic solid which is highly soluble in water, ethanol, and methanol. The modification yields a sweetener with a glycosidic linkage significantly more resistant to acid or enzymatic hydrolysis than the parent compound, sucrose.

Sucralose is noncaloric, therefore it does not contribute calories when added to foods. Sucralose has been demonstrated to be noncariogenic, making it beneficial for use in noncariogenic applications such as chewing gum.

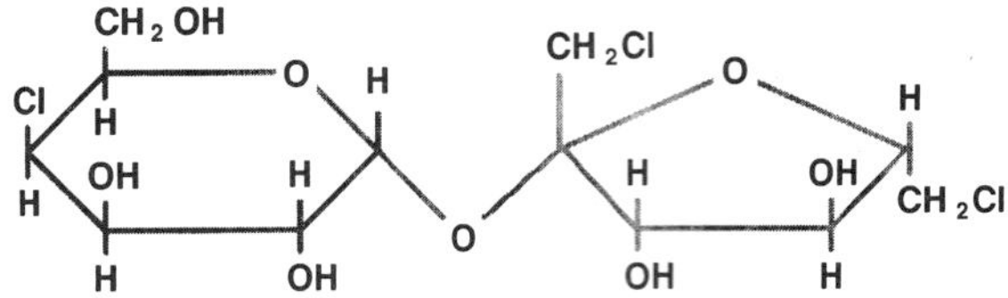


Figure 1 Sucralose.

## SENSORY CHARACTERISTICS OF SUCRALOSE

### Sweetness Intensity

The sweetness factor curves for sucralose and three other high-potency sweeteners—aspartame, saccharin, and acesulfame-K—are shown in Figure 2. Sucrose is the sweetness reference used in the construction of these curves. The sweetness factor curves express the number of times a compound is sweeter than an isosweet concentration of sucrose. The curves demonstrate a general decreasing sweetness concentration for each high-potency sweetener relative to increasing sucrose concentration. Sucralose displayed the greatest sweetness factor or highest potency among the sweeteners examined. Sucralose is approximately 750 times sweeter than sucrose at a concentration equisweet to a 2% sucrose

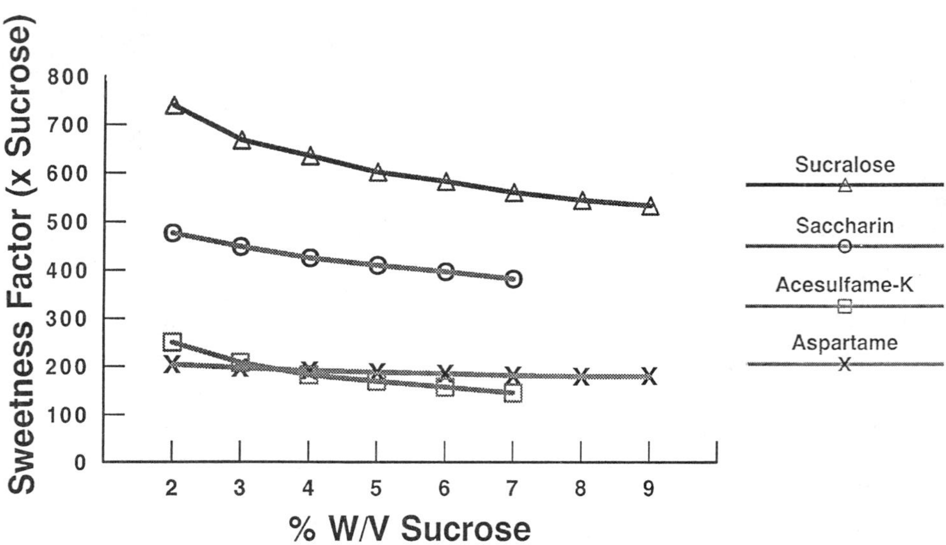


Figure 2 Sweetness factors.

solution. At 9% sweetness equivalence, sucralose is approximately 500 times sweeter than sucrose. As a general guideline, sucralose is regarded as being 600 times sweeter that sucrose (2).

Because of the intense sweetness of sucralose, an extremely small amount is necessary to achieve the desired sucrose isosweetness level. As an example, approximately 20 mg sucralose per liter is needed to equal the sweetness of a 2% sucrose solution. At 5% sucrose equivalency, a level at which a normal cup of coffee is typically sweetened, approximately 83 mg sucralose per liter would be required. At 9% equivalency, close to where many diet beverages are sweetened, approximately 170 mg/liter would be required for isosweetness. It is important to note that the actual use levels of sucralose will be influenced by characteristics such as the pH, temperature, and viscosity of the food or beverage system in which sucralose will be incorporated.

## Temporal Qualities

An understanding of the temporal properties of sucralose as they relate to other sweeteners is of interest to food and beverage formulators. The temporal profiles of the sweeteners relate to the onset of sweetness, time to maximum intensity, and rate of sweetness decay. At an aqueous isosweet concentration, the time to maximum sweetness intensity is similar for sucrose and the other high-potency sweeteners tested (Table 1). Deviations from sucrose become apparent when studying the rate of decline in sweetness with individual high-potency sweeteners. Under the test conditions used, sucralose and aspartame had a similar but slower rate of sweetness decay than sucrose. It should be remembered that different temporal properties may emerge as a result of the concentration of the sweetener used as well as other formulation issues.

**Table 1** Temporal Sweetness Characteristics of Sucralose

| | Maximum intensity (0–12 Scale) | Time to maximum sweetness (sec) | Sweetness duration time (Sec) |
|---|---|---|---|
| Sucrose | 7.6 | 4.1 | 66.1 |
| Sucralose | 7.8 | 5.0 | 75.4 |
| Aspartame | 7.5 | 6.2 | 76.7 |
| Saccharin | 7.5 | 3.1 | 77.2 |
| Acesulfame-K | 7.5 | 4.9 | 77.4 |

### Taste Quality

Flavor attributes of isosweet concentrations of sucralose and sucrose in aqueous solution were identified and quantified using a sip and spit procedure. The sweeteners were profiled at a moderate sweetness level of 5%. The panelists evaluated a 10-ml sample by holding the sample in their mouth for 5 seconds and then expectorating. During this time, intensity estimates were made of several taste attributes using an unstructured line scale. Sweet and nonsweet aftertaste were evaluated 60 seconds after expectoration. The experimental series was repeated twice and the scores averaged across replicates.

Several of the more important flavor attributes are presented in Figure 3. This comparison demonstrates that sucralose is a high quality sweetener with a flavor profile which is similar to its sugar origin. At a 90% confidence interval, no significant difference was found between the intensity of the flavor attributes evaluated in either sweetener. Bitterness was very low in both sucralose and sucrose. The rate or sharpness of sweetness onset and persistence of sweet and nonsweet aftertaste were similar under the conditions of this test. These flavor attributes are impacted by the food or beverage system in which they are studied or the concentration of the sweetener evaluated (5).

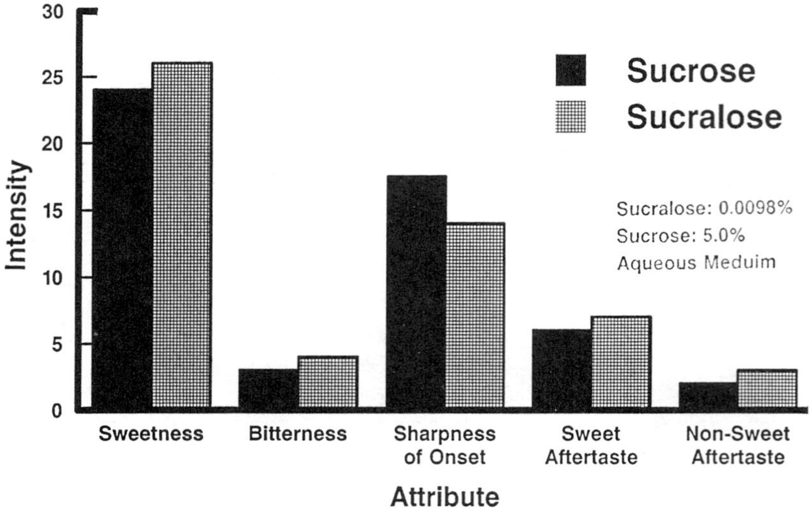


**Figure 3** Flavor profile comparison of sucralose and sucrose.

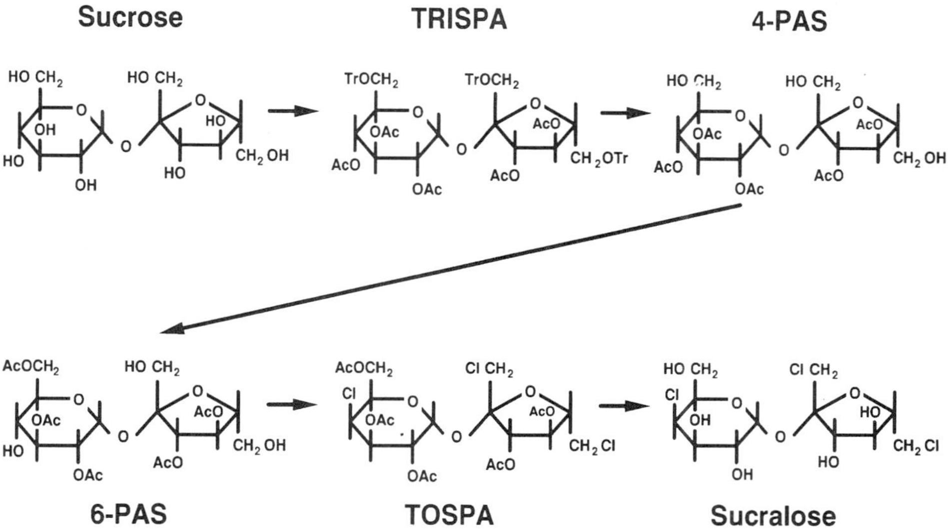


**Figure 4** Synthesis of sucralose from sucrose.

## CHEMICAL AND STABILITY CHARACTERISTICS OF SUCRALOSE

### Synthesis of Sucralose

The synthesis of sucralose (Fig. 4) begins with sucrose as the starting material. The three primary hydroxyl groups of the sucrose molecule are blocked as ethers. This is followed by the blocking of the five secondary hydroxyl groups as acetates. This yields 6,1′,6′-tritrityl 2,3,4,3′,4′ penta acetyl sucrose (TRISPA). The second step in the synthesis is the deblocking of the three primary hydroxyls resulting in the formation of 2,3,4,3′,4′-penta-*o*-acetyl sucrose (4-PAS). Next the acetate group from the 4 position of the glucose moiety is migrated to the 6 position of the glucose moiety, yielding 2,3,6,3′,4′-penta-*o*-acetyl sucrose (6-PAS). The fourth step in this synthesis involves the chlorination of 6-PAS at the exposed hydroxyl groups, giving 4,1′,6′-trichloro-4,1′,6′-trideoxygalactosucrose penta acetate (TOSPA). During this step the glucose moiety is converted to a galactose moiety. The final step in the synthesis of sucralose is the removal of the acetates by hydrolysis, yielding 1,6-dichloro-1,6-dideoxy-*b*-D-fructofuranosyl-4-chloro-4-deoxy-*a*-D-galactopyranoside, which is commonly known as sucralose (6).

## Physicochemical Characteristics of Sucralose

Physicochemical characteristics of sucralose that have particular relevance to the potential utility of this sweetener in food and beverage applications have been evaluated. Physicochemical properties, such as solubility, are important to understand because they impact how a food ingredient can be used in the food manufacturing environment.

### *Solubility*

Solubility was measured in a thermostatically controlled Wheaton jacketed glass vessel. The temperature range studied was 20–60°C. As Figure 5 shows, sucralose is freely soluble in water and ethanol. As the temperature increases, so does the solubility of sucralose in each of the solvents.

The solubility of sucralose in water ranges from 28.3 g/100 ml at 20°C to 66 g/100 ml at 60°C. The solubility of sucralose in ethanol was found to range from 9.5 g/100 ml at 20°C to 18.9 g/100 ml at 60°C. These data demonstrate that sucralose will be easy to use in food operations and that special techniques to solubilize the sweetener will not have to be developed (7).

### *Dynamic Viscosities*

The dynamic viscosities of aqueous sucralose solutions were measured from 20° to 60°C and from 10 to 50% w/w concentrations within the

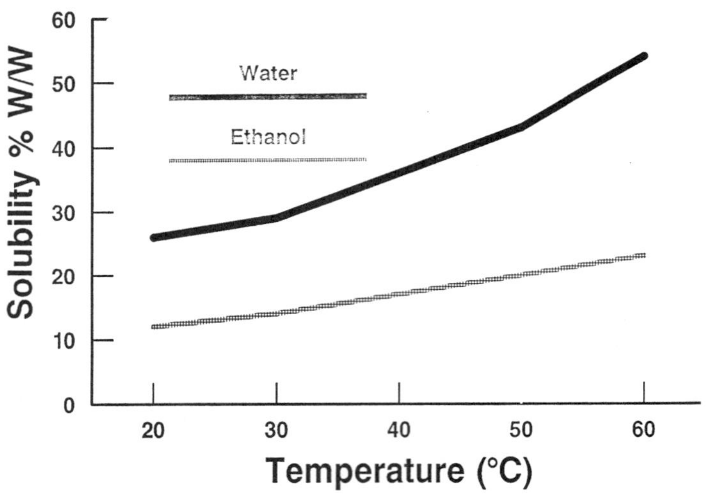


**Figure 5** Solubility of sucralose.

solubility range. A Rheomat 30 viscometer was used and was fitted with a concentric cylinder head suitable for solutions of low viscosity. The dynamic viscosities of distilled water and 10% w/w aqueous sucralose solutions at different shear rates are shown in Figure 6. Variations in viscosity over the four shear rates of 512, 691, 939, and 1280 $sec^{-1}$ were insignificant, demonstrating that the sucralose solutions were Newtonian in behavior.

Dynamic viscosities of aqueous sucralose solutions measured at 939 $sec^{-1}$ are shown in Figure 7. Comparison of these data with available

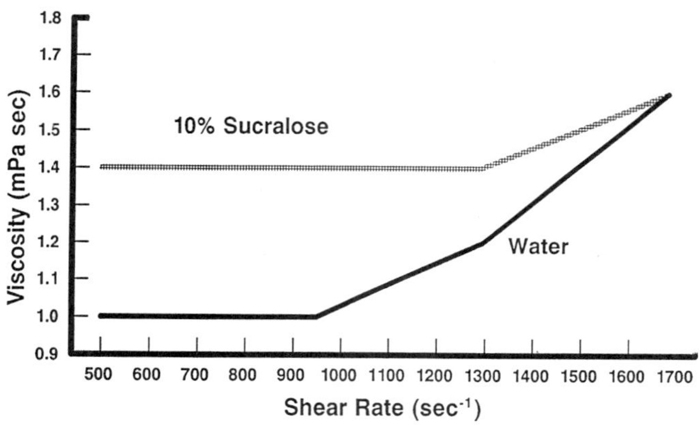


**Figure 6** Dynamic viscosity of water and aqueous 10% W/W sucralose solution at 20°C.

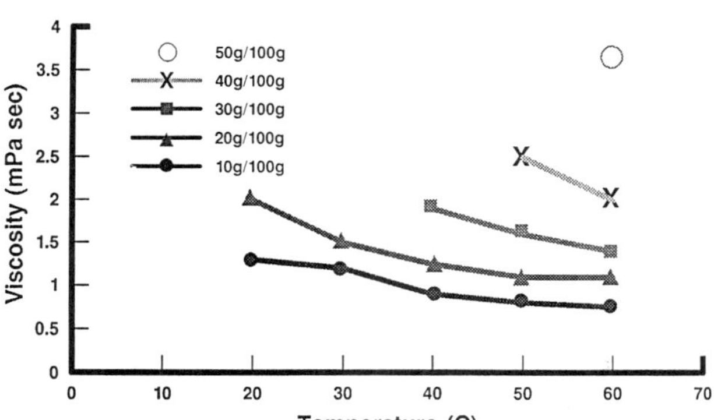


**Figure 7** Dynamic viscosity of aqueous sucralose.

data for the viscosity of sucrose solutions shows that sucralose and sucrose have very similar viscosities.

The comparatively low viscosities observed for sucralose solutions and the Newtonian behavior show that, at normal usage levels of sucralose, viscosity will not create mixing or dispersion problems in normal food unit operations or processes.

### *Surface Tension*

The surface tension of aqueous sucralose solutions held at 20°C was measured with a Kruss model K8600 tensiometer, using the ring method. The surface tension of dilute sucralose solutions, 0.1% and 1%, exhibited negligible lowering of surface tensions. The reduction was 0.8 milliNewtons per meter (mN/m) to 3.1 mN/m, respectively. This negligible reduction in surface tension implies that sucralose is not a surfactant and it should not cause foaming in liquid products.

### *Refractive Index*

Although valid and precise high-performance liquid chromatography (HPLC) and gas chromatography (GC) analytical methods for the determination of sucralose are available, more rapid and simpler techniques

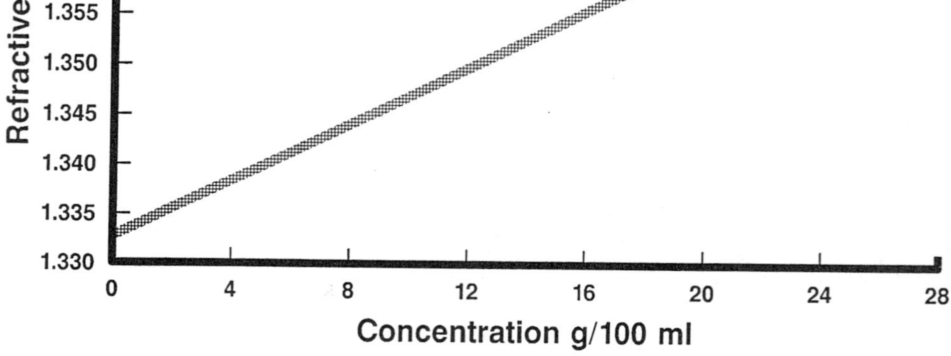


**Figure 8** Refractive index of aqueous sucralose solutions.

may be useful for food manufacturers for quality assurance and quality control purposes.

The use of refractive index as a simple means of determining the concentration of sucralose in simple aqueous solutions was evaluated with the use of a Bellingham and Stanley model 60/95 Abbe refractometer. Aqueous solutions of sucralose were prepared at 5, 10, 15, 20, and 25% and tested at 20°C.

As Figure 8 shows, the refractive index of the sucralose solutions is linear ($r^2 = 0.999$) with respect to concentration over the range of the concentrations studied. These results demonstrate the potential use of refractive index measurements for the determination of sucralose concentration in simple solutions.

These physicochemical characteristics demonstrate that sucralose will function well within typical unit operations and processes that are used in the food industry (7).

## Chemical Interactions

In the process of chlorination of the sucrose molecule to synthesize sucralose, the primary reaction sites on the molecule are occupied, and this reduces the chemical reactivity of the compound.

The reactivity of sucralose was evaluated in a model system study where 1% solutions of sucralose were prepared with 0.1% of one of the following;

- Bases
  - Niacinamide
  - Monosodium Glutamate
- Oxidizing and reducing agents
  - Hydrogen peroxide
  - Sodium metabisulfite
- Aldehydes and ketones
  - Acetaldehyde
  - Ethyl acetoacetate
- Metal salt
  - Ferric chloride

These classes of compounds and the specific compounds were selected as typical food system chemicals, and they are representative of many compounds found in food formulations. These solutions were held at 40°C for 7 days, at pH 3, 4, 5, or 7, and analyzed for sucralose content by an HPLC technique. Table 2 shows the sucralose retention in each of the solutions. As can be seen from the data, there are no chemi-

**Table 2** Sucralose Interaction Study: Sucralose Retention (%) after 7 Days at 40°C

| Sample | pH 3.0 | pH 4.0 | pH 5.0 | pH 7.0 |
|---|---|---|---|---|
| Control | 100 | 99.8 | 98.9 | 99.5 |
| Sucralose & hydrogen peroxide | 100 | | 99.7 | |
| Sucralose & sodium metabisulfite | 99.9 | | 99.9 | |
| Sucralose & acetaldehyde | 100 | | 100 | |
| Sucralose & ethyl acetoacetate | 98.8 | | 100 | |
| Sucralose & ferric chloride | 95.9 | | 98.0 | |
| Sucralose & niacinamide | | 100 | | 100 |
| Sucralose & monosodium glucamate | | 99.8 | | 100 |

cal interactions of concern. These data demonstrate that sucralose is relatively inactive chemically, and therefore chemical interactions of sucralose with food ingredients in food systems is not likely to occur.

### Stability of the Ingredient

Stability of the dry pure sucralose has been evaluated to determine the shelf life and proper storage and distribution conditions for the ingredient. Due to the physicochemical characteristics and the inherent stability of the molecule in aqueous solution, two forms of the ingredient, dry and a liquid concentrate, are possible product forms.

#### *Stability of Dry Neat Sucralose*

The stability of dry neat sucralose has been measured at several temperatures to assess the shelf life and the storage conditions required to properly store and distribute this form of the ingredient. The first indication of the breakdown of neat sucralose is the development of a mild discoloration (tan/pink), which is accompanied by a very small release of HCl. The development of color occurs prior to the detection of a loss of sucralose by HPLC analysis. The mild discoloration is a quality issue and proper storage of dry sucralose is necessary to assure that sucralose will be of the proper quality for its incorporation in food products.

Dry neat sucralose was stored at 75°F (23.8°C), 86°F (30°C), 104°F (40°C), and 122°F (50°C), and monitored for the development of color. The time to first discoloration was 18 months at 75°F, 3 months at 86°F, 3 weeks at 104°F, and less than one week at 122°F. As these data show, dry neat sucralose has sufficient shelf life if the product is properly stored and handled at appropriate temperatures. Further, sucralose can be abused by improper storage conditions. Abuse of the ingredient is a quality issue which requires effective materials management techniques to assure the full shelf life of the ingredient.

### *Stability of Sucralose Liquid Concentrate*

The solubility and inherent aqueous stability of sucralose provided the opportunity to examine the feasibility of a sucralose liquid concentrate for use by the food industry. A liquid product form provides many opportunities for the industrial customer with regard to operational issues in the utilization of sucralose in the manufacturing environment.

To demonstrate the feasibility, a 25% (w/w) sucralose solution with a dual preservative system of 0.1% sodium benzoate and 0.1% potassium sorbate, and a citrate buffer (0.02 M) to maintain a pH of 4.4 can be manufactured using a cold fill process. Product in this form has been manufactured, and its stability characteristics have been evaluated. No chemical changes or physical changes have been found in product stored for greater than one year at 68°F (20°C) and 104°F (40°C). No loss of sucralose, sodium benzoate, or potassium sorbate was analytically detectable, the solution maintained its clarity, no color development occurred, and there was no change in pH. These data can be used to predict a shelf life of at least 2 years for the liquid sucralose concentrate product. Triple challenge microbiological studies on fresh and aged liquid sucralose concentrate have clearly demonstrated that the liquid ingredient will not be a vector for microbiological contamination when good manufacturing practices are observed.

This product form provides the end user with an ingredient system that is extremely stable, chemically and microbiologically, and compatible with most food unit operations.

## Stability of Sucralose in Model Systems

The use of sucralose in food products was first evaluated by the use of model food systems. The model food systems are simplified to reduce the complexity of food analysis, food interactions and to provide a milieu in which the inherent stability of a food ingredient can be understood.

### *Model Dry Food Systems*

Two model dry systems were used to evaluate the stability of sucralose in dry food systems: sucralose dry blended with maltodextrin, and sucralose co-spray dried with maltodextrin.

Sucralose was dry blended with maltodextrin in the following ratios 75:25, 50:50, and 25:75 (sucralose:maltodextrin). As the amount of sucralose was decreased the time to discoloration increased by factors of 1.5, 2, and 8, respectively, when compared to neat sucralose. The dispersion of sucralose in dry mix, typically at usage levels of less than 1%, will markedly increase the stability of sucralose. This provides sufficient stability of sucralose in dry applications.

To determine the appropriateness of using sucralose in spray-drying operations and in spray-dried products, sucralose was spray-dried with maltodextrin in 50:50 and 25:75 sucralose:maltodextrin ratios. As the amount of sucralose decreased, the time to first discoloration was increased by factors of 10 and 30 times, respectively. This remarkable increase in stability was unexpected. These data demonstrate that the use of sucralose in spray-drying and in spray-dried applications is appropriate, and that co-spray-drying sucralose and maltodextrins may be a means to develop a more stable dry industrial ingredient form.

Both of these studies demonstrate that sucralose can be used in dry food applications with no expectation of discoloration when the food products are handled in normal food distribution systems.

*Model Aqueous Food Systems*

Simple aqueous solutions systems studies and radiolabeled studies in beverages and baked goods were performed to demonstrate the inherent stability of sucralose in liquid and aqueous food systems.

The aqueous stability of sucralose was determined over various pHs, temperatures, and times in model system studies.

Figure 9 shows the stability data for sucralose solutions stored at 100°C. After one hour there was 2% loss of sucralose at pH 3, 5, and 7.

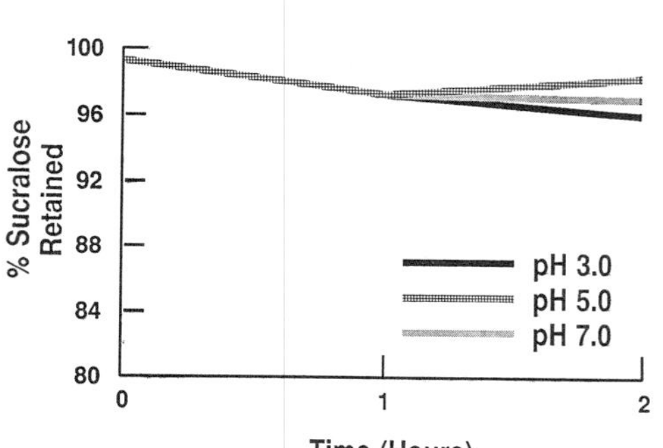


**Figure 9** Aqueous stability of sucralose, effect of pH at 100 degrees centigrade.

After 2 hours there was 2% loss at pH 5, a 3% loss at pH 3, and a 4% loss of sucralose at pH 7. These data were gathered using an HPLC technique that has a coefficient of variation of ± 2%.

Since a loss of 10–15% of a sweetener is necessary before a loss of sweetness can be detected, these losses of sucralose will not be detected by consumers as a loss of sweetness (8). These data show that sucralose will withstand heat-processing conditions that are typical of many food processes, and that sucralose will be able to withstand the rigors of rework or reprocessing.

Figure 10 shows the stability of sucralose solutions stored at 30°C for up to one year at pH 3.0, 4.0, 6.0, and 7.5. Again an HPLC technique that has a coefficient of variation ± 2% was used. There was no measurable loss of sucralose after one year of storage at pH 4.0, 6.0, and 7.5. At pH 3.0 there was less than 4% loss of sucralose after one year of storage (9). This level of stability will provide food manufacturers and consumers with food products that have a consistent level of sweetness throughout their shelf life.

The stability data in model aqueous systems can be used to predict the efficacy of using sucralose in aqueous food systems. Sucralose can be used to formulate high quality food products that will maintain their high quality and level of sweetness throughout the food-processing and distribution system. This will provide the consumer the benefit of high quality products that are consistent in their sweetness profile and level.

These studies and all the stability studies on sucralose demonstrate that sucralose, in aqueous systems, will slowly break down to its component moieties, 4-chlorogalactose (4-CG) and 1,6-dichlorofructose (1,6-DCF), as shown in Figure 11. This breakdown is dependent upon pH, temperature, and time.

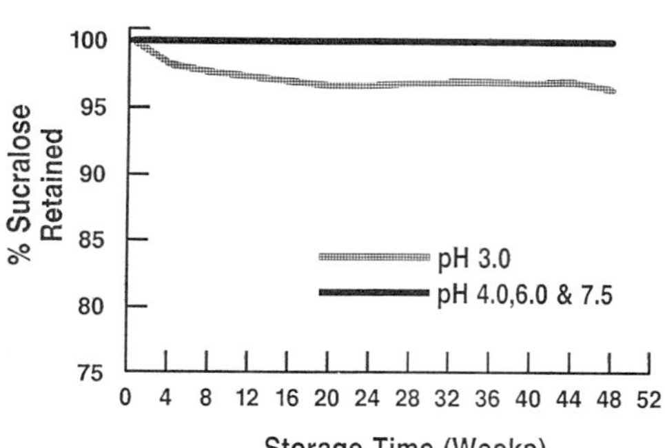


**Figure 10** Aqueous stability of sucralose, effect of pH at 30 degrees centigrade.

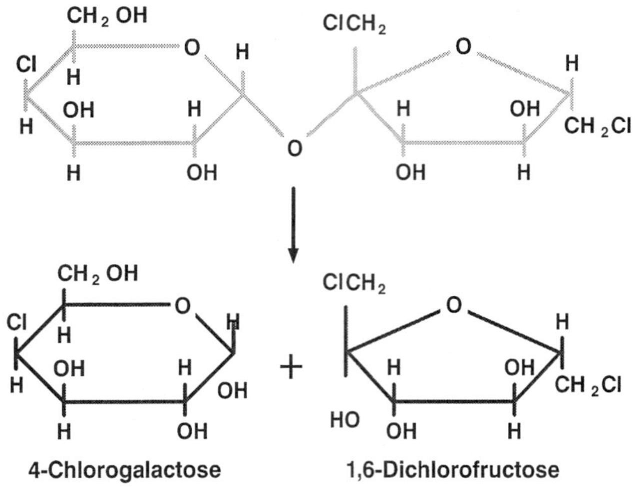


**Figure 11** Acidic hydrolysis of sucralose.

To fully delineate the mechanism of breakdown in food systems, a $^{36}Cl$ aqueous stability study was conducted. Radiolabeled sucralose was prepared with a specific activity of 1.027 microcuries ($\mu c$) per milligram and a radiochemical purity of 99%. This was then formulated into glycine/HCl buffers at pH 2.5, 3.0, and 3.5 or in a cola carbonated soft drink formulated at pH 3.0 and 3.5. Sucralose was added at 200 parts per million concentration. Buffers and colas were stored at 30°C or 40°C for one year and sampled at 0, 8, 16, 26, and 52 weeks.

Samples of the buffers and colas were analyzed using thin-layer chromatography and liquid scintillation counting. The amounts of sucralose, 4-chlorogalactose, and 1,6-dichlorofructose were calculated after correcting for the amount of 4-chlorogalactose and 1,6-dichlorofructose present in the radiolabeled sample analyzed on day 0.

The pattern of radioactivity of samples from the buffers and colas was consistent with the pattern of radioactivity that was found in the $^{36}Cl$ sucralose sample analyzed at day 0. No new compounds were separated or identified. There was no evidence of the presence of inorganic chloride in the buffer or cola products.

Figure 12 shows the stability data of the pH 3.0 buffered solution, and Figure 13 shows the stability data of the pH 3.0 cola. Both samples were stored for one year at 40°C. the`loss of sucralose in both cases indicated here is consistent with the stability data discussed earlier. The

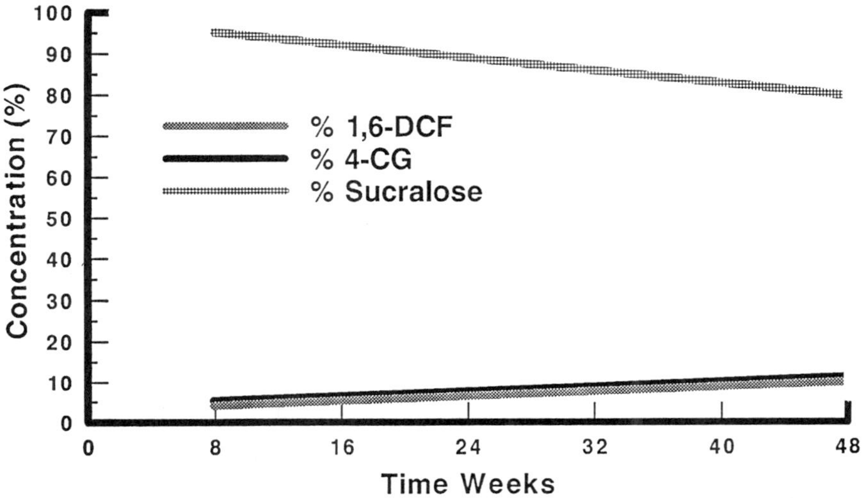


**Figure 12** Aqueous stability of $^{36}$Cl sucralose. Buffered pH 3.0 solution at 40 degrees C storage.

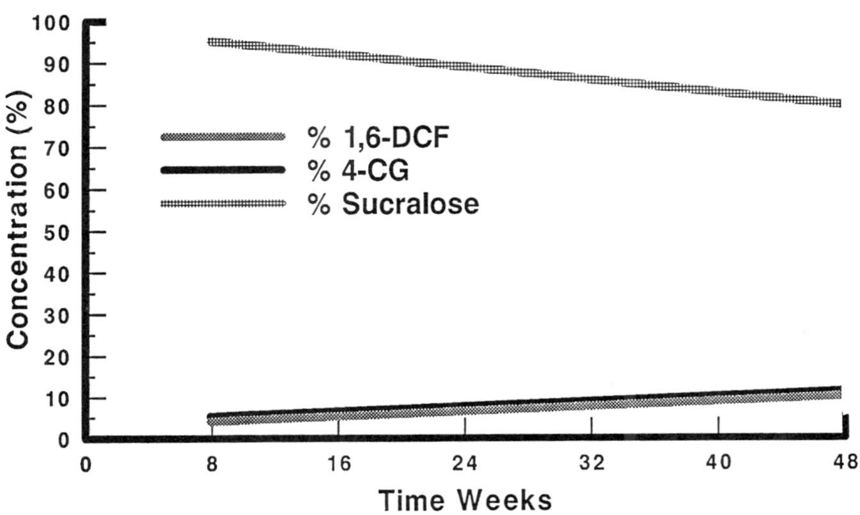


**Figure 13** Aqueous stability of $^{36}$Cl sucralose. pH 3.0 cola carbonated soft drink at 40 degrees C storage.

increase in 4-chlorogalactose and 1,6-dichlorofructose is consistent with the knowledge that sucralose only breaks down to these two products.

Further, the similarity in the data between the cola at pH 3.0 and the buffered solution at pH 3.0 is indicative of the lack of interaction of sucralose with other food components. This is predicted by the interaction study cited above and is probably due to the lack of reactive sites on the sucralose molecule.

This data from the $^{36}Cl$ study was then used to determine whether all the sucralose could be accounted for as the hydrolysis product and sucralose. Mass balance, for the purpose of this study, is defined as the percent of sucralose accounted for by sucralose and the hydrolysis product and is shown in Table 3. These data show that all of the sucralose was accounted for as sucralose, 4-chlorogalactose, or 1,6-dichlorofructose, and there is no indication of any other sucralose breakdown products at these pHs. No other compounds were formed, and the 4-chlorogalactose and 1,6-dichlorofructose did not interact or break down further (1).

A second radiolabeled stability study was performed in baked goods. The objective of this study was to demonstrate the stability of sucralose in a variety of common baked goods prepared under typical baking conditions. For this study, $^{14}C$-labeled sucralose was used to minimize the difficulties associated with the recovery and detection of low levels of sucralose in the presence of other carbohydrates found in baked goods. Cakes, cookies, and graham crackers were selected for this study since they represent a cross section of common bakery ingredients and typical baking conditions encountered in the baking industry.

The baked goods were prepared under simulated baking time and temperature conditions. The sucralose was recovered by aqueous extraction, and the extracts were cleaned up by a methanol precipitation prior to TLC analysis.

**Table 3** % Mass Balance in $^{36}Cl$ Stability Study, at 40°C Storage

| Solution | 8 Weeks | 16 Weeks | 26 Weeks | 52 Weeks |
|---|---|---|---|---|
| Aqueous Buffer | | | | |
| pH 2.5 | 99.8 | 98.9 | 98.8 | 100.0 |
| pH 3.0 | 99.9 | 99.9 | 99.6 | 100.1 |
| pH 3.5 | 100.2 | 100.0 | 100.0 | 100.1 |
| Cola | | | | |
| pH 3.0 | 99.9 | 102.0 | 99.8 | 100.8 |
| pH 3.5 | 100.0 | 99.7 | 100.1 | 100.1 |

The total radioactivity expected for 100% recovery of the $^{14}C$-labeled material from each baked product was calculated and was compared to the actual level of radioactivity recovered in the aqueous extract. This comparison demonstrated that essentially 100% of the sucralose radioactivity that was added to each formulation prior to baking was recovered by the extraction of the baked goods.

With the demonstration of complete recovery of the radiolabeled material, the extracts were subjected to thin-layer chromatography using two solvent systems independently as eluents. The solvent systems were selected because of their ability to separate sucralose from its potential breakdown products. The amount of radioactivity was quantitatively determined. The quantity and distribution of radioactivity contained in each baked product extract was compared to the quantity and distribution of a sucralose standard solution.

It is significant to note that for each baked product, no peaks other than sucralose could be detected by TLC analysis. The distribution of radioactivity corresponded well with the distribution of radioactivity in the sucralose standard solution.

The most significant findings of this study were that no products other than sucralose were found under the conditions of this experiment, and the TLC distribution of radioactivity from the baked goods extracts corresponded almost exactly with that of the sucralose standard. These findings demonstrate that sucralose is a sweetener that is suitable for its intended use as a high potency sweetener in baked goods (10).

These model system stability studies clearly demonstrate that sucralose is a remarkably stable ingredient that will have a wide range of applicability. Further, it is unlikely that stability will in any way restrict the use of sucralose in food applications.

## Analytical Methodology

Significant efforts have been made to develop analytical methods for the recovery and analysis of sucralose from food products. To date more than 50 methods to determine the sucralose level in food products have been developed. HPLC has been shown to be the most effective analytical method for precision and accuracy in the analysis of sucralose.

As with much of food analysis, extraction methodology is the key to acceptable recoveries. The solubility of sucralose in water and in alcohol facilitate the selective extraction of sucralose from food products. When required, additional sample clean-up and concentration of sucralose is achieved using solid phase extraction (SPE) techniques. A nonpolar C-18 SPE cartridge has been most effective in a majority of the applications evaluated. An Alumina SPE cartridge has been most effective for dairy products.

Reversed-phase HPLC using a C-18 column with a mobile phase of 70% water, 30% methanol is used in most analyses. Occasionally acetonitrile is used in place of methanol. Injection volumes typically in range of 100–150 $\mu$l are used, but they vary with the concentration of sucralose in the sample being evaluated.

Although there are several options regarding detection, we have found that refractive index is the most appropriate. Refractive index provides a detection system that has fewer problems with interference and can be used effectively when temperature is controlled (11).

## UTILIZATION OF SUCRALOSE IN FOOD APPLICATIONS

### Food Categories Included in U.S. Food Additive Petition

Due to the excellent physicochemical characteristics and the remarkable stability of sucralose a petition, broad in scope, was developed and submitted to the U.S. Food and Drug Administration (FDA) in February 1987. This petition covered many traditional categories of use for high-potency sweeteners and several categories that can be viewed as new opportunities for the use of high-intensity sweeteners. Table 4 lists the Code of Federal Regulation (CFR) categories for which sucralose use has been petitioned. These categories are broad in nature and are not product specific, allowing the food technologist the freedom to develop new products that will be covered by the regulatory category as defined in the CFR (12). The data submitted in the sucralose food additive petition clearly demonstrate that sucralose has the stability requirements for the effective use of the sweetener in all of the petitioned categories.

These data suggest that there are many other applications for which sucralose's stability, physicochemical, and high quality taste characteris-

**Table 4** Application Categories in Sucralose Food Additive Petition (U.S.)

| | |
|---|---|
| • Baked Goods and Baking Mixes | • Fruits and Water Ices |
| • Beverages and Beverage Bases | • Gelatins and Puddings |
| • Chewing Gum | • Jams and Jellies |
| • Coffee and Tea | • Milk Products |
| • Confections and Frostings | • Processed Fruits and Fruit Juices |
| • Dairy Products Analogs | • Sugar Substitutes |
| • Fats and Oils (Salad Dressings) | • Sweet Sauces, Toppings, and Syrups |
| • Frozen Dairy Desserts and Mixes | |

**Table 5** Suggested Sucralose Usage Levels in Selected Applications

| Product category | Recommended sucralose usage level (ppm) | Product category | Recommended sucralose usage level (ppm) |
|---|---|---|---|
| Frozen desserts | | Beverages | |
| Ice cream type | 100–150 | Ready to drink | 150–250 |
| Ice milk type | 75–150 | Powdered (as consumed) | 200–250 |
| Sherbet type | 100–200 | Carbonated | 175–225 |
| Water ice | 150–250 | Fountain syrup | 1700–1900 |
| Baked goods | | Syrups | |
| Cake | 500–600 | Pancake | 1000–1500 |
| Cookies | 300–350 | Chocolate | 1000–1300 |
| Apple pie filling | 175–225 | Fruit products | |
| Brownies | 200–300 | Jams and jellies | 400–850 |
| Graham cracker | 200–300 | Canned fruit | 250–300 |
| Cheese cake | 200–250 | Powdered desserts | |
| Dairy products | | Instant pudding | 250–350 |
| Plain yogurt | 100–150 | Chewing gum | 500–550 |
| Fruit yogurt | 175–225 | Pourable salad dressing | 100–150 |
| Flavored milk | 60–100 | | |

tics will be very compatible for the successful inclusion of this sweetener in new food products. To date, all the data available show that sucralose can be used in virtually all food applications.

### Typical Sucralose Usage Levels

Extensive prototype development in support of commercialization of sucralose has demonstrated the utility of sucralose in high quality low-calorie/reduced-calorie food and beverage products. Table 5 lists suggested usage levels for sucralose in a range of applications. The actual usage level in specific formulations will vary depending upon sweetness level desired, other ingredients used in the formulation and flavor systems utilized.

## SAFETY AND REGULATORY

### Safety

One hundred scientific studies have been conducted to establish the safety of sucralose for human consumption. There is no evidence that sucralose or its by-products cause any untoward effects. Sucralose is non-toxic and does not hydrolyze or dechlorinate following ingestion. The animal studies clearly demonstrate the overall safety of sucralose even under life-long, high-dose test conditions that would exaggerate any health effects. Studies included evaluation of animals that were exposed to sucralose from conception throughout normal life span and with amounts that far exceed the probable maximum human consumption.

In addition the breakdown products of sucralose were subjected to almost the same level of testing as sucralose, including a separate cancer study.

Metabolism studies indicate that the dog, rat, mouse, rabbit, and man metabolize sucralose similarly. Therefore, the results of the safety studies conducted on sucralose in animals can be extrapolated to man with confidence.

Sucralose was determined to be nontoxic following acute exposure in rodents up to 16,000 mg/kg body weight. This represents about 16,000 times the projected average intake of sucralose, or the consumption of approximately 4000 gallons of carbonated soft drinks per day.

Sucralose and its breakdown products demonstrated no carcinogenic activity, as was shown in three lifetime studies in animals. Two of these studies, a mouse study and a rat study, used doses of up to 3% of the diet as sucralose. No adverse effects were seen in rats or mice at doses equivalent to 1500 mg/kg body weight/day, or approximately 1500 times the projected average human consumption.

A two-generation reproduction study in rats and teratology and genetic toxicity studies showed that sucralose and the hydrolysis products had no effect on fertility or reproduction and no teratological effects. Specific studies in rats and monkeys at doses hundreds of times the maximum amount of anticipated human consumption have shown that sucralose and the hydrolysis products had no effect on the nervous system.

The safety of sucralose in man has been evaluated in a number of clinical studies. In one study, daily doses of 500 mg, which is equivalent in sweetness to two-thirds of a pound of sugar a day, were fed to 77 volunteers. There were no physical, biochemical, or hematological changes in the volunteers in this study or any of the other clinical studies conducted. In addition, sucralose was found to have no effect on blood glucose, blood fructose, and/or insulin secretion in healthy animals.

An independent panel of 16 internationally recognized scientific experts was asked to review the sucralose toxicological database and projected human intake data. They were to provide a detailed review on the adequacy of the toxicological data and evaluate the safety of sucralose for human consumption. The panel included experts in toxicology, oncogenicity, clinical toxicology, genetic toxicology, metabolism, biochemistry, reproductive toxicology, physiology, nutrition, hematology, pediatrics, risk assessment, neurotoxicology, and immunology. The expert panel on sucralose concluded that, based on the evaluation of the data presented to them and regulatory agencies, and on the discussions and deliberations of the entire panel, "sucralose is safe for its intended purpose of use" (13).

## Cariogenicity

Sprague-Dawley Rats infected with *Streptococcus mutans* were provided with sucrose or sucralose along with a diet that contained no additional sucrose. After 35 days the level of infection of *S. mutans* was determined and the sulcal caries scored. Rats red sucralose had a similar number of total viable flora, however, the level of *S. mutans* of the total viable flora was decreased by up to 20-fold when compared to the sucrose control. Sucralose-consuming rats had less than 50% of the sulcal caries evident in the sucrose control.

Rats were desalivated and inoculated with *S. mutans* and *Actinomyces viscosus* and fed diets that contained either 56% sucrose or sucralose at 93 mg/100 g. The number of coronal lesions did not differ significantly among groups, but the severity of the lesions was significantly lower after 35 days of feeding the sucralose diet when compared to the sucrose control.

Substitution of sucralose for sucrose resulted in substantially fewer lingual and proximal lesions. Although similar levels of root surface exposures were achieved in the test and control diet, only rats fed sucralose remained free of root surface caries. These data show that sucralose is noncariogenic (14).

In a second study it was demonstrated that where sucralose was the sole source of carbon, the growth of 10 strains of oral bacteria and plaque could not be supported. When sucralose was incorporated in a liquid medium containing glucose or sucrose, all organisms tested displayed similar acid production when compared to a control without sucralose. Sucralose was shown, using $^{14}C$-labeled sucralose to inhibit the formation of the glucan and fructan polymers found in plaque. These data show that sucralose is not metabolized by oral bacteria and that sucralose has a noncompetitive and reversible inhibitory action on the enzymes necessary to synthesize glucan and fructan polymers (15). The results of these two studies demonstrate that sucralose is noncariogenic and will not contribute to the development of caries.

### Regulatory Status

Petitions to receive regulatory approval of sucralose for use in a variety of food applications have been filed separately in the United States, Canada, the United Kingdom, the European Economic Community, Denmark, Australia, and with the Joint FAO/WHO Expert Committee on Food Additives (JECFA). In June 1990, JECFA assigned a permanent ADI for sucralose of 15 mg/kg body weight/day.

## PATENT STATUS

Sucralose has an extensive patent portfolio that covers its use, methods of synthesis, and other patentable issues. Patents regarding sucralose were first issued in the United Kingdom in 1979, Canada in 1981, and the United States in 1982.

Some of the significant patents include:

Crystalline form, U.S. Patent 4,343,934 issued 4/10/82
Synthesis, U.S. Patent 4,363,869 issued 12/7/82
Use, U.S. Patent 4,435,440 issued 3/6/84
Synergy, U.S. Patent 4,495,170 issued 1/22/85

## CONCLUSION

Sucralose is a high quality, high-potency sweetener that is derived by the selective chlorination of sucrose. The resulting product is extraordinarily stable with excellent physicochemical characteristics that will permit the use of this product in a wide variety of food and beverage applications. The safety database has shown sucralose and the hydrolysis products to be safe for its intended uses for human consumption. These characteristics provide the food and beverage industry with a unique opportunity to improve existing low-calorie products and develop totally new low-calorie/reduced-calorie product applications that will meet the ever-growing consumer demand for good tasting, high quality, low-calorie food and beverages.

## REFERENCES

1. L. Hough and R. A. Khan, *Trends Biochem. Sci. 3:* 61–63 (1978).
2. G. A. Miller, *Changing Food Technology 2,* (M. Kroger and A. Freed, ed.), Technomic Publishing Co., Lancaster, PA 1989, pp. 107–126.
3. M. R. Jenner, *Progress in Sweeteners,* (T. Grenby, ed.) Elsevier, London, 1989, pp. 121–141.

4. S. G. Wiet and P. K. Beyts, *Psychosensory Characteristics of the Sweetener Sucralose,* 49th Annual Meeting of the Institute of Food Technologists, New Orleans, LA, 1988.
5. S. G. Wiet and P. K. Beyts, *Sensory Characteristics of Sucralose,* International Conference, Sweeteners: Carbohydrates and Low Calorie, American Chemical Society, Agricultural and Food Chemistry Division, Los Angeles, CA, 1988.
6. G. A. Miller and M. R. Jenner, *The Chemical and Stability Characteristics of Sucralose,* International Conference, Sweeteners: Carbohydrates and Low Calorie, American Chemical Society, Agricultural and Food Chemistry Division, Los Angeles, CA, 1988.
7. M. R. Jenner and A. Smithson, *Physico-chemical properties of the sweetener sucralose", J. Food Sci., 54:* 1646–1649 (1989).
8. Anon., Food Additive Petition 7A3987, 1987, pp. 739–754.
9. Anon., Food Additive Petition 7A3987, 1987, pp. 623–731.
10. R. L. Barndt and G. Jackson, *Food Technology, 44* (1): 62–66 (1990).
11. V. M. Mulligan, M. E. Quinlan, and M. R. Jenner, *General Procedures for the Analytical Determination of the High Intensity Sweetener, Sucralose, in Food Products,* 49th Annual Meeting of the Institute of Food Technologists, New Orleans, LA, 1988.
12. Anon., Code of Federal Regulations, the Office of the Federal Register, National Archives and Records Administration, 21 CFR 170.3(n), 1988, pp. 6–8.
13. Anon., Food Additive Petition 7A3987, 1987, pp. A-000755–A-000783.
14. (W. H. Bowen, D. A. Young, and S. K. Pearson, *J. Dental Research* (in press).
15. D. A. Young and W. H. Bowen, *J. Dental Research* (in press).

# 11
# Less Common High-Potency Sweeteners

A. Douglas Kinghorn *University of Illinois at Chicago, Chicago, Illinois*
César M. Compadre *University of Arkansas, Little Rock, Arkansas*

## INTRODUCTION

In this chapter, recent progress on the study and utilization of several less commonly encountered naturally occurring and synthetic sweet substances will be described. Several of the natural sweeteners discussed have served as template molecules for extensive synthetic modification. Greater emphasis will be provided for those compounds that have commercial use as high-potency sweeteners or as flavoring agents in one or more countries, and priority will also be given to those sweet compounds for which there is recent published information. Following the format of the earlier version of this chapter (1), substances that modify the sweet taste response will also be mentioned.

## NATURALLY OCCURRING COMPOUNDS

### Glycyrrhizin

Glycyrrhizin, which is also known as glycyrrhizic acid, is an oleanane-type triterpene glycoside whose use as a sweetener has been covered in several recent reviews (1–4). This compound, a diglucuronide of the aglycone, glycyrrhetinic acid (Fig. 1), is extracted from the roots of licorice (*Glycyrrhiza glabra* L., Fabaceae) and other species in the genus *Glycyrrhiza*. In its natural form, the compound occurs in the plant in yields of 6–

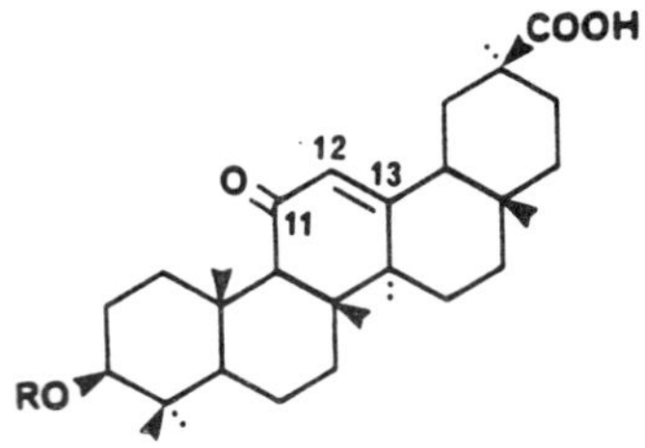

| | R |
|---|---|
| Glycyrrhizin | β-glcA²-β-glcA |
| Apioglycyrrhizin | β-glcA²-β-api |

**Figure 1** Structures of glycyrrhizin and apioglycyrrhizin. (GlcA = S-glucuronopyranosyl; api = S-apiofuranosyl.)

14% w/w as a mixture of various metallic salts. Well-established procedures are available for the extraction and purification of glycyrrhizin from the plant. Conversion of glycyrrhizin to ammoniated glycyrrhizin, the fully ammoniated salt of glycyrrhizin, results in a more water-soluble compound that is reasonably stable at elevated temperatures. Glycyrrhizin has been rated as approximately 50–100 times sweeter than sucrose, although it has a slow onset of sweet taste and a long aftertaste. Ammoniated glycyrrhizin has similar hedonic properties to glycyrrhizin. Its sweetness intensity, which is about 50 times sweeter than sucrose, is increased in the presence of sucrose (1–4). Attempts to improve the sensory parameters of glycyrrhizin by synthetically modifying its carbohydrate moieties have not so far led to any sweeter or more pleasant-tasting analoges (2,3). However, apioglycyrrhizin (Fig. 1), a glycyrrhizin analog recently isolated from the roots of *Glycyrrhiza inflata* Batal, has been found to be about twice as sweet as the parent compound (5). Periandrins I–IV are additional naturally occurring oleanane-type triterpene glycosides that were isolated by Hashimoto and colleagues from the roots of *Periandra dulcis* L. (Fabaceae) (Brazilian licorice). These compounds are 90–100 times sweeter than sucrose, but occur in the plant in low yields and are somewhat difficult to purify from bitter substances with which they co-occur (2,4).

*Glycyrrhiza* root extracts that are constituted by at least 90% w/w pure glycyrrhizin, are widely used in Japan for sweetening and flavoring foods, beverages, medicines, cosmetics, and tobacco (1,6,7). In 1987, extracts of *Glycyrrhiza* species containing glycyrrhizin were estimated to have a share of the Japanese high-potency sweetener market (27%) that was second only to *Stevia rebaudiana* sweeteners, with a sales volume of 2 billion yen (6). Roots of *Glycyrrhiza* species for the Japanese market are produced mainly in the People's Republic of China, the Soviet Union, Afghanistan, and Pakistan. In addition, extracts of *Glycrrhiza* roots are imported into Japan from the United States and the People's Republic of China (6). Ammoniated glycyrrhizin is included in the GRAS (Generally Recognized as Safe) list of approved natural flavoring agents by the U.S. Food and Drug Administration. There are many applications of this compound as a flavorant, flavor modifier, and foaming agent. While it is very useful for incorporation into confectionery and dessert items, ammoniated glycyrrhizin is only employed in carbonated beverages that do not have too low a pH, since this substance tends to precipitate at pH levels below 4.5 (1).

Glycyrrhizin exhibits a number of other biological activities in addition to its sweetness, as exemplified by its antiallergic, antiinflammatory, antitussive, and expectorant actions. Unfortunately, the widespread use of glycyrrhizin and ammoniated glycyrrhizin by humans has been shown to lead to pseudoaldosteronism, which is manifested by hypertension, edema, sodium retention and mild potassium diuresis (2,7,8). The 11-oxo-$\Delta^{12,13}$- functionality in ring C of the aglycone of glycyrrhizin has been attributed as the part of the molecule responsible for this untoward activity (7,9). The Ministry of Health in Japan has issued a caution stipulating that glycyrrhizin should be limited to less than 200 mg/day when used in drug formulations (8). Glycyrrhizin, at a level of 0.5–1%, has been shown to inhibit in vitro plaque formation mediated by *Streptococcus mutans,* a cariogenic bacterial species. As a consequence, it has been suggested that glycyrrhizin is suitable for wider use as a vehicle and sweetener for medications employed in the oral cavity (9).

## Mogrosides

Mogrosides IV and V are sweet curcurbitane-type triterpene glycoside constituents of the dried fruits of the Chinese plant *lo han kuo.* This species, a member of the family Cucurbitaceae, was accorded the binomial *Momordica grosvenorii* Swingle in 1941, which was then changed to *Thladiantha grosvenorii* (Swingle) C. Jeffrey in 1979. However, new evidence has indicated that the name of this plant should be *Siraitia grosvenorii* (Swingle) C. Jeffrey. Chemical studies on the sweet principles of this plant did not begin until the 1970s (2,4).

The structures of mogrosides IV and V (Fig. 2) were established by Takemoto and colleagues after extensive chemical and spectroscopic studies (10–12). The major sweet principle of *lo han kuo* fruits is mogro-

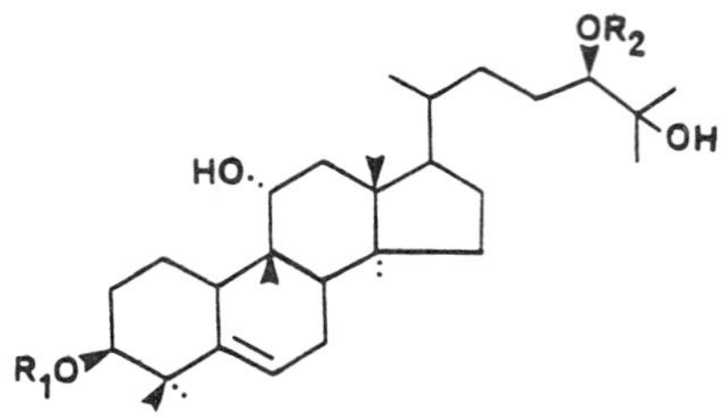


| | $R_1$ | $R_2$ |
|---|---|---|
| Mogroside IV | $\beta$-glc$^{6}$-$\beta$-glc | $\beta$-glc$^{2}$-$\beta$-glc |
| Mogroside V | $\beta$-glc$^{6}$-$\beta$-glc | $\beta$-glc$^{2}$-$\beta$-glc<br>\|6<br>$\beta$-glc |

**Figure 2** Structures of mogroside IV and mogroside V. (Glc = S-glucopyranosyl.)

side V, which occurs in concentration levels of over 1% w/w (2). Mogroside V is a polar compound, since it contains five glucose residues, which readily permit its extraction into either water or 50% ethanol. Aqueous solutions containing mogroside V are stable when boiled (1). Mogroside V was rated as being 256 times sweeter than sucrose by a human taste panel, when a 0.02% w/v solution was tested (10). However, this compound has been found to possess a lingering, licorice-like aftertaste (1,2). Three other sweet cucurbitane glycosides have been isolated from two species in the genus *Hemsleya,* although none of these compounds was ranked as more than 80 times sweeter than sucrose, when tested at a concentration of 0.025% w/w (13,14).

Extracts made from *lo han kuo* fruits have long been used by local populations in Kwangsi province in southern China for the treatment of colds, sore throats, and minor stomach and intestinal complaints (2). Preparations made from *lo han kuo* are available in Chinese medicinal herb stores in several countries including the United States. It was estimated that the domestic demand of *lo han kuo* fruits containing mogroside V for food and medicinal products in Japan in 1987 was two metric tons, representing a sales volume of 40 million yen (6).

Safety studies on mogroside V have not been extensive to date. The compound has been found to be nonmutagenic, when tested in a forward mutation azaguanine assay using *Salmonella typhimurium* strain

TM677 (2,4). Mogroside V produced no mortalities when administered by oral intubation at doses up to 2 g/kg body weight in acute toxicity studies in mice, and an aqueous extract of *lo han kuo* fruits exhibited an $LD_{50}$ in mice of > 10 g/kg body weight (2,4). There appear to have been no adverse reactions among human populations who have ingested aqueous extracts of *lo han kuo* fruits, which would be expected to contain substantial quantities of mogroside V. Therefore, *lo han kuo* extracts containing mogroside V might well be worthy of wider application for sweetening purposes in the future, because of their apparent safety, in addition to favorable sensory, stability, solubility, and economic aspects.

## Phyllodulcin

Phyllodulcin is produced from its naturally occurring glycoside form by enzymatic hydrolysis when the leaves of *Hydrangea macrophylla* Seringe var. *thunbergii* (Siebold) Makino (Saxifragaceae) and other species in this genus are crushed or fermented. Phyllodulcin (Fig. 3) is a dihydroisocoumarin and was initially isolated in 1916 and structurally characterized in the 1920s. In 1959, this sweet compound was found to have 3*R* stereochemistry (1–4). Phyllodulcin has been detected in the leaves of *H. macrophylla* subsp. *serrata* var. *thunbergii* in levels as high as 2.36% w/w (2). Several patented methods are available for the purification of phyllodulcin. In one such procedure, after initial extraction from the plant with methanol or ethanol, hydragenol (a nonsweet analog of phyllodulcin) and pigment impurities were removed after pH manipulations and extraction with chloroform. Phyllodulcin was then selectively extracted in high purity at pH 10 with a non-polar solvent (15). The relative sweetness of phyllodulcin has been variously reported as 400 and 600–800 times sweeter than sucrose, although the compound exhibits a delay in sweetness onset and a licorice-like aftertaste (16). There have been extensive attempts to modify the phyllodulcin structure in order to produce compounds with improved sensory characteristics, which have recently been subjected to detailed review (3). As a result of such investigations, it has been established that the 3-hydroxy-4-methoxyphenyl unit of phyllodul-

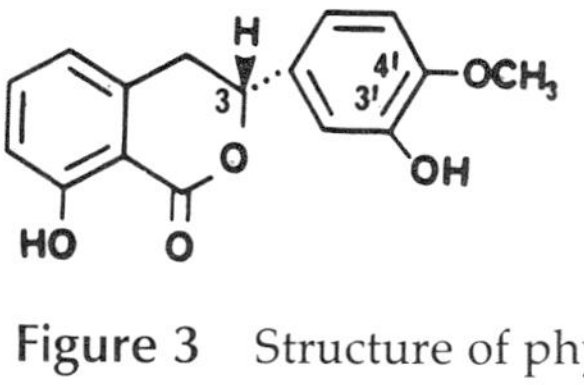


**Figure 3** Structure of phyllodulcin.

cin must be present for the exhibition of a sweet taste, but the phenolic hydroxy group and the lactone function can be removed without losing sweetness (3,4,16). To date, the phyllodulcin derivatives that have been produced synthetically seem to have limitations in terms of their water solubility, stability, and/or sensory characteristics (3,4,16).

The fermented leaves of *H. macrophylla* var. *thunbergii* ("Amacha") are used in Japan to produce a sweet tea that is consumed at *Hamatsuri*, a Buddhist religious festival (1,2). A 1987 estimate indicated that the demand for extracts of *Hydrangea* species containing phyllodulcin was 1 metric ton, with a value of 15 million yen (6). Pure phyllodulcin has been found to be nonmutagenic in a forward mutation assay, and also not acutely toxic for mice, when administered by oral intubation at up to 2 g/kg body weight (2). The low solubility in water and the sensory limitations of phyllodulcin that have been referred to would seem to limit the prospects of this compound being more widely utilized as a sweetening agent in the future.

## Sweet Proteins

### *Thaumatin*

Thaumatins I and II are the major sweet proteins that have been obtained from the arils of the fruits of the West African plant *Thaumatococcus daniellii* (Bennett) Benth. (Marantaceae). Thaumatin I has a relative sweetness intensity of between 1600 and 3000, when compared with sucrose on a weight basis. Thaumatin protein (which is known by the trade-name of Talin® protein) was comprehensively reviewed by Higginbotham in the first edition of *Alternative Sweeteners* (17), in terms of botany, production, biochemistry, physical characteristics, sensory parameters, sweetness synergy with other substances, applications (including flavor potentiation and aroma enhancement effects), safety assessment, cariogenic evaluation, and regulatory status. Since a further detailed review of thaumatin by van der Wel has also appeared in the literature (18), this subject is not covered in depth in the present chapter.

Talin® protein was initially permitted as a natural food additive in Japan in 1979 (17). In 1987, there was an estimated Japanese demand of 200 kg of thaumatin, which was valued at a price of 350,000 yen per kilogram (6). Despite the fact that Talin® protein has been approved as a sweetener in the United Kingdom and Australia and is being similarly reviewed in several other countries (17), it now appears that the major future use of this product will be as a flavor enhancer. In the United States, Talin® protein was accorded GRAS status as a flavor adjunct for chewing gum in 1984 (17).

*Monellin*

The sweet protein isolated from the fruits of another African plant, *Dioscoreophyllum cumminsii* (Stapf) Diels (Menispermaceae), has been called "monellin," a substance composed of two polypetide chains, of molecular weight 11,000 daltons (1). "Monellin" is actually monellin 4, one of five sweet proteins, named monellins 1–5, that have been isolated from this plant. Present procedures, in which buffered aqueous extracts of *D. cumminsii* fruit pulps are fractionated, enable 3–5 g of protein to be purified per kg of fruit, with a sweetness relative to 7% w/v sucrose of 1500–2000 times (18). Monellin is costly to produce, and its plant of origin is difficult to propagate. In addition, the compound has a slow onset of taste, along with a persistent aftertaste, and its sweet effect is both thermolabile and pH sensitive (1,2,18). No toxicological data are yet available for monellin (18). Therefore, unless new formulations can be produced to limit these disadvantages, it looks as if monellin will not be developed in the near future as a high-intensity sweetener and will have greater utilization as a molecular tool for the laboratory investigation of the sweetness sensation (1,2,18).

*Other Sweet Proteins*

Mabinlins I and II are sweet proteins that are produced in the seeds of *Capparis masaikai* Lévl (Capparidaceae), a plant which grows in Yunnan province in the People's Republic of China. The molecular weights of mabinlins I and II are 11,600 and 10,400 daltons, respectively, and both proteins are composed of single polypeptide chains, and have some 80 amino acid residues in common. These sweet albumins were found to occur in a combined yield of 13% w/w of the dry weight of the defatted seeds, with mabinlin I being the major sweet component. Mabinlins I and II were treated with 50% aqueous acetone, and then purified by chromatography over cellulose. The compounds are reported to be less sweet than thaumatin and monellin, although similar in sweetness quality. The sweet taste of mabinlin II persists on incubation at 80°C for 48 hours at pH 6, although the sweetness of the less stable mabinlin I is lost after 30 minutes when stored under the same conditions. Children are reported to chew the seed meal of *C. masaikai* because of its sweet taste and because it imparts a sweet taste to water drunk later. No pharmacological studies have been performed on either the mabinlins or on *C. masaikai* seed meal, although the latter is used in traditional Chinese medicine to treat sore throats and as an antifertility agent (19).

Pentadin is a sweet protein that was recently reported as a constituent of the fruits of the African plant *Pentadiplandra brazzeana* Baillon (Pentadiplandraceae). The compound has an estimated molecular weight of 12,000 daltons and was obtained from the plant by water extraction, ultrafiltration, and gel filtration. Pentadin has been estimated to have a sweetness potency about 500 times that of sucrose on a weight basis. Using a rhesus monkey, 0.1% pentadin gave a steeper onset and more rapid stimulatory-response decline than did 0.02% thaumatin and monellin. Further studies are planned to more thoroughly characterize pentadin (20).

## Miscellaneous Highly Sweet Plant Constituents

### *Hernandulcin*

Hernandulcin (Fig. 4) is a bisabolane sesquiterpene that was isolated as a minor constituent in a petroleum ether-soluble extract of the aerial parts of the herb *Lippia dulcis* Trev. (Verbenaceae), collected in Mexico. This plant was known to be sweet by the Aztec people, according to the Spanish physician Francisco Hernández, who wrote a monograph entitled "Natural History of New Spain" between 1570 and 1576. The *L. dulcis* sweet constituent was named in honor of Hernández, and was rated as 1000 times sweeter than sucrose on a molar basis when assessed by a taste panel (21). Racemic hernandulcin was synthesized by a directed aldol condensation from two commercially available ketones (21,22), and the naturally occurring (6*S*,1′*S*)-diastereomer of this compound was produced in the laboratory from (*R*)-limonene (23). When the functional groups of hernandulcin were modified, no sweet compounds resulted (24). However, this investigation enabled the conclusion to be made that, in addition to the C-1′ hydroxyl and the C-1 carbonyl groups of hernandulcin, which represent the AH and B groups in the Shallenberger model of sweetness, the C-4′–C-5′ double bond appears to be a third functionality necessary for the exhibition of a sweet taste by this compound (22,24). Racemic hernandulcin was not mutagenic and not acutely toxic for mice at the doses tested in preliminary safety studies (21,22). Unfortunately, the high sweetness potency of hernandulcin is marred by an unpleasant aftertaste and a somewhat bitter-tasting nature (21).

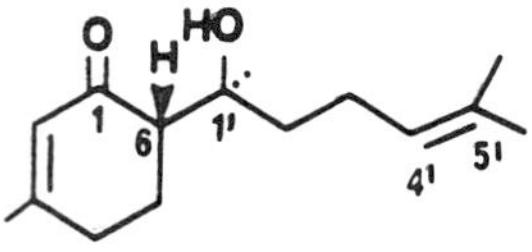


**Figure 4** Structure of hernandulcin.

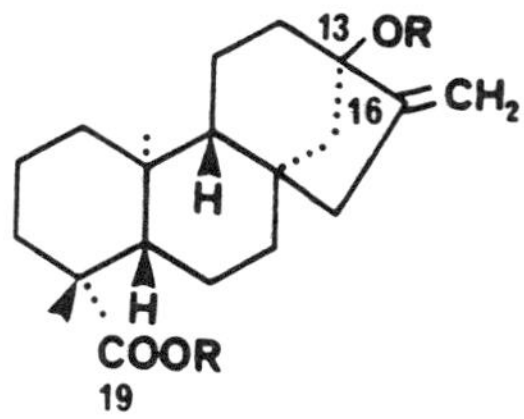


| | R |
|---|---|
| Rubusoside | β-glc |

**Figure 5** Structure of rubusoside. (Glc = S-glucopyranosyl.)

### *Rubusoside*

Tanaka and co-workers have determined the *ent*-kaurene diterpene rubusoside (Fig. 5) to be responsible for the sweet taste of the leaves of *Rubus suavissimus* S. Lee (Rosaceae), which is indigenous to southern regions of the People's Republic of China (2,4). Rubusoside is extractable from the plant with hot methanol and occurs in high yield in the leaves of *R. suavissimus* ( > 5% w/w) and at lower concentration levels in the fruits of this plant (25). When evaluated at a concentration level of 0.025%, rubusoside was rated as possessing 114 times the sweetness of sucrose, although its quality of taste sensation was marrred by some bitterness (2,4). Several analogs of rubusoside have been produced by treatment with the cyclodextrin glucosyltransferase produced by *Bacillus megaterium* strain No. 5, and some of these products proved to be sweeter and more pleasant-tasting than rubusoside itself (26).

A sweet tea called *Tian-cha*, prepared from the leaves of *R. suavissimus*, is consumed as a summer beverage in the Guangxi Autonomous Region of the People's Republic of China. Also, during festivals, local populations mix aqueous extracts of this plant with rice in order to make cakes. In recent years, teas made from the leaves of *R. suavissimus* have been used in folk medicine to treat diabetes, hypertension, and obesity. In an acute toxicity test on rubusoside, the $LD_{50}$ was established as about 2.4 g/kg body weight, when administered orally to mice. Subsequently, in a subacute toxicity study, rubusoside was incorporated into the diet of mice for 60 days at a dose of one-tenth of its $LD_{50}$, and no distinct toxicity nor side effects were observed (27). However, the aglycone of rubusoside is steviol (*ent*-13-hydroxykaur-16-en-19-oic acid), which has been shown to be mutagenic in a bacterial forward mutation assay when

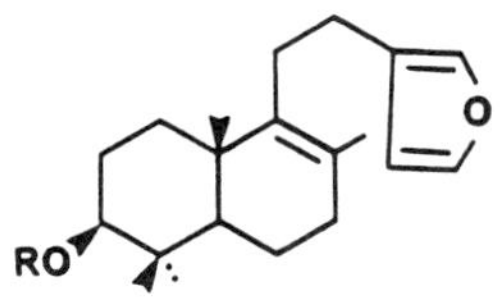

| | R |
|---|---|
| Baiyunoside | β-glc²-β-xyl |
| Phlomisoside I | β-glc²-α-rha |

**Figure 6** Structures of baiyunoside and phlomisoside I. (Glc = S-glucopyranosyl; xyl = S-xylopyranosyl; rha = S-rhamnopyranosyl.)

metabolically activated (28). The mutagenicity of steviol is discussed more fully in Chapter 9.

*Baiyunoside*

Baiyunoside (Fig. 6) is a labdane-type diterpene glycoside, based on the aglycone (+)-baiyunol, that was first isolated in 1983 by Tanaka and coworkers from a plant used in Chinese medicine, namely, *Phlomis betonicoides* Diels (Labiatae) (2,4,29). This butanol-soluble compound was found to be about 500-fold sweeter than sucrose and to possess a lingering aftertaste lasting more than one hour. Phlomisoside I (Fig. 6) is a further sweet constituent of *P. betonicoides,* although its sensory characteristics have not been detailed (29). Synthetic routes are available for both (±)- and (+)-baiynunol (30,31). A general glucosylation procedure has been developed for baiyunol and has not only been used to produce baiyunoside synthetically, but offers a general method for the preparation of baiyunoside glycosidic analogs, in the hope that one or more might exhibit better sensory properties than the parent compound (32). No safety studies appear to have been performed thus far on baiyunoside.

*Steroidal Saponins*

Osladin (Fig. 7) is a steroid saponin constituent of the fern *Polypodium vulgare* L. (Polypodiaceae), which was isolated and structurally characterized by Herout and co-workers in 1971 (1–4). The stereochemistry of the aglycone of osladin was established by Havel and Cerny in 1975, although the configuration of the S-rhamnopyranosyl moiety at C-26 of this glycoside has not yet been determined (2,4). While osladin has been rated as 3000 times as sweet as sucrose, its hedonic taste characteristics are not described in the literature.

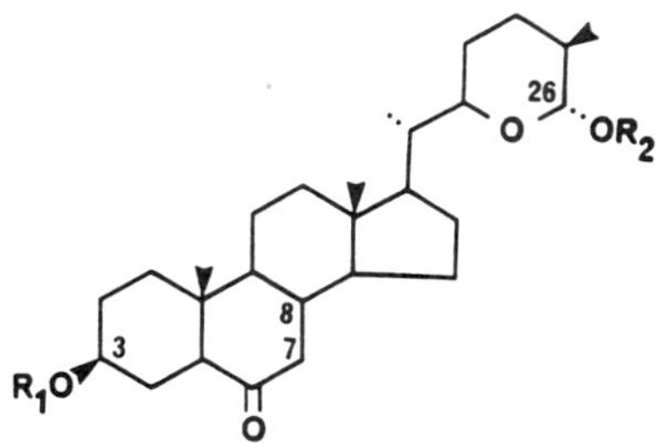

| | $R_1$ | $R_2$ | Other |
|---|---|---|---|
| Osladin | β-glc²-α-rha | rha | -- |
| Polypodoside A | β-glc²-α-rha | α-rha | $\Delta^{7,8}$ |
| Polypodoside B | β-glc | α-rha | $\Delta^{7,8}$ |

**Figure 7** Structures of some sweet steroidal saponins. (Glc = D-glucopyranosyl; rha = L-rhamnopyranosyl.)

In recent work, two sweet compounds, polypodosides A and B (Fig. 7), which are closely related in structure to osladin, have been isolated from the rhizomes of *Polypodium glycyrrhiza* D.C. Eaton (Polypodiaceae) 33,34). This plant, known by the common name of "Licorice fern," is a North American species that is native to the Pacific northwest. Polypodoside A, the major sweet constituent of *P. glycyrrhiza,* has a neohesperidose (2-*O*-α-S-rhamnopyranosyl-β-S-glucopyranose) unit attached to C-3 of the aglycone, as well as a C-26 affixed S-rhamnopyranosyl unit (with a β configuration). The aglycone of polypodoside A (polypodogenin) corresponds to the $\Delta^{7,8}$-derivative of the aglycone of osladin. Polypodoside A was found to be nonmutagenic and not acutely toxic for mice, when dosed by oral intubation at up to 2 g/kg body weight. In subsequent sensory tests using a small human taste panel, polypodoside A was assessed as having 600 times the sweetness intensity of a 6% w/v sucrose solution, but also revealed a licorice-like off-taste and a lingering aftertaste (33). Polypodoside B, an isolate similar structurally to polypodoside A, except that it has a glucose unit affixed to C-3, was found to be only slightly sweet (34). Comparison of polypodosides A and B with other compounds based on the same aglycone has indicated that polypodogenin glycosides must be bisdesmosidic in order to exhibit a sweet taste (34). The potential of polypodoside A for commercialization is marred by its relative insolubility in water, its sensory characteristics, and difficulties in collecting *P. glycyrrhiza* rhizomes (33). It is probable that osladin will suffer from similar limitations, in addition to its extremely low concentration levels in its plant of origin, as has already been noted (1).

### *Dihydroflavonol Sweeteners*

Naturally occurring flavonoids are generally regarded as either bitter- or neutral-tasting compounds, although a number of sweet or bittersweet dihydrochalcone glycosides have been reported, including glycyphyllin, phloridzin, and trilobatin (2,4). Recently, a new class of sweet flavonoids was discovered, namely, the dihydroflavonol sweeteners. In an initial report, the sweetness and astringency of the stembark of *Glycoxylon huberi* Ducke (Sapotaceae) was correlated with the presence of four dihydroflavonol constituents, although the individual compounds were not assessed for the presence or absence of sweetness (35).

The known compound dihydroquercetin 3-acetate (Fig. 8) was isolated as a sweet constituent of the young shoots of *Tessaria dodoneifolia* (Hook. & Arn.) Cabrera (Asteraceae). Dihydroquercetin 3-acetate-4′-methyl ether (Fig. 8), a synthetic derivative of this sweet plant constituent, was determined as possessing about 400 times the sweetness potency of a 3% w/v sucrose solution, and exhibited no bitterness, although it had a somewhat slow onset of sweet taste. By comparison, dihydroquercetin 3-acetate was rated as about 80 times sweeter than 3% w/v sucrose. Both these quercetin derivatives were found to be nonmutagenic and not acutely toxic for mice. The aerial parts of *T. dodoneifolia* are used in Paraguay in folk medicine as an emmenogogue, and the plant is referred to colloquially as *Kaá hê-é* (sweet herb) (36).

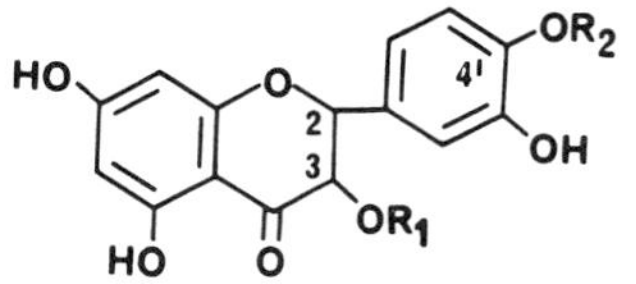


| | $R_1$ | $R_2$ | |
|---|---|---|---|
| Quercetin 3-acetate | acetyl | H | (2*R*,3*R*) |
| Quercetin 3-acetate-4′-methyl ether | acetyl | $CH_3$ | |
| Neoastilbin | α-rha | H | (2*S*,3*S*) |

**Figure 8** Structures of some sweet dihydroflavonol derivatives. (Rha = S-rhamnopyranosyl.)

Neoastilbin (Fig. 8) is a further dihydroflavonol that has been reported to possess a sweet taste. This compound is a rhamnoside derivative of taxifolin that is produced by the leaves of *Engelhardtia chrysolepis* Hance (Juglandaceae), a species which grows in southern parts of the People's Republic of China. The sweetness intensity of neoastilbin has not been reported. The leaves of *E. chrysolepis* are used as sweet tea by local populations where they are available (37).

## SYNTHETIC COMPOUNDS

### Oximes

Perillartine (Fig. 9), the *α-syn*-oxime of perillaldehyde, has been known to be highly sweet since 1920, and is reported to be up to 2000 times sweeter than sucrose (3,38,39). In contrast, perillaldehyde itself (Fig. 9), a major constituent of the volatile oil of *Perilla frutescens* (L.) Britton (Labiatae), is only slightly sweet. Perillartine is used commercially in Japan as a replacement for maple syrup or licorice for the sweetening of tobacco, but more widespread use of this compound for sweetening purposes has been restricted by a limited solubility in water, an appreciably bitter taste, as well as a menthol-licorice off-taste that accompanies sweetness (2,3,38,39).

The intense sweetness and structural simplicity of perillartine have promoted the synthesis of numerous analogs (3,38,39). This work has not only led to a better understanding of the functional groups in compounds of the oxime class that confer sweetness and bitterness, but has also led to the development of several improved sweet compounds. One of the most promising of such derivatives is SRI oxime V (Fig. 9). This compound is 450 times sweeter than sucrose on weight basis and exhibits much improved water solubility when compared with perillartine.

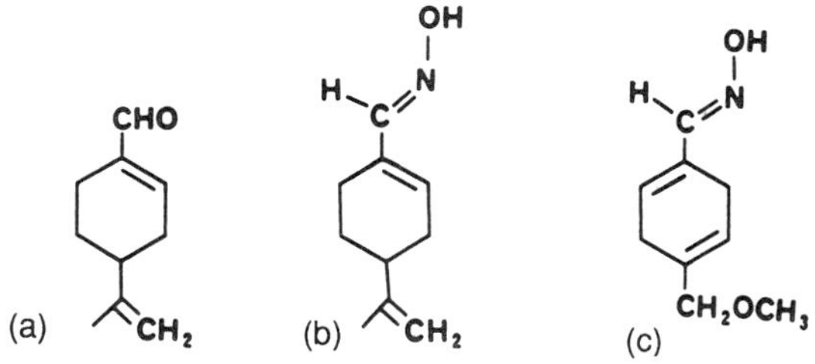


**Figure 9** Structures of (a) perillaldehyde, (b) erillartine, and (c) SRI oxime V.

SRI oxime V has no undesirable aftertaste and is stable above pH 3 (1–3,38–40). This substance was shown not to be a bacterial mutagen in the Ames assay, and to exhibit an $LD_{50}$ of > 1 g/kg body weight in the rat after a single oral dose (40). The compound is readily absorbed and metabolized, with excretion nearly quantitative within 48 hours after administration to the rat, dog, and rhesus monkey. The major metabolites of SRI oxime V were found to be products resulting from oxidation of either the methoxymethyl or the aldoxime moieties, as well as those occurring after thioalkylation and glucuronidation (40). Subchronic toxicity tests on this compound conducted in rats with a diet containing 0.6% SRI oxime V for 8 weeks revealed no apparent toxic effects. It has been suggested that SRI oxime V shows such promise as an artificial sweetener that a chronic toxicity test is warranted (40).

### Urea Derivatives

Dulcin (*p*-ethoxyphenylurea) has been known to be sweet for over a century. The compound is about 200 times sweeter than sucrose and was briefly marketed as a sucrose substitute in the United States. Commercial use of this compound was discontinued after it was found to be toxic to rats at a low dose. Dulcin has also been found to be mutagenic (1,39).

Another group of sweet ureas of more recent interest are the carboxylate-solubilized *p*-nitrophenyl derivatives, which were discovered by Peterson and Müller (3,39). Suosan, the sodium salt of *N*-(*p*-nitrophenyl)-*N'*-($\beta$-carboxyethyl)-urea (Fig. 10), is representative of this series and has been reported as about 350 times sweeter than sucrose, although it has significant bitterness. Other compounds in this class are even sweeter than suosan (3,39,42). Structure–sweetness relationships have been investigated for the sweet-tasting arylureas (3,42,43).

### *Miscellaneous Compounds*

Tryptophan Derivatives The sweetness of derivatives of the amino acid tryptophan was discovered by Kornfeld and his co-workers in 1968, when it was observed that racemic 6-trifluoromethyltryptophan has an

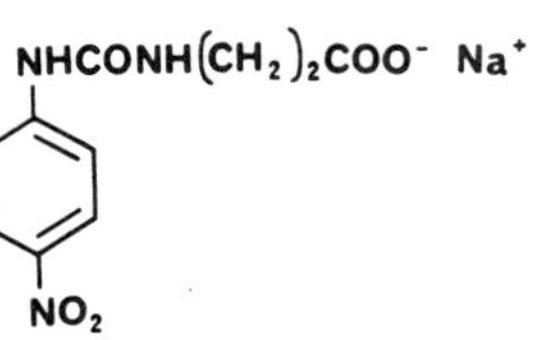


**Figure 10** Structure of suosan.

intensely sweet taste (3,44). Additional studies demonstrated that these compounds are sweet when in the S form, with 6-chloro-S-tryptophan being some 1000 times sweeter than sucrose. The S form of this compound is tasteless but has been found to possess antidepressant activity (44).

It was reported by Finley and Friedman that racemic *N'*-formyl and *N'*-acetyl derivatives of kynurenine, an intermediate in the metabolism of tryptophan, are approximately 35 times sweeter than sucrose and elicit an immediate sweet taste on contact with the tongue (38,44,45). More recently, the 6-chloro derivative of kynurenine [3-(4-chloroanthraniloyl)-SS-alanine] (Fig. 11), has been reported to be 80 times sweeter than sucrose and to possess no significant aftertaste or off-flavor (45).

Trihalogenated Benzamides 2,4,6-Tribromo benzamides that are substituted in the C-3 position by a carboxyalkyl or a carboxylalkoxy group are intensely sweet. For example, 3-(3-carbamoyl-2,4,6-tribromophenyl) propionic acid (Fig. 12) was rated as 4000 times sweeter than sucrose, although this compound has a slow onset and a slight lingering aftertaste, as well as some bitterness (3,46). Within this compound class, the intensity of sweet taste depends upon the chain length of the carboxyalkyl or carboxyalkoxy group (3,46). The acute toxicities of several tribromo benzamides have been determined in mice, and the results were comparable with analogous data obtained for saccharin and cyclamate (46).

(Phenylguanidino)-acetic Acid Derivatives Nofre and co-workers have recently reported the synthesis of a series of (phenylguanidino)- and {[1-(phenylamino)ethyl]amino}-acetic acid derivatives, with sweetness intensities between 30 and 200,000 times that of a 2% sucrose solution (47).

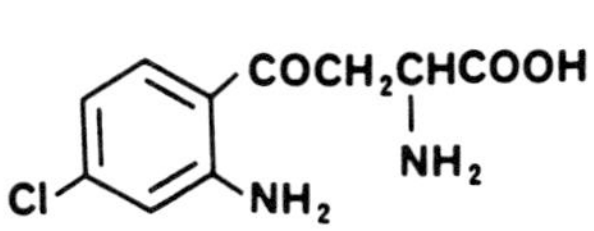


**Figure 11** Structure of 3-(4-chloroanthraniloyl)-SS-alanine.

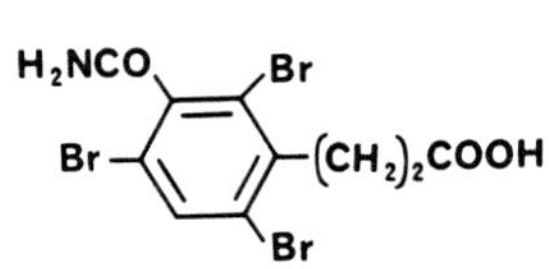


**Figure 12** Structure of 3-(3-carbamoyl-2,4,6-tribromophenyl)propionic acid.

These compounds appear to represent some of the most potent sweeteners discovered to date.

## SWEETNESS MODIFYING SUBSTANCES

### Sweetness Inducers and Enhancers

#### *Miraculin*

Miraculin is a basic glycoprotein constituent of the fruits of *Richardella dulcifica* (Schumach. & Thonning) Baehni [formerly *Synsepalum dulcificum* (Schumach. & Thonning) DC.] (Sapotaceae) (miracle fruit), that has the propensity of making sour or acidic materials taste sweet (1,2). Although miracle fruit concentrates are no longer commercially available in the United States as a food additive (1), efforts have continued in the structural characterization of this substance. Miraculin is single polypeptide of 191 amino acid residues, with a carbohydrate content of 13.9% and an overall molecular weight of 24,600 daltons. The complete amino acid sequencing of miraculin was recently carried out by Kurihara and coworkers (48).

#### *Other Sweetness Enhancers*

A number of plant constituents have sweetness inducing properties, including cynarin, chlorogenic acid, and caffeic acid (2). Arabinogalactin (larch gum) is capable of enhancing the sweetness potency and taste qualities of saccharin, cyclamate, and protein sweeteners such as thaumatin and monellin (1).

### Sweetness Inhibitors

#### *2-(4-Methoxyphenoxy)propanoic Acid*

A series of aralkylcarboxylic acid salts, inclusive of the sodium salt of 2-(4-methoxyphenoxy)propanoic acid (Fig. 13) have been found to in-

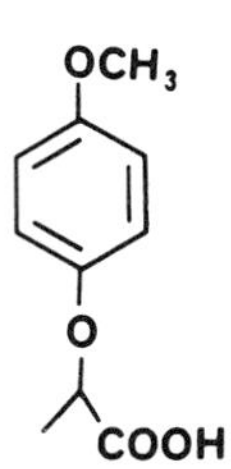


**Figure 13** Structure of 2-(4-methoxyphenoxy)propanoic acid.

hibit the sweetness of sucrose (49). Following the analysis of roasted Colombian Arabica coffee beans by gas chromatography/mass spectrometry and high-performance liquid chromatography, 2-(4-methoxyphenoxy)propanoic acid was found to be a naturally occurring constituent (50). This compound (as its sodium salt) is now commercially available as a flavoring agent intended to modulate high sweetness, under the tradename of Cypha®, and has been accorded Generally Recognized as Safe (GRAS) status, for use up to levels of 150 ppm (51).

*Arylalkylketones*

Several arylalkylketones and arylcycloalkylketones have also been discovered to inhibit the sweet taste of sucrose as well as other bulk and intense sweeteners. One such compound, the commercially available 3-(4-methoxybenzoyl)propionic acid (Fig. 14), is capable of reducing the sweetness intensity of 40% w/v aqueous sucrose by over a sixfold margin, when present at a 2% w/v concentration at pH 7, compared with when absent from the formulation. This compound and its analogs are recommended for use in soft freeze puddings, infused vegetables, and other food products (52).

*Triterpene Glycoside Sweetness Inhibitors*

Considerable recent progress has been made in the characterization of plant-derived triterpene glycoside sweetness inhibitors. Gymnemic acid has been known for some time to be a mixture of variously acylated glucuronide derivatives of the oleanane-type aglycone, gymnemagenin, found in the leaves of *Gymnema sylvestre* R. Br. (Asclepiadaceae) (2,4). Two groups have now purified six discrete sweetness inhibitors (Fig. 15) and several inactive analogs from this plant source (53–55), which should be of great utility in future research on the physiology of sweetness. The most abundant antisweetness factor found was gymnemic acid II, and it has been found that the more highly acylated substances are more potent as sweetness inhibitors (53–55).

While the suppressive effect of gymnemic acid affects taste sensations other than sweetness, ziziphin, a constituent of *Ziziphus jujuba* P. Miller (Rhamnaceae), has a selective inhibitory action on sweet taste. Ziziphin (Fig. 16) has been isolated and fully characterized by Kurihara

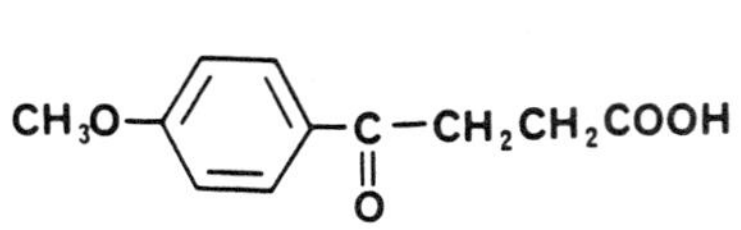


**Figure 14** Structure of 3-(4-methoxybenzoyl)propionic acid.

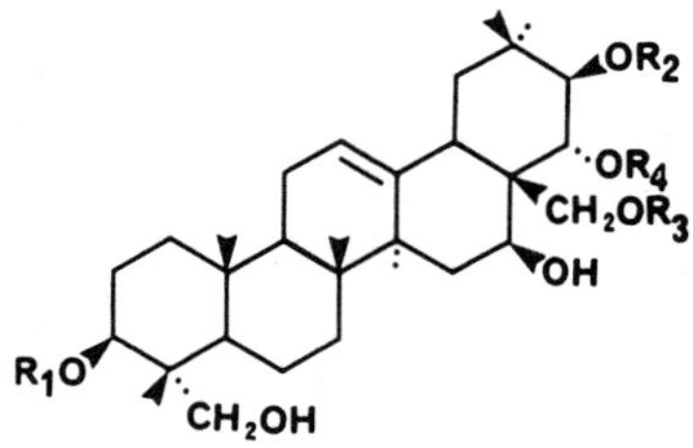


| | $R_1$ | $R_2$ | $R_3$ | $R_4$ |
|---|---|---|---|---|
| Gymnemic acid I | glcA | tiglyl | acetyl | H |
| Gymnemic acid II | glcA | 2-methylbutyryl | acetyl | H |
| Gymnemic acid III | glcA | 2-methylbutyryl | H | H |
| Gymnemic acid IV | glcA | tiglyl | H | H |
| Gymnemic acid V | glcA | tiglyl | H | tiglyl |
| Gymnemic acid VI | glc$^3$-glcA | tiglyl | H | H |

**Figure 15** Structures of gymnemic acids I–VI. (GlcA = S-glucuronopyranosyl; glc = S-glucopyranosyl.)

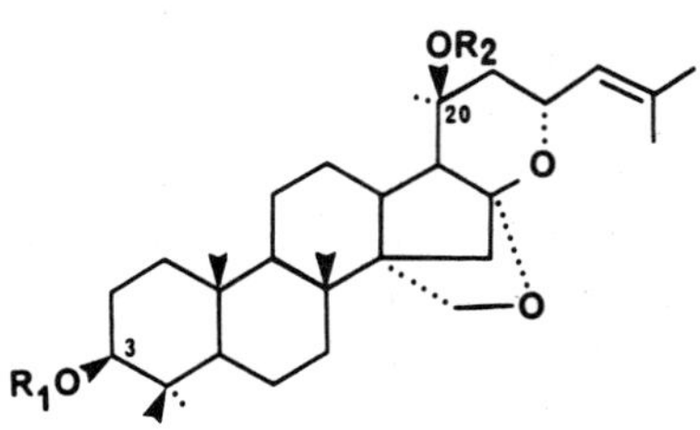


| | $R_1$ | $R_2$ |
|---|---|---|
| Zizyphin | α-rha$^4$-α-ara | α-rha-2-Ac-3-Ac |

**Figure 16** Structure of ziziphin. (Rha = S-rhamnopyranosyl; ara = S-arabinopyranosyl; rha-2-Ac-3-Ac = (2,3-di-*O*-acetyl)-*α*-S-rhamnopyranosyl.)

and colleagues and is apparently the only constituent of the plant to manifest a sweetness inhibitory effect (56). Hodulcin, from the leaves of *Hovenia dulcis* Thunb. (Rhamnaceae), which is as yet incompletely characterized, is a further triterpene glycoside that has been found to possess a selective inhibitory effect on sweet taste. However, it has been been established that hodulcin is based on a different aglycone than either gymnemic acids I–VI or ziziphin (57).

## REFERENCES

1. L. O'Brien Nabors, and G. E. Inglett, A review of various other alternative sweeteners, in *Alternative Sweeteners* (L. O'Brien Nabors and R. C. Gelardi, eds.), Marcel Dekker, New York, 1986, pp. 309–323.
2. A. D. Kinghorn, and D. D. Soejarto, Sweetening agents of plant origin, *CRC Crit. Rev. Plant Sci., 4:* 79–120 (1986).
3. H. van der Wel, A. van de Heijden, and H. G. Peer, Sweeteners, *Food Rev. Internat., 3:* 193–268 (1987).
4. A. D. Kinghorn and D. D. Soejarto, Intensely sweet compounds of natural origin, *Med. Res. Rev., 9:* 91–115 (1989).
5. I. Kitagawa, M. Sakagami, F. Hashiuchi, J. L. Zhou, M. Yoshikawa, and J. Ren, Apioglycyrrhizin and araboglycyrrhizin, two new sweet oleanene-type triterpene oligoglycosides from the root of *Glycyrrhiza inflata, Chem. Pharm. Bull., 37:* 551–553 (1989).
6. Anonymous, High sweeteners—Market size 7.2 billion yen: Stevia occupies 41%, but future gains will be made by aspartame, *Food Chemicals* (*Tokyo*) No. 6: 19–26 (1988).
7. Hikino, H., Recent research on Oriental medicinal plants, in *Economic and Medicinal Plant Research,* Vol. 1 (H. Wagner, H. Hikino, and N. R. Farnsworth, eds.), Academic Press, London, 1985, pp. 53–85.
8. J. D. Higginbotham, Recent developments in non-nutritive sweeteners, in *Developments in Sweeteners— 2* (T. H. Grenby, K.J. Parker, and M.G. Lindley, eds.), Applied Science Publishers, London, 1983, pp. 119–155.
9. R. Segal, S. Pisanty, R. Wormser, E. Azaz, and M. N. Sela, Anticariogenic activity of licorice and glycyrrhizin I: Inhibition of *in vitro* plaque formation by *Streptococcus mutans. J. Pharm. Sci., 74:* 79–81 (1985).
10. T. Takemoto, S. Arihara, T. Nakajima, and M. Okuhira, Studies on the constituents of Fructus Momordicae. I. On the sweet principle, *Yakugaku Zasshi, 103:* 1151–1154 (1983).
11. T. Takemoto, S. Arihara, T. Nakajima, and M. Okuhira, Studies on the constituents of Fructus Momordicae. II. Structure of sapogenin, *Yakugaku Zasshi, 103:* 1155–1166 (1983).
12. T. Takemoto, S. Arihara, T. Nakajima, and M. Okuhira, Studies on the constituents of Fructus Momordicae. III. Structure of mogrosides, *Yakugaku Zasshi, 103:* 1167–1173 (1983).

13. R. Kasai, K. Matsumoto, R.-L. Nie, T. Morita, A. Awazu, J. Zhou, and O. Tanaka, Sweet and bitter cucurbitane glycosides from *Hemsleya carnosiflora, Phytochemistry, 26:* 1371–1376 (1987).
14. R. Kasai, K. Matsumoto, R.-L. Nie, J. Zhou, and O. Tanaka, Glycosides from Chinese medicinal plant, *Hemsleya panacis-scandens,* and structure-taste relationship of cucurbitane glycosides, *Chem. Pharm. Bull., 36:* 234–243 (1988).
15. Yuki Gosei Kogyo Co. Ltd., High purity phyllodulcin production, *Jpn. Kokai Tokkyo Koho 81, 96,677,* 4 pp. (1981); through *Chem. Abstr., 95:* 167369a (1981).
16. A. Arnoldi, L. Camarda, and L. Merlini, Synthesis and taste of some 2-phenylbenzodioxanes, *J. Agric. Food Chem., 34:* 339–344 (1986).
17. J. D. Higginbotham, Talin protein (thaumatin), in *Alternative Sweeteners* (L. O'Brien Nabors and R. C. Gelardi, eds.), Marcel Dekker, New York, 1986, pp. 103–134.
18. H. van der Wel, The sweet proteins, in *Developments in Food Proteins-4,* (B.J.F. Hudson, ed.), Elsevier Applied Science, London, 1986, pp. 219–245.
19. H. Zhong, The sweet proteins from the seeds of *Capparis masaikai* Lévl, in *Proceedings of the München-Shanghai Symposium on Peptide and Protein Chemistry,* Schloss Ringberg am Tegernsee, FRG, June 9–10, 1986, pp. 109–113.
20. H. van der Wel, G. Larson, A. Hladik, C. M. Hladik, G. Hellenkant, and D. Glaser, Isolation and characterization of pentadin, the sweet principle of *Pentadiplandra brazzeana* Baillon, *Chemical Senses, 14:* 75–79 (1989).
21. C. M. Compadre, J. M. Pezzuto, A. D. Kinghorn, and S. K. Kamath, Hernandulcin: An intensely sweet compound discovered by review of ancient literature, *Science, 227:* 417–419 (1985).
22. C. M. Compadre, R. A. Hussain, R. L. Lopez de Compadre, J. M. Pezzuto, and A. D. Kinghorn, The intensely sweet sesquiterpene, hernandulcin: Isolation, synthesis, characterization, and preliminary safety evaluation, *J. Agric. Food Chem., 35:* 273–279 (1987).
23. K. Mori and M. Kato, Synthesis of (6*S*,1′*S*)-hernandulcin, a sweetener, and its stereoisomers, *Tetrahedron, 42:* 5895–5900 (1986).
24. C. M. Compadre, R. A. Hussain, R. L. Lopez de Compadre, J. M. Pezzuto, and A. D. Kinghorn, Analysis of the structural features responsible for the sweetness of the sesquiterpene, hernandulcin, *Experientia, 44:* 447–449 (1988).
25. W.-H. Chou, T. Oinaka, F. Kanamaru, K. Mitzutani, F.-H. Chen, and O. Tanaka, Diterpene glycosides from leaves of Chinese *Rubus chingii* and fruits of *R. suavissimus,* and identification of the source plant of the Chinese folk medicine "Fu-pen-zi", *Chem. Pharm. Bull., 35:* 3021–3024 (1987).
26. M. Darise, K. Mizutani, R. Kasai, O. Tanaka, S. Kitahata, S. Okada, S. Ogawa, F. Murakami, and F.-H. Chen, Enzymic transglucosylation of rubusoside and the structure-sweetness relationship of steviol-bisglycosides, *Agric. Biol. Chem., 48:* 2483–2488 (1984).
27. W.-K. Xu, Y.-K, Li, L.-S. Meng, and D.-Z. Huang, *Sweet Tea,* Chinese Forestry Publishing House, Beijing, People's Republic of China, 1984.

28. C. M. Compadre, R. A. Hussain, R. L. Lopez de Compadre, J. M. Pezzuto, and A. D. Kinghorn, Mass spectral analysis of some derivatives and in vitro metabolites of steviol, the aglycone of the natural sweeteners, stevioside, rebaudioside A, and rubusoside, *Biomed. Environ. Mass Spectrom., 15:* 211–222 (1988).
29. T. Tanaka, O. Tanaka, Z.-W. Lin, and J. Zhou, Sweet and bitter principles of the Chinese plant drug, Bai-yun-shen: Revision of the assignment of the source plant and isolation of two new diterpene glycosides, *Chem. Pharm. Bull., 33:* 4275–4280 (1985).
30. M. Nishizawa, H. Yamada, and Y. Hayashi, Total synthesis of (±)-baiyunol, *Tetrahedron Lett., 27:* 3255–3256 (1986).
31. K. Mori and M. Komatsu, Synthesis of (+)-baiyunol, the diterpene aglycone of the sweet glycoside baiyunoside, *Tetrahedron, 43:* 3409–3412 (1987).
32. H. Yamada, and M. Nishizawa, Total synthesis of baiyunoside by a novel 2′ discriminated glycosylation, *Tetrahedron Lett., 28:* 4315–4318 (1987).
33. J. Kim, J. M. Pezzuto, D. D. Soejarto, F. A. Lang, and A. D. Kinghorn, Polypodoside A, an intensely sweet constituent of the rhizomes of *Polypodium glycyrrhiza, J. Nat. Prod., 51:* 1166–1172 (1988).
34. J. Kim and A. D. Kinghorn, Further steroidal and flavonoid constituents of the sweet plant, *Polypodium glycyrrhiza, Phytochemistry, 28:* 1225–1228 (1989).
35. H. Jacquemin, A. Boissonnat, G. Faugeras, F. Tillequin, and P. Delaveau, Flavonoides de *Glycoxylon huberi* Ducke, *Ann. Pharm. Fr., 43:* 521–525 (1985).
36. N. P. D. Nanayakkara, R. A. Hussain, J. M. Pezzuto, D. D. Soejarto, and A. D. Kinghorn, An intensely sweet dihydroflavonol derivative based on a natural product lead compound, *J. Med. Chem., 31:* 1250–1253 (1988).
37. R. Kasai, S. Hirono, W.-H. Chou, O. Tanaka, and F.-H. Chen, Sweet dihydroflavonol rhamnoside from leaves of *Engelhardtia chrysolepis*, a Chinese folk medicine, Hung-qi, *Chem. Pharm. Bull., 36:* 4167–4170 (1988).
38. G. A. Crosby, New Sweeteners, *Crit. Rev. Food Sci. Nutr., 7:* 297–323 (1976).
39. G. A. Crosby and R. E. Wingard, Jr., A survey of less common sweeteners, in *Developments in Sweeteners—1* (C. A. M. Hough, K. J. Parker, and A. J. Vlitos, eds.), Applied Science Publishers, London, 1979, pp. 135–164.
40. T. J. Vernanzi and C. A. Vernanzi, *Ab initio* molecular electrostatic potentials of perillartine analogues: Implications for sweet-taste receptor recognition, *J. Med. Chem., 31:* 1879–1885 (1988).
41. C. Mitoma, E. M. Acton, J. I. DeGraw, and D. W. Thomas, Metabolic and toxicologic study of an artificial sweetener, oxime V, *Drug Chem. Toxicol., 8:* 195–206 (1985).
42. J. M. Tinti, C. Nofre, and A. M. Peytavi, Interaction of suosan with the sweet taste receptor, *Z. Lebensm.-Unters. Fortsch., 175:* 266–268 (1982).
43. M. R. Caijolo, F. Lelj, T. Tancredi, P. A. Temussi, and A. Tuzi, Interaction of conformationally flexible agonists with the active site of sweet taste. A study of arylureas, *J. Med. Chem., 26:* 1060–1065 (1983).
44. B. Crammer and R. Ikan, Properties and synthesis of sweetening agents, *Chem. Soc. Rev., 6:* 431–465 (1977).
45. K. Kawashima, H. Itoh, N. Yoneda, K. Hagio, T. Moriya, and I. Chibata,

An intensely sweet analogue of kynurenine: 3-(4-Chloroanthraniloyl)-DL-alanine, *J. Agric. Food Chem., 28:* 1338–1340 (1980).

46. H. Gries, W. Muetzel, H. D. Belitz, H. Wieser, I. Krause, and W. Stempfl, Trihalogenated benzamides—a new class of artificial sweeteners, *Z. Lebensm.-Unters. Forsch., 176:* 376–378 (1983).
47. C. Nofre, J. M. Tinti, F. Chatzapoulos-Ouar, Preparation of (phenylguanidino)- and {[1-(phenylamino)ethyl]amino}acetic acids as sweeteners, *Eur. Pat. Appl. EP 241,395,* 116 pp. (1987); though *Chem. Abstr., 109:* 190047k (1988).
48. S. Theerasilp, H. Hitotsuya, S. Nakajo, K. Nakaya, Y. Nakamura, and Y. Kurihara, Complete amino acid sequence and structure characterization of the taste-modifying protein, miraculin, *J. Biol. Chem., 264:* 6655–6659 (1989).
49. M. G. Lindley, Method of inhibiting sweetness, *Eur. Pat. Appl. EP 84302496.9,* 26 pp. (1984).
50. E. B. Rathbone, G. D. Patel, R. W. Butters, D. Cookson, and J. L. Robinson, Occurrence of 2-(4-methoxyphenoxy)propanoic acid in roasted coffee beans: analysis by gas-liquid chromatography and by high-performance liquid chromatography, *J. Agric. Food Chem., 37:* 54–58 (1989).
51. F. LaBell, Sweetness reduction improves flavor delivery, functionality, *Food Processing* (May): 74, 76 (1989).
52. R. E. Barnett, Foodstuffs containing sweetness inhibiting agents, *U. S. Pat. 4,544,565,* 10 pp. (1985).
53. K. Yoshikawa, K. Amimoto, S. Arihara, and K. Matsuura, Structure studies of new antisweet constituents from *Gymnema sylvestre, Tetrahedron Lett., 30:* 1103–1106 (1989).
54. M. Maeda, T. Iwashita, and Y. Kurihara, Studies on taste modifiers. II. Purification and structure determination of gymnemic acids, antisweet active principle from *Gymnema sylvestre* leaves, *Tetrahedron Lett., 30:* 1547–1550 (1989).
55. K. Yoshikawa, K. Amimoto, S. Arihara, and K. Matsuura, Gymnemic acid V, VI, and VII from Gur-ma, the leaves of *Gymnema sylvestre* R. Br., *Chem. Pharm. Bull., 37:* 852–854 (1989).
56. Y. Kurihara, K. Ookubo, H. Tasaki, H. Kodama, Y. Akiyama, A. Yagi, and B. Halpern, Studies on the taste modifiers. I. Purification and structure determination of sweetness inhibiting substance in leaves of *Ziziphus jujuba, Tetrahedron, 44:* 61–68 (1988).
57. L. M. Kennedy, L. R. Saul, R. Sefecka, and D. A. Stevens, Hodulcin: selective sweetness-reducing principle from *Hovenia dulcis* leaves, *Chem. Senses, 13:* 529–543 (1988).

# II
# CALORIC ALTERNATIVES

# 12

# Crystalline Fructose

**Thomas F. Osberger** *Food Industry Consultant, Upland, California*

## INTRODUCTION

One of the first things taught to students of organic chemistry is that sucrose is a disaccharide that can be readily hydrolyzed to provide equimolar amounts of two monosaccharides, D-glucose and D.-fructose. Annual U.S. consumption of D-glucose, or dextrose, to use its more familiar synonym, exceeds two billion pounds. Although it would seem simple to obtain both fructose and glucose from the hydrolysis of sucrose, it was only in the mid-1970s that sucrochemical advances enabled commercial quantities of pure fructose to become available in the United States.

It was certainly not the case that food scientists and members of the medical profession were unaware of the increased sweetening ability and unique metabolic properties of pure fructose. However, it made no sense to expend research time and money to formulate special foods and diets utilizing pure fructose without assurance that industrial quantities (i.e., truckloads) of pure fructose would be available.

## MANUFACTURE OF FRUCTOSE

Serious efforts to produce pure fructose in the United States date from the early 1950s and the issue of patent number 3,050,444 to Holstein and Holsing. The method described in this patent involves the inversion of sucrose by enzymatic action or acid hydrolysis followed by enzyme oxidation of the invert sugar to produce gluconic acid. Subsequent crystallization of sodium gluconate permits pyrogen-free levulose of high purity to be crystallized ultimately from the balance of the original invert syrup.

Earlier attempts to commercially extract fructose centered upon the use of the storage carbohydrate inulin, which is found in the bulbs of plants such as dahlias, Jerusalem artichokes, and the Hawaiian Ti plant. Through the controlled hydrolysis of the polyfructan, inulin, and the disruption of the $\beta$ (2→ 1) glycosidic linkages, the fructofuranose units, which overwhelmingly predominate, are freed and converted to the more stable fructopyranose anomer. Despite considerable efforts, these processes, which originally held considerable promise, could never provide fructose at a competitive price in the quantities required for use as an ingredient in large-scale food processing.

Meanwhile, in Europe, primarily in Finland, Germany, and France, the results of activities in sucrochemical research and development aimed at finding the best source for fructose had gone full circle. The pioneering work of Dubrufant in France in 1847 was reappraised, and sucrose again became the preferred source of raw material for fructose.

During the 1960s, industrial quantities of crystalline fructose began to be manufactured in Europe. The process involved the inversion of sucrose to split the glycosidic bond and release the simple sugars fructose and glucose from their disaccharide union. This constituted the simple part of the process. The subsequent separation and purification of the sugars by ion-exclusion techniques, followed by carefully controlled crystallization of the fructose, caused the total manufacturing time to exceed one week. In each country, except for one, the crystallization was accomplished from alcohol solutions. Only in Finland was crystallization done from distilled water.

As the worldwide demand for fructose escalated, the need to increase capacity and find less expensive ingredients became critical. Because the manufacture of pure crystalline fructose is intensely energy dependent, economies based on raw material were required to partially offset the skyrocketing energy costs. Considerable research time and effort was expended in order to reduce the cost of production and make it reasonably competitive with sucrose. Unfortunately, most of these efforts were based on textbook chemistry, which was divorced from the realities of chemical processing experience. These, in turn, resulted in the erroneous assumption that vast quantities of pure crystalline fructose priced at or below commodity sweeteners (i.e., sucrose; high-fructose corn syrup, HFCS) could be produced via the application of space-age genetic engineering techniques, which would result in an enzyme capable of 100% isomerization of glucose to fructose in a single step (1).

In 1981, the world's largest facility for making pure crystallization fructose came on stream in Thomson, Illinois, on the banks of the Mississippi River. Located in the heart of the corn belt, the American Xyrofin plant used liquid dextrose as the starting raw material. After purification and enzymatic isomerization, the traditional Finnish techniques were employed to produce a particularly high-quality crystalline fructose. Production time was reduced to about 5 days.

Initially, as a result of an intense marketing effort to educate the medical community, the food-processing industry, and selected consumers about the metabolic and sweetening advantages of crystalline fructose, sales were strong and the fructose production capabilities at the American Xyrofin plant were utilized to the fullest.

After several years, however, crystalline fructose sales reached a plateau, primarily because of its price relative to alternative natural and synthetic sweeteners. In addition, shipments began to arrive in significant quantities from Europe and Asia at prices sharply below those of the domestic producer. Further production of crystalline fructose was deemed to be economically unfeasible. Today the Thomson, Illinois facility is engaged primarily in making top grade Xylose.

Fortunately, for those companies who were committed to use of crystalline fructose in their products and to an even larger number of firms who had evaluated crystalline fructose but had heretofore been deterred from its use because of its high cost, the leaders of the corn sweetener industry decided the time was ripe to begin large-scale manufacture of crystalline fructose and pure fructose solutions.

To long-time observers of the corn wet-milling industry, this decision appeared to be long overdue. In view of their access to low-cost raw materials, their acknowledged expertise in producing high-quality high-fructose corn syrups and the developmental work in finding ways to enable crystalline fructose to be profitably positioned as a commodity sweetener, A. E. Staley Manufacturing Company and Archer Daniels Midland were logical candidates to move to the forefront of crystalline fructose production in the United States.

This has proven to be the case, and reductions in process time, coupled with economies of scale, have enabled selling prices to be reduced to around $0.36/lb. At this price, crystalline fructose is able to compete with sucrose and dextrose on a sweetening per dollar basis. This is indeed a dramatic reduction in price, and reports indicate that a great deal of new and renewed evaluation of this sweetener is being

**Table 1** Typical Physical Properties

| | |
|---|---|
| Appearance | White crystalline powder, forming anhydrous needle-shaped crystals |
| Empirical formula | $C_6H_{12}O_6$ |
| Molecular weight | 180.16 |
| Melting point | 102°–105°C |
| Density | 1.60 g/cm$^3$ |
| Bulk density (loose) | 0.8 g/cm$^3$ |
| Caloric value | 3.7 cal/g |
| Loss on drying (70°C for 4 hr in vacuum oven) | Less than 0.2% |
| Residue on ignition | Less than 0.5% |
| Heavy metals | Less than 5 ppm |
| Arsenic | Less than 1 ppm |
| Chloride | Less than 0.018% |
| Sulfate | Less than 0.025% |
| Calcium and magnesium | Less than 0.005% |
| Hydroxymethylfurfural | Less than 50 ppm |
| Glucose | Less than 0.1% |
| Assay (dry basis) | 98.0–102.0% |
| pH in aqueous solution (1 g/10 ml) | 5.0–7.0 |

State-of-the-art instrumentation is now available to permit rapid, precise measurements of fructose purity utilizing liquid chromatography. The reader is referred to the Sweetener Group of A. E. Staley Manufacturing Company of Decatur, IL, for this methodology.

done by food processors who are anxious to utilize the acknowledged capabilities of crystalline fructose.

Table 1 shows the composition of a typical lot of fructose, USP, FCC,* produced in the United States.

## EXISTENCE IN NATURE

Levulose, D-fructose, and fruit sugar are synonyms used to designate the ketohexose considered to be the sweetest sugar found in nature. Historically, the most available form of naturally occurring fructose has been the numerous varieties of honeys which are harvested throughout the world. According to Reay Tannahill, "until the end of the Middle

*USP = *The United States Pharmacopeia XX,* 21st revision, U.S. Pharmacopeial Convention, Rockville, MD, 1984; FCC = *Food Chemicals Codex III,* National Academy of Sciences, Washington, DC.

Table 2 Constituents of Honey

| | |
|---|---|
| Fructose | 40% |
| Glucose | 35% |
| Water | 18% |
| Other saccharides | 4% |
| Other substances | 3% |

Ages, honey was everywhere the sweetener par excellence" (3). Earliest recorded references to substances that provided a sweet taste were primarily concerned wtih honey. Thousands of years before the principal sugar of honey would be isolated and characterized as fructose, the superior sweetening ability of the as yet unidentified fructose, as present in wild honey, was recognized. Honey as an ingredient in sweet foods was popular in Rome before the birth of Christ, and honey was referred to as "chief among all sweet things." Biblical references to honey are found throughout both the New and Old Testaments.

There is no single quantitative list of constituents that is able to define all types of honey; however, Table 2 gives representative values which adequately approximate those found in most commercial honeys.

During the period prior to the discovery of insulin, honey was generally considered to be a preferred sweetener for diabetics. In view of our present knowledge about the components of honey and the metabolic superiority of fructose compared to other sugars, one can speculate that most of the sweetening advantages formerly attributed to honey were the result of the predominance of fructose in honey. Recent clinical studies by Akgun and Ertel (4) and Olefsky and Crapo (5) confirm the efficacy and suitability of fructose in the diets of some diabetics but cast considerable doubt on the overall merit of honey because of the presence of glucose and other high glycemic response–inducing saccharides. The characterization of fructose as "fruit sugar" stems from its significant presence in fruits and berries. Table 3 shows the percentages of fructose on the basis of the entire fruit and in relation to the total fruit solids (6).

## FRUCTOSE SWEETNESS

The property of fructose that most distinguishes it from all other edible carbohydrates is its high degree of sweetness. Without this attribute, the other distinctive characteristics of fructose (i.e., metabolic superiority, rapid solubility) would rarely serve as an inspiration to food technologists or pharmaceutical chemists to cause them to formulate new products with this remarkable sweetener.

**Table 3** Fructose Content of Fruits

| Fruit | Percent fructose in fruit | Percent fructose in total solids |
|---|---|---|
| Apple | 6.04 | 37.8 |
| Blackberry | 2.15 | 14.1 |
| Blueberry | 3.82 | 24.0 |
| Currant | 3.68 | 20.8 |
| Gooseberry | 3.90 | 26.3 |
| Grape | 7.84 | 41.0 |
| Pear | 6.77 | 49.9 |
| Raspberry | 4.84 | 17.2 |
| Sweet cherry | 7.38 | 32.9 |
| Strawberry | 2.40 | 25.4 |

Sweetness evaluation is a subjective estimate of the organoleptic perception, by a trained observer, of the sweetness of a substance. Usually, sucrose is used as a reference sweetener against which other compounds are compared. Moskowitz describes four major measures of sweetness, namely (a) threshold measures, (b) equal-sweetness matches, (c) category or rating scales, and (d) ratio scales (6).

Equal-sweetness matches are probably most frequently employed by food technologists in their efforts to achieve a market-dictated sweetness level through the use of an alternate sweetener. From such comparisons, a ratio is obtained that can provide the technologist with data on which to base amendments and revisions of existing formulations.

Reported sweetness multiples for fructose usually fall in the range of about 1.2–1.8 times that of sucrose. However, it cannot be overemphasized that the relative sweetness of fructose is dependent on the anomeric state of the fructose at the time the estimation of sweetness is made.

In the crystalline state, only the $\beta$-D-fructopyranose anomer exists; however, upon dissolution in water, rapid mutarotation occurs, resulting in the formation of three and possibly four fructose tautomers (7). The extent of this mutarotation can be determined by measuring the optical rotation of a solution of pure fructose in water. Other more sophisticated techniques, including gas-liquid chromatography and nuclear magnetic resonance, have been employed to determine that at 22°C the tautomeric equilibrium concentrations of 20% fructose in $D_2O$ are as illustrated in Figure 1 (8).

The temperature, pH, and concentration of the solution are the most important environmental conditions that influence the magnitude of perceived sweetness of freshly prepared solutions of fructose in water.

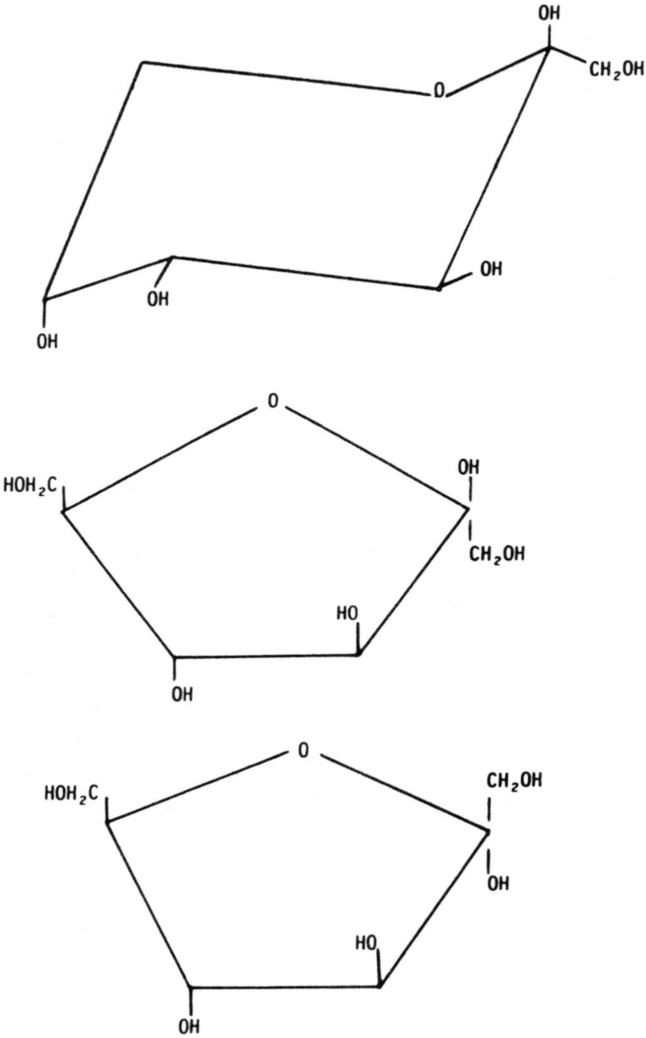


β-D-Fructopyranose

73%

β-D-Fructofuranose

21%

α-D-Fructofuranose

6%

**Figure 1** Cyclic tautomers of fructose. (From Ref. 8.)

Of these, only temperature exerts a significant effect on the mutarotational behavior of fructose in solution and on the transformation from the sweetest form, β-D-fructopyranose, to an equilibrium state in which other less sweet tautomeric forms of fructose are present (7,9).

In order to determine the reason for the differences of sweetness reported by some investigators, nuclear magnetic resonance (NMR) determinations were conducted at different temperatures (10). Loss in

**Table 4** The Tautomeric Equilibrium of Fructose at Different Temperatures

| Temperature (°C) | α-D-Fructo-furanose (%) | β-D-Fructo-furanose (%) | β-D-Fructo-pyranose (%) |
|---|---|---|---|
| 20 | 7 | 24 | 69 |
| 40 | 7 | 31 | 62 |
| 60 | 9 | 33 | 58 |
| 80 | 11 | 38 | 51 |

sweetness appears to bear a direct relationship to the shift in tautomeric equilibrium as temperature is increased, as shown in Table 4.

According to Shallenberger, "Based upon the correlation between fructose sweetness as related to its mutarotational behavior, it was deduced that the furanoses were nearly devoid of sweet taste" (7). This deduction, based primarily on conformational considerations, has not, however, been verified by a trained sweetness taste panel. Nevertheless, it is evident that an increase in the furanose anomer at the expense of the pyranose form will cause a reduction in perceived sweetness.

In any case, in order to derive the maximum sweetness from a given quantity of fructose, crystals of pure fructose should be consumed before being allowed to undergo mutarotation after being put into solution. In practice, the degree of loss of sweetness caused by this partial change to other less sweet furanose tautomers can be minimized through the use of cold solutions and slightly acid conditions. Experience has shown that citrus-flavored beverage bases, sweetened with pure crystalline fructose and containing the usual amounts of acidulants, can realize a reduction of up to 50% of usual sweetener calories. Conversely, one of the least efficient uses of fructose is in hot coffee, in which the mutarotation to the furanose forms diminishes the sweetness to the point that it is isosweet with sucrose. One further point to note in regard to fructose sweeteners is that the equilibrium state of the anomers is determined by the temperature at the time of consumption. Thus, cakes made with pure crystalline fructose will taste sweeter after they have been allowed to cool than they will if tasted just out of the oven.

## METABOLISM OF FRUCTOSE

Absorption of fructose in humans begins in the gastrointestinal tract, from which it is absorbed into the intestinal epithelial cells. The exact mechanism involved is uncertain; however, the process is probably one of facilitated diffusion rather than that of active transport (11). After absorption, which occurs at a rate slower than that of sucrose or glucose, it is primarily and rapidly metabolized in the liver. Significantly, its entry into liver cells and subsequent phosphorylation is insulin independent (12). Following cleavage by liver aldolase, the resulting trioses can be utilized for gluconeogenesis and glycogenesis or the synthesis of triglycerides, or they can enter the glycolytic pathway. The ultimate fate of these trioses is dependent on the metabolic state of the individual. Figure 2 illustrates the pathways involved in fructose metabolism.

In normal persons and in well-regulated diabetics, glycogenesis predominates, and only insignificant amounts of fructose fragments will be channeled toward glucose formation. Ultimately, liver glycogen is converted to glucose and released into the blood, at which time some insulin is required for phosphorylation and subsequent utilization in peripheral tissues. However, this conversion and release will be in response to lowered levels of blood sugar and will not cause the severe postprandial plasma glucose peaks and resulting hypoglycemic valleys that are often associated with ingestion of glucose and sucrose. Olefsky and Crapo have demonstrated that postprandial insulin and plasma glucose fluctuations are minor, following oral ingestion of 50 g of fructose in a liquid test meal that also contained 20 g each of fat and protein (5).

## COMPARATIVE METABOLIC ADVANTAGES

Earlier work by Dr. Bohannon and associates at the University of California resulted in the conclusion that "fructose ingestion causes less marked variation from baseline of the blood glucose, insulin, glucagon, and growth hormone levels than does sucrose or glucose in normal human beings" (13). Figures 3 and 4 illustrate these comparative effects.

The increased awareness of the harmful side effects of diabetes and the necessity for keeping blood sugar levels of diabetics as close to normal without inducing severe hypoglycemia have led diabetologists and food technologists to renew investigations into the use of fructose as a preferrred sweetener for diabetics (14).

Sorbitol has been widely accepted as a sugar substitute suitable for use in well-managed diabetic diets (15). After being absorbed, it is rapidly converted to fructose (see Fig. 2) through the enzymatic action of sorbitol dehydrogenase, and its metabolism is identical to that of ingested fructose. The important difference to note is that the perception of sweetness from sorbitol is only about one-third that from an equal weight of fructose (6). Therefore, dietetic foods can be sweetened much more efficiently with fructose in comparison to sorbitol, and, once absorbed, the two sweeteners will have the same metabolism. In addition,

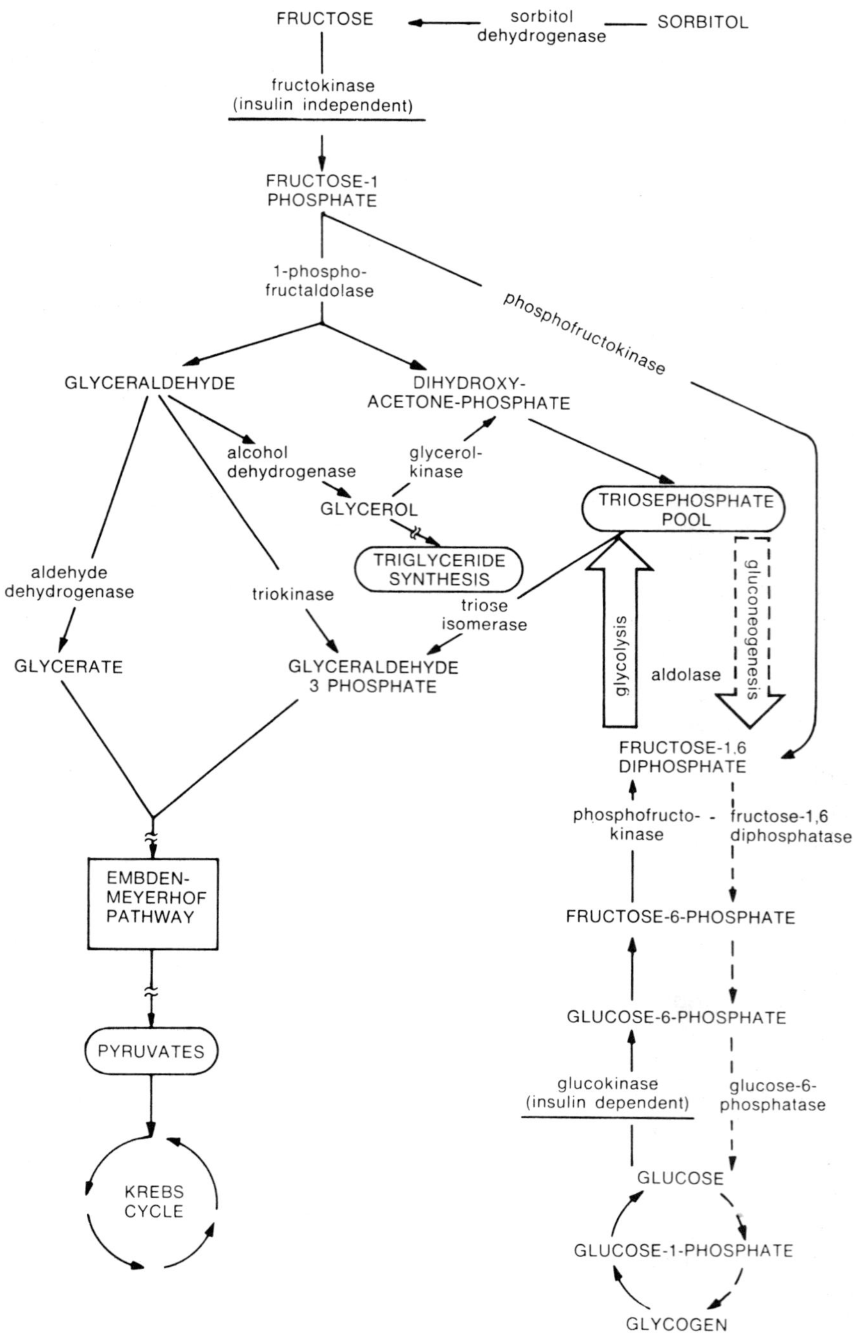


**Figure 2** Metabolic pathways of fructose in the liver.

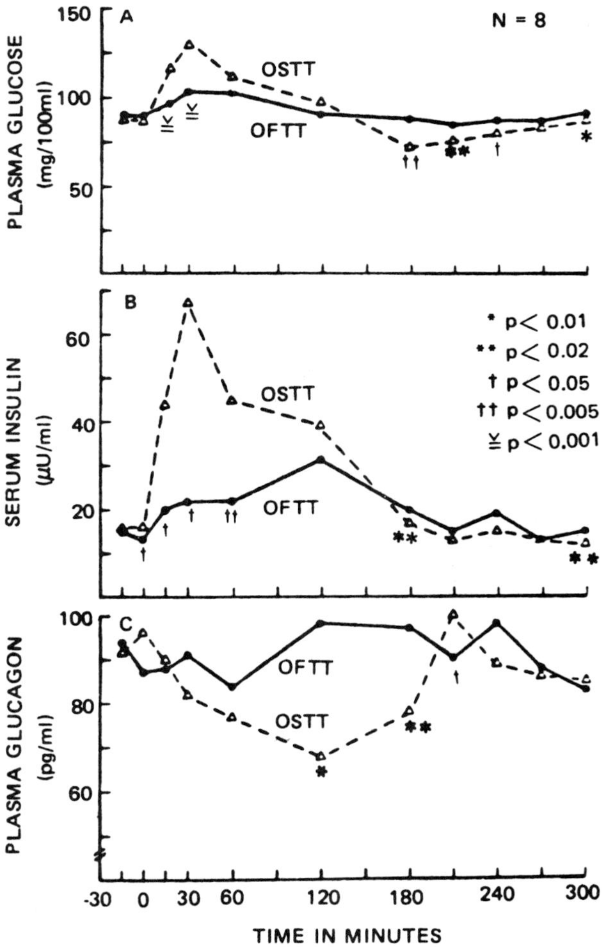


**Figure 3** Comparison of glucose, insulin, and glucagon responses to oral sucrose versus oral fructose (OSTT = oral sucrose tolerance test; OFTT = oral fructose tolerance test).

the possibility of osmotic diarrhea, frequently encountered by the consumption of as little as 30 g of sorbitol, is obviated by the use of pure fructose (16).

Perhaps diabetic exchange values might be profitably amended to reflect the Glycemic Index (see below) in addition to carbohydrate calories. Food technologists should be cognizant of the differences in physiological response to various carbohydrates when formulating foods in-

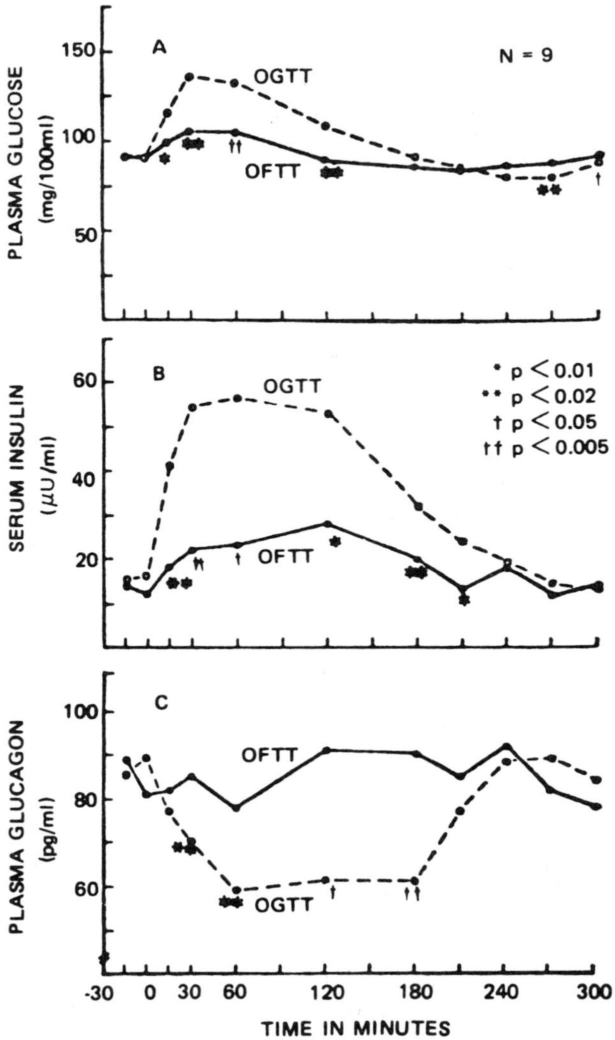


**Figure 4** Comparison of glucose, insulin, glucagon responses to oral glucose versus oral fructose (OGTT = oral glucose tolerance test; OFTT = oral fructose tolerance test).

tended for sale to customers in a specific segment of the market (i.e., diabetics, hypoglycemics, obese). Through the judicious use of pure crystalline fructose in special dietary foods, it may be possible to create good-tasting foods that do not cause severe changes in blood sugar and

Table 5 Glycemic Index

| | |
|---|---|
| Glucose | 100 |
| Sucrose | 59 ± 10 |
| Fructose | 20 ± 5 |
| Maltose | 105 ± 12 |
| Apples | 39 ± 3 |
| Raisins | 64 ± 11 |
| White bread | 69 ± 5 |

*Source:* From Ref. 17.

insulin levels. In this way, the likelihood of maintaining a more healthful dietary regime will be increased.

## GLYCEMIC EFFECT OF FRUCTOSE

Jenkins et al. determined the effect of 62 commonly eaten foods and sugars fed to groups of human volunteer subjects. They constructed a Glycemic Index, which is defined as "the area under the blood glucose response curve for each food expressed as a percentage of the area after taking the same amount of carbohydrate as glucose" (17). Table 5 illustrates some of their findings expressed as a number relative to glucose, which is assigned a value of 100.

These glycemic values suggest that simple carbohydrate exchanges based on carbohydrate content may not, in fact, be an accurate predictor of physiological response. A clinical comparison concerning the effect of orally administered fructose, sucrose, and glucose in subjects with reactive hypoglycemia was reported in *Diabetes Care.* The investigators determined that the use of 100-g loads of pure fructose as a sweetener in cakes and beverages or by itself resulted in a significantly reduced glycemic effect, as indicated by markedly less severe glucose and insulin responses. The authors concluded that "fructose may thus prove useful as a sweetening agent in the dietary treatment of selected patients with reactive hypoglycemia" (18).

## COMPARISON OF THE PROPERTIES OF PURE CRYSTALLINE FRUCTOSE AND ISOMERIZED CORN SYRUPS

As a result of the similarity of nomenclature and a tendency for otherwise accurate publications and technologists to employ specific terms in a nonspecific fashion, considerable confusion exists regarding the difference between pure crystalline fructose and the family of isomerized corn syrups usually referred to as high-fructose corn syrup or HFCS (19,20). This is regrettable because the sweetening and metabolic differences between pure fructose and some of the members of the HFCS grouping can be so significant that serious flavor, physiological, and regulatory problems may result from improper label declarations, which create the impression that these products are one and the same (21).

Fructose is clearly defined in USP and FCC, and these definitions exclude any label description of an HFCS ingredient as "fructose." In order to employ the term "fructose," the sweetener must conform to the compendial descriptions.

Any comparisons of pure crystalline fructose with the isomerized corn syrups should take into account not only the percentage of fructose, but also the quantity and quality of other syrup solids which frequently exhibit a significant negative effect on sweetness and flavor as well as on the glycemic response (4). It has been found that the dextrose, maltose, and other saccharides, which constitute as much as 52% of the solids in some isomerized syrups, can seriously affect delicate flavor systems. These same nonfructose saccharides can also upset the metabolic-superiority rationale of otherwise well-formulated dietary supplements. Therefore, it is imperative that the composition of pure fructose and fractional fructose syrups be considered before being used in a processed food. Table 6 provides a quick comparison of the primary differences among these products.

Since the appearance of the first edition of *Alternative Sweeteners* in 1986, significant processing and marketing changes have occurred in the fructose industry. As a result, solutions of pure crystalline fructose in water, made by remelting crystalline fructose, have been made available.

This form of pure fructose should appeal to food and pharmaceutical users who are primarily engaged in the manufacture of products requiring liquid sweeteners. Incremental costs of shipping and storing liquid fructose can probably be offset by savings in weighing and mixing.

As regards the labeling of liquid fructose, in view of the fact that, unlike the usual types of HFCS, liquid fructose is a 77% solution of compendial grade crystalline fructose in water, its presence in a liquid product can be declared simply as "fructose."

## STORAGE AND HANDLING

One of the most important attributes of crystalline fructose is that it is a high-purity, free-flowing solid, which can be blended in all proportions with all types of dry ingredients without any special equipment or pro-

**Table 6** Comparison of Pure Crystalline Fructose with Isomerized Corn Syrups

| Properties | Pure crystalline fructose | 42% HFCS | 55% HFCS | 90% HFCS | Liquid fructose |
|---|---|---|---|---|---|
| Physical state | Crystalline | Liquid | Liquid | Liquid | Liquid |
| Chemical description | Ketohexose, pure monosaccharide | Mixture of dextrose, fructose, maltose, isomaltose, and higher saccharides | Mixture of fructose, dextrose, and higher saccharides | Mixture of fructose, dextrose, and higher saccharides | Pure fructose, water |
| Percent fructose[a] | 99.5 | 29.8 | 42.4 | 69.3 | 77.0 |
| Percent moisture[a] | 0.2 | 29 | 23 | 23 | 23.0 |
| Range of relative sweetness (sucrose = 100)[a] | 130–180 | 90–95 | 95–100 | 100–106 | 110–120 |
| Should be declared as | Fructose or levulose | Corn sweetener or corn syrup | Corn sweetener or corn syrup | Corn sweetener or corn syrup | Fructose, water |
| Shipping containers | 25-kg bags | Drums, tank cars | Drums, tank cars | Drums, tank cars | Drums, tank cars |
| Typical products that are best made with this sweetener | Dietetic and reduced-calorie foods; dry mixes requiring extraordinary sweetness | Salad dressings, baked goods, james and jellies | Soft drinks, canned fruits | Soft drinks, liquid dietetic foods | Soft drinks, dietetic foods and beverages |

[a]All calculations are based on the commercially available form of the sweetener in question.

cessing techniques. This is in sharp contrast to all of the isomerized corn syrups which, because they are liquid, require storage tanks, pipelines, metering devices, and temperature controls. Crystalline fructose alone can be warehoused and blended as a simple dry ingredient. The most frequently used type of crystalline fructose has a particle size in the range of 0.35–0.45 mm. Milled or powered forms of crystalline fructose are also available. However, because of the increased surface area resulting from the milling of fructose, it is advisable to incorporate about 2% of silicon dioxide in order to ensure free-flow characteristics.

For those applications requiring pure crystalline fructose to be used only a few days each month (i.e, to make dietetic versions of regular products), definite logistic and economic advantages can result from having pure crystalline fructose available in easily stored bags. Good manufacturing practices dictate that when products containing crystalline fructose are exposed to the atmosphere during processing and packaging, the relative humidity be kept below 60% and the ambient temperature not exceed 25°C.

One of the most frequently heard apprehensions about pure crystalline fructose concerns handling problems that may result from its being hygroscopic. The ability of fructose to function as a humectant is a particular asset in keeping cakes, bread, and other food products fresh and moist. However, if the moisture level of the air in contact with the surface of the fructose crystal increases to the extent that dissolution of the highly soluble fructose commences, then subsequent lowering of the humidity in the adjacent air will cause recrystallization to occur. This "drying effect" will frequently result in hardening and lumping problems of varying degrees of severity.

The failure to prevent exposure of crystalline fructose to high humidity atmosphere for even short periods of time (i.e., 15 minutes) has probably caused most of the processing problems encountered with fructose. The author has seen batches exceeding several tons of food and pharmaceutical ingredients being blended with crystalline fructose rendered irreversibly unworkable by the failure to observe the precautions of conducting all processing and packaging operations in an atmosphere under 60% R.H. and with scrupulously dry equipment.

Another indication that the product has been unduly exposed to a humid atmosphere is the presence of a "wet feel" which causes the fructose to bridge and be unsuitable for volumetric filling. In most applications, crystalline fructose in this condition can only be salvaged by being used in small percentages (under 10%) in blends where other ingredients such as starches and anticaking agents can mitigate the textural deficiencies of the "wet" crystalline fructose. Table 7 shows this

**Table 7** The Effect of Relative Humidity on the Moisture Absorption of Crystalline Fructose at 20°C[a]

| Relative humidity (%) | Days exposed | | | | | | | | |
|---|---|---|---|---|---|---|---|---|---|
| | 1 | 2 | 5 | 7 | 9 | 12 | 16 | 20 | 26 |
| 33 | 0.02 | 0.01 | 0.02 | 0.02 | 0.01 | 0.01 | 0.01 | 0.01 | 0.01 |
| 48 | – | 0.03 | 0.03 | 0.03 | 0.03 | 0.03 | — | — | — |
| 55 | — | 0.06 | 0.06 | 0.06 | 0.06 | 0.05 | — | — | — |
| 60 | 0.10 | 0.12 | 0.13 | 0.13 | 0.13 | 0.13 | 0.13 | 0.13 | 0.13[b] |
| 66 | 0.32 | 0.52 | 1.12 | 1.51 | 1.87 | 2.45 | 3.16 | 3.84 | 4.89 |
| 72 | 0.68 | 1.33 | 3.34 | 4.66 | 6.01 | 7.91 | 10.3 | 12.2 | 14.7 |
| 76 | 0.91 | 1.78 | 4.40 | 6.14 | 7.86 | 10.3 | 12.8 | 14.9 | 17.7 |
| 82 | 1.27 | 2.49 | 6.14 | 8.67 | 11.0 | 13.7 | 16.9 | 19.7 | 23.8 |

[a]The amount of moisture absorbed is indicated as a percentage of dry weight.
[b]When moisture content exceeds 0.7%, fructose becomes increasingly wet and sticky. Processing conditions should be controlled to achieve values above this lline.

effect and establishes humidity limits which must be observed in order to avoid processing problems (22).

## PRACTICAL APPLICATION OF CRYSTALLINE FRUCTOSE

Crystalline fructose can be effectively used as a sweetener in any food. However, in view of its higher/lb. cost relative to the usual natural sweeteners, it is worthwhile to construct a model of a potential food-processing customer who could profitably use pure crystalline fructose. Such a customer should:

1. Make a product that is sweetened with nutritive or nonnutritive substances. These include sucrose, glucose, lactose, whey powder, honey, milk powder, sorbitol, mannitol, saccharin, aspartame, acesulfame, or a combination of these sweeteners.
2. Have a dietetic, functional, or organoleptic reason for using fructose as a replacement of all or part of other sweeteners.
3. Be convinced that the presence of crystalline fructose in the list of ingredients will enable the product to have an improved image and to command a premium price.

Some examples of food products that successfully utilize pure crystalline fuctose are:

1. Dietetic cake mixes
2. Desserts (gelatins and puddings)

3. Candies—dietetic and regular*
4. Gums—athletic types
5. Tabletop sweeteners—packets, boxes, etc.
6. Frozen desserts
7. Diet or reduced-calorie beverages, alone or in combination with artificial sweeteners
8. Energy drinks to improve athletic performance
9. Dietary meal replacers
10. Powdered beverage bases
11. Starch functionality improvement

In addition to the increased natural sweetness and flavor-enhancing attributes of fructose, food products that are sold in powdered form, intended for subsequent dissolving in water, have been found to be particularly well served by the inclusion of crystalline fructose as the primary sweetener. Figure 5 shows the solubility of three sugars as a function of temperature and illustrates the superiority of crystalline fructose.

Some manufacturers of sweetened powdered mixtures are faced with the problem of how to achieve an acceptable level of sweetness while keeping their use rate significantly below normal levels. Often this requirement arises when single serving packets or a dispenser are used to deliver the powdered mix.

Artificial sweeteners like saccharin, aspartame, and acesulfame-K have frequently been found lacking when used alone for this purpose. However, in combination with fructose a more acceptable sweetness together with improved mouth-feel can be achieved.

Nearly all attempts at making hard candies with fructose have been unsuccessful, primarily because boiled candy made from it will not set properly. Recently, a hard candy sweetened exclusively with fructose has come on the market in Italy; however, this candy also contains about 44% maltodextrin (25). Thus, its potential as a dietetic confection appears to be limited.

---

*The most successful applications by candymakers have been in making premium-quality chocolate or carob coatings used to enrobe energy bars. Approximately 35% of these coatings are pure fructose. The resulting increased sweetness permits a thinner coating to be applied without reducing total sweetness. One particularly important process change must be observed when making fructose coatings: Temperatures must not exceed 120°F, otherwise agglomeration of crystals and concomitant severe glazing of pumps and pipelines will result (24).

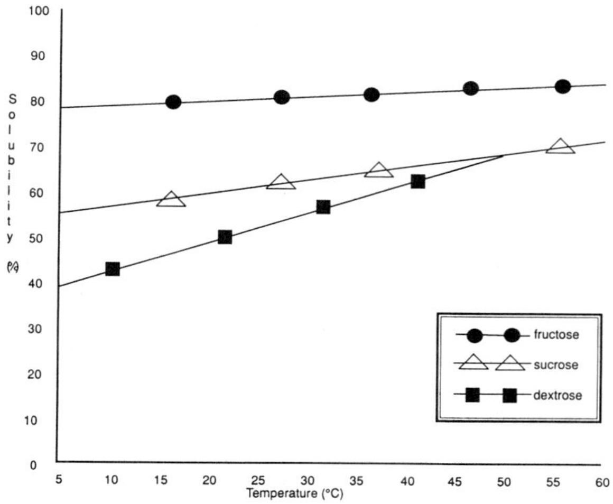


**Figure 5** Solubility of sucrose, dextrose, and fructose at different temperatures.

## CALORIE-CONTROLLING POSSIBILITIES

The reduction of sweetener calories in foods, while maintaining overall quality and acceptability, can be achieved through the judicious use of crystalline fructose to replace other carbohydrate sweeteners (26). However, it is not simply a case of plugging in a factor to determine the quantities of fructose that are required. All of the expertise of a food technologist is required to formulate such a reduced-calorie product, because in the majority of instances textural and body changes will result from the decreased amount of sweetener employed. The following examples are intended to show what adjustments of ingredients and procedures might be required when pure crystalline fructose is used as the key ingredient in making a reduced-calorie product that possesses most of the characteristics of its full-calorie counterpart.

### Reduced-Calorie Cakes

With the emergence of pure crystalline fructose as a commercial food ingredient, a number of new product possibilities were noted by alert food technologists. Of particular importance was the fact that regulations published by the Food and Drug Administration defined a class of dietetic foods that could be labeled as being "reduced-calorie" foods by virtue of containing at least 33⅓% fewer calories than their full-calorie counterparts (27). Because crystalline fructose can, under optimum conditions, be as much as 70% sweeter than an equal weight of sucrose, and because sugar is the major dry ingredient in cakes, it seemed likely that crystalline fructose might be employed to great advantage in developing reduced-calorie cakes and cake mixes.

Layer cakes have long been one of the most preferred desserts in the American diet. However, the high percentage of sugar contained in such cakes has often caused cake lovers to resist the urge to partake of this dessert.

The search for a method to produce reduced-calorie cakes had not been particularly rewarding up to this time. Most efforts had been concerned with utilizing extra-sweet natural or artificial sweeteners in order to reduce the usual cake sweeteners such as granulated sugar, dextrose, invert sugar, and corn syrup.

In most cases, the replacement of the sweetener by a lesser amount of a more potent substitute was negated by the necessity to incorporate additional quantities of flour or shortening. Some supposedly reduced-calorie cakes actually contained more calories because of this added fat content. Attempts to use noncaloric bulking agents to make up for the lost solids content were not particularly successful.

In 1980, Dr. Simon Jackel reported on work done at the Quality Bakers of America Research Laboratory (28), which gave strong evidence that palatable reduced-calorie cakes could be made using pure crystalline fructose as the exclusive sweetener in conjunction with increased levels of surfactants. However, although promising, these efforts were not commercially successful. The problems with cakes made according to these "reduced-caloric" formulations were that:

1. They did not exhibit the rise and texture of a full-calorie cake.
2. Their eating qualities were inferior to those of a regular cake.

Finally, in 1984, a joint program was begun by Hoffmann-La Roche and American Xyrofin to develop a series of fructose-sweetened, reduced-calorie cake mixes which would overcome the deficiencies cited above and enable a commercial baker or homemaker to produce a truly good cake with a 33⅓% calorie reduction.

Of critical importance to this renewed effort by Roche and American Xyrofin was the development by the National Starch and Chemical Corporation of the "Matrix System." This is a balanced blend of emulsifiers, modified food starch, and guar gum, which is sold under the trademarked name, N-Flate.

According to National Starch,

> When the Matrix System is used in a cake, it improves the textural properties by controlling the batter viscosity and providing a well-

balanced emulsified system. This results in cakes with a larger volume and a finer, more uniform cell structure. The moisture retention is also greater in cakes made with the Matrix System. This usually relates to better eating quality, and better shelf stability.

It was found at Roche that the Matrix System (i.e., N-Flate) could be used in combination with crystalline fructose to produce a cake that was virtually free of shortening or vegetable oil. Because fats and oils provide the largest amount of calories of any food (9kcal/g), it became apparent that crystalline fructose, which is the sweetest natural sugar found in nature, might be successfully used in conjunction with the Matrix System to make special dietary cake mixes. Significant reductions in calories would result from the reduced sweetener carbohydrates and the absence of shortening.

The only problem remaining was to determine if cakes made from such mixes would have the visual and taste appeal that are so frequently missing from foods designed for persons on restricted diets.

An exhaustive series of baking trials was undertaken at the laboratory and test kitchen of American Xyrofin. More than 100 different cake mix formulations were made, from which cakes were baked and subjected to test panels.

The three formulations shown in Table 8 represent the mixes that were used to make the cakes receiving the highest scores. It is expected that larger bakeries will find it most economical to purchase the individual ingredients called for in these formulations, whereas smaller establishments may prefer to purchase complete mixes from a bakery supply house.

Water activity greatly affects the performance of various types of starches (i.e., cook up, pregelatinized and granular instant). The water activity, in turn, is significantly affected by carbohydrates, including fructose.

Compared to systems involving starch and a sweetener such as sucrose or dextrose, those containing fructose demonstrate lower gelatinization temperatures and increased final viscosities. These altered characteristics are due to the effect of fructose on the starch molecule and the change in "water mobility" brought about by the starch–fructose combination. Table 9 charts the difference in the loss of birefringence by starch in several solutions of sugars.[29]

Because of textural and volume requirements, reformulations are probably among the most difficult ones that might be encountered. In most instances, particularly where the final food product is liquid, the substitution of fructose for all other carbohydrate sweeteners can be

Table 8 Reduced-Calorie Cake Mixes

| Ingredient | White cake (%) | Yellow cake (%) | Chocolate cake (%) |
|---|---|---|---|
| Krystar® crystalline fructose | 47.28 | 47.28 | 44.90 |
| Hi-ratio bleached cake flour | 38.75 | 38.75 | 37.72 |
| N-Flate (National Starch) | 7.71 | 7.71 | 7.52 |
| Henningsen type P-20 egg white solids | 3.44 | 3.44 | 3.35 |
| Baking powder | 2.24 | 2.24 | 1.49 |
| Givaudan FD-9993 vanilla powder | 0.58 | 0.58 | 0.57 |
| De Zaan 11SB cocoa powder | — | — | 4.17 |
| Baking soda | — | — | 0.28 |
| Roche 1% CWS dry $\beta$-carotene | — | —0.15 | — |

Directions for use: Blend 16 oz (454 g) of reduced-calorie cake mix with 10 oz of water for 30 sec at low speed, followed by 5 min whipping at high speed. Transfer the heavy batter into two 9-inch round cake pans (or one 9½ × 13½ rectangular pan) which have been sprayed with a low-calorie, nonstick spray. Bake at 325°F 30–32 min.

Table 9 Temperature Range for Loss of Birefringence by Starch in Bleached Commercial Cake Flour in Several Sugar Solutions and % Loss Birefringence

| Sugar Solutions (% w/w) | Sucrose 2% (°C) | Sucrose 50% (°C) | Sucrose 98% (°C) | Glucose 2% (°C) | Glucose 50% (°C) | Glucose 98% (°C) | Fructose 2% (°C) | Fructose 50% (°C) | Fructose 98% (°C) |
|---|---|---|---|---|---|---|---|---|---|
| None | 55 | 58.5 | 62.5 | 55 | 58.5 | 62.5 | 55 | 58.5 | 62.5 |
| 10 | 58 | 61 | 65 | 58.5 | 61 | 64.5 | | | |
| 20 | 59 | 63.5 | 69.5 | 61 | 64 | 67.5 | 59 | 63 | 67 |
| 30 | 66 | 70 | 74 | 66 | 68 | 72 | | | |
| 40 | 73 | 76 | 79 | 68.5 | 73 | 76 | 67 | 71 | 74 |
| 50 | 82 | 84 | 87 | 77 | 79 | 83 | | | |
| 57 | 90 | 91.5 | 94 | 82 | 84 | 86.5 | 78 | 80 | 83 |
| 60 | 93.5 | 94.5 | 96.5 | 85 | 86.5 | 90.0 | 81 | 83 | 85 |
| 62 | | | | 88 | 89.5 | 91.5 | 84 | 86 | 88.5 |
| 65 | 98 | 101 | 104 | 90 | 91.5 | 94 | 85 | 87 | 90 |
| 70 | 104.5 | 106 | Boiled | 95 | 97.5 | 100.5 | 89 | 91 | 94.5 |
| 73 | | | | | | | 91.5 | 94.5 | 97 |
| 80 | | | | | | | 99 | 103 | 105 |

done primarily on the basis of sweetness; moreover, the clean, quick sweetness of pure fructose will enable characterizing flavors, particularly those from citrus and berry sources, to be more truly perceived. The formulas shown in Tables 10, 11, and 12 illustrate this concept.

## Fructose–Saccharin Combinations

It is not surprising, in our increasingly health- and diet-conscious society, that two sweeteners which have been prominently in the news are pure crystalline fructose and saccharin: One is the sweetest sugar found

**Table 10** Reduced-Calorie Lemonade Mix

| | |
|---|---|
| Krystar® crystalline fructose | 93.0 lb |
| Citric acid (anhydrous fine granular) | 5.45 lb |
| Ascorbic acid | 0.42 lb |
| Carrageenan (Viscarin #402) | 0.10 lb |
| Riboflavin | 0.0016 lb |
| Syloid #244 | 0.50 lb |
| Permaseal clouding agent FD-9208-B | 0.20 lb |
| Naturalseal lemon flavor FD-8949-D | 0.33 lb |

Usage: Mix 8 oz of above product with sufficient water to make 1 gal of reduced-calorie lemonade. Chill and serve.

**Table 11** Low-Calorie Fructose Lemonade Miz[a]

| | |
|---|---|
| Fructose + saccharin blend (98:1) | 89.1 lb |
| Citric acid (anhydrous fine granular) | 9.2 lb |
| Ascorbic acid | 0.7 lb |
| Carrageenan (Viscarin #402) | 0.16 lb |
| Riboflavin | 0.0025 lb |
| Permaseal clouding agent FD-9208-D | 0.32 lb |
| Naturalseal lemon flavor FD-8949-D | 0.53 lb |

[a]Usage: Mix 5 oz of above product with sufficient water to make 1 gal of low-calorie lemonade. Chill and serve.

**Table 12** Reduced-Calorie Gelatin Dessert

| | |
|---|---|
| Crystalline fructose | 79 lb, 8 oz |
| 300 Bloom gelatin | 14 lb |
| Fumaric acid | 3 lb, 8 oz |
| Sodium citrate | 2 lb |
| Syloid No. 244 | 1 lb |
| Spray-dried color and flavor | As required |

in nature, whereas the other is the sweetest nonnutritive substance permitted to be used to sweeten foods.

A study of the scientific and food industry literature and the nearly universal taste experience leads one to the conclusion that foods sweetened with saccharin do indeed exhibit a bitter or metallic aftertaste. Numerous attempts have been made to diminish or mask these off-flavors through the use of flavoring substances or other sweeteners. The results have not been very successful, and many persons who would like to reduce their consumption of sweetener calories are unable to do so because they simply cannot tolerate the taste of saccharin.

Sweetness research in the United States and Finland has provided encouragement to those food processors who are willing to accept a reduction of about 80% in sweetener calories in return for being able to make products with superior flavor characteristics (30).

The key to this low calorie–good taste breakthrough is the combination of crystalline fructose with rigidly controlled quantities of saccharin. Organoleptic trials have demonstrated that crystalline fructose can be dry-mixed with up to 1.0% saccharin to provide a sweetener with sweetening potential equal to as much as five times that of an equal weight of sucrose. Equally important is the fact that in many products, the usually encountered saccharin aftertaste is diminished to the point where it is not perceived adversely.

Federal regulations, which formerly imposed severe restrictions on the kinds of food products that could employ blends of nutritive and nonnutritive sweeteners, now permit combinations of crystalline fructose and saccharin "to be used in a special dietary food which is useful in reducing or maintaining caloric intake of body weight" (21 CFR section 105.66).

Table-top sweeteners are encountered in almost every restaurant in the country in the form of small packets of a mixture that is usually about 94% dextrose and 4% saccharin, with 2% anticaking agents. Such a mixture is about 10–12 times sweeter than an equal weight of sucrose. This "sweetness" is different from that imparted by natural sweeteners and is further burdened with a bitter aftertaste, which is cumulative. However, the caloric reduction, which approaches 90%, is excellent.

A superior tabletop sweetener can be formulated with crystalline fructose, a maximum of 1% calcium saccharin, and an anticaking agent. Because this blend is about five times sweeter than sucrose, a packet weighing 0.9 g is needed to provide the same sweetness as a full teaspoon of sucrose.

There is a significant flavor difference between a product blend of crystalline fructose and saccharin and one made with dextrose (or lactose) and saccharin. The best way to determine the flavor superiorty of fructose–saccharin is to conduct simple taste tests using the quantities of each sweetener described above in iced tea, in lemonade, and on a breakfast cereal.

Presweetened, low-calorie beverages in a variety of flavors can be made using 0.6–1.0% saccharin, in combination with 98.0–98.4% crystalline fructose. As the percentage of saccharin is increased, the sweetening multiple becomes greater; however, at the 1% level, some slight bitter notes may be detected by persons who are sensitive to saccharin. A simple method for using such fructose–saccharin combinations is to make a blend as follows:

| | |
|---|---|
| Crystalline fructose | 98 lb |
| Calcium saccharin | 1 lb |
| Silicon dioxide | 1 lb |
| | 100 lb |

This sweetening blend can then be used to make a low-calorie form of a full-calorie beverage base via removal of all of the sugars (sucrose, dextrose, corn syrup solids) in the original formula and replacement with one-fifth by weight of the above fructose–saccharin blend. The reduction in calories results from the fact that the new beverage base is used in much smaller quantities in the low-calorie beverage. Table 12 shows how the above combination can be employed to make a low-calorie version of an excellent powdered lemonade mix.

One further note is in order concerning the use of fructose–saccharin combinations. Usually the sweetness of fructose tends to be diminished in a beverage such as hot cocoa. Nevertheless, the 98:1 sweetener combination is uniquely effective in hot chocolate mixes, wherein it enables a reduced-calorie hot chocolate beverage to be made with flavor characteristics comparable to its full-calorie counterpart. Table 13 offers a proven formula for a 100-lb batch of this type of beverage mix.

### Fructose–Aspartame Combinations

In 1981, when aspartame received final, albeit limited, clearance for use in food products, it became the task of food technologists to determine precisely how to use this sugar substitute. Contrary to often expressed

**Table 13** Reduced-Calorie Hot Chocolate Mix

| | |
|---|---|
| Fructose–saccharin blend (98:1) | 30.2 lb |
| Dutch cocoa | 26.0 lb |
| Instantized nonfat dry milk | 43.0 lb |
| Vanillin | 0.8 lb |
| Total | 100 lb |

opinion, aspartame does not exhibit a sweetness profile exactly identical to that of sucrose. Some subtle organoleptic differences can be perceived, with the significance of these differences being dependent on the specific conditions under which the aspartame is employed.

It has been found by Hyvonen (31) in Finland, and by Johnson (32) in the United States, that combinations of aspartame and crystalline fructose can be used to achieve a synergistic sweetening effect and to minimize any lingering nonsweet flavors of aspartame. Reports of a recent study conducted at the University of Missouri indicate a sweetness multiple of 133 for aspartame, compared to sucrose, in an orange-flavored beverage (33).

Using magnitude-estimation procedures which compared various concentrations of aspartame combinations to a reference solution of 10% sucrose in water, Hyvonen et al. concluded that "where the ratio of the component sweetness were nearly 50:50, the amount of synergism noted was 83.5%" (30).

These results indicate that in some food systems, an organoleptically determined mixture of aspartame could be used to dramatically reduce sweetener calories without adversely affecting cost or overall flavor. It appears that the possibilities of judiciously blending such nutritive sweeteners to achieve the maximum ratio of benefits (metabolic, flavor, calorie-sparing) to cost offers great challenge and opportunity to enterprising food technologists and companies willing to encourage and support their research endeavors.

## FRUCTOSE MARKETING

Crystalline fructose has been available as an industrial food ingredient in the United States since 1975. After a tentative beginning, this unique sweetener is now afforded consideration by food technologists whenever a sweetener is required for a particularly food. Awareness of fructose has greatly increased among industry professionals and the general public as a result of a barrage of newspaper articles, books, and sales efforts by fructose manufacturers (34).

Most crystalline fructose sold in the United States is produced domestically. Prior to 1981, the main sources of supply were plants in Finland and Germany. Other fructose producers are located in the Common Market countries, Japan, Taiwan, and Israel. Current worldwide capacity for crystalline fructose exceeds demand, following shortages in the early 1980s. However, the largest use of crystalline fructose is in the United States, where the creative energies of imaginative food technologists have created uses far beyond those found in Europe, where the largest application is as a table-top sweetener. It seems likely that the worldwide demand for fructose will continue to increase.

However, the current U.S. marketing rationale for the use of fructose as a natural sweetener emphasizes the functionality of fructose and downplays the nutritional aspects. This approach has been disconcerting to those "true believers" who maintain that crystalline fructose is not just another sweetener, but rather is a unique sugar with proven advantages for health conscious consumers and for the entire food-processing industry.

## REFERENCES


1. Cetus Fructose Project threatened as Social ends involvement. *Eur. Chem. News*, June 14 (1982).
2. Y. Tsuzuki, J. Yamazaki, and K. Kagami, The specific rotation of fructose, *J. Am. Chem. Soc., 72:* 1071 (1950).
3. R. Tannahill, in *Food in History*, Stein & Day, Briarcliff Manor, NY, 1973.
4. S. Akgun and II. Ertel, Plasma glucose and insulin after fructose and high fructose corn syrup meals in subjects with non-insulin dependent diabetes mellitus, *Diabetes Care, 4:* 464 (1981).
5. J. M. Oleksky and P. Crapo, Fructose—its characteristics, physiology and metabolism, *Nutr. Today*, July 10 (1980).
6. H. Moskowitz, in *Sugars in Nutrition* (H. Sipple and K. McNutt, eds.), Academic Press, Inc., New York, 1974, pp. 37–64.
7. R. S. Shallenberger, *Pure and Applied Chemistry*, Pergamon Press Ltd., London, 1978.
8. L. Hyvonen, P. Varo, and P. Koivistonen, Tautomeric equilibria of D-glucose and D-fructose. *J. Food Sci., 42:* 657 (1977).
9. L. Hyvonen, in *Varying Relative Sweetness*, University of Helsinki EKT-546, 1980.
10. R. Scott, Hoffmann-La Roche, Internal Report, Nutley, N.J., 1980 (unpublished).
11. R. Crane, Intestinal absorption of sugars, *Physiol. Rev., 40:* 789 (1060).

12. H. Mehnert, in *Handbuch des Diabetes Mellitus,* J. F. Lehmanns Verlag, Munich, 1971.
13. N. V. Bohannon, J. H. Karam, and P. H. Forsham, Endocrine responses to sugar ingestion in man, *J. Am. Diet. Assoc., 76:* 555 (1980).
14. J. Talbot and K. Fisher, The need for special foods and sugar substitutes by individuals with diabetes mellitus, *Diabetes Care, 1:* 231 (1978).
15. B. Dwivedi, in *Carbohydrates and Health* (L. Hood, ed.), AVI Publishing Co., Westport, CT.
16. R. Ylikahri and R. Pelkonen, in *Carbohydrate Sweeteners in Food and Nutrition* (P. Koivistonen and L. Hyvonen, eds.), Academic Press, London, 1980.
17. D. J. A. Jenkins, T. M. S. Wolever, and R. H. Taylor et al., Glycemic index of foods: A physiological basis for carbohydrate exchange, *Am. J. Clin. Nutr., 34:* 362 (1981).
18. P. Crapo, J. Scarlett, O. Koltermann, et al., The effects of oral fructose, sucrose and glucose in subjects with reactive hypoglycemia, *Diabetes Care, 5:* 512 (1982).
19. G. Inglett, A review of natural, sweet polyhydric alcohols, proteins, and synthetic sweeteners, *Food Technol. 3:* 37 (1981).
20. J. Lukasick, The future of fructose, *Rev. World,* Aug. (1982).
21. M. Burros, A case of sweet deception; *The Washington Post,* Feb. 21, 1980.
22. Research Report, Finnish State Technical Research Center #A-6759-73, Helsinki, 1973.
23. Technical Data Sheet, No. 1, 88MS04, A. E. Staley Sweetener and Commodities Group, Decatur, Il.
24. M. Zimmermann, Fructose in the manufacture of dietetic chocolate, *Manufact. Confectioner, 39:* (1974).
25. *Sweet Topic* No. 35, Xyrofin, Baar, Switzerland, 1982.
26. T. Osberger and H. Linn, in *Low Calorie and Spec. Diet Foods,* CRC Press, West Palm Beach, FL, 1978, pp. 115–123.
27. S. Gardner, *Special Dietary Foods in Federal Reigster, 43:* 185 (1978).
28. S. Jackel, New research makes reduced calorie cakes possible, *Bakery Production and Marketing,* July, 1980.
29. M. M. Bean, W. T. Yamazak, and D. H. Donelson, Wheat starch gelatinization in sugar solutions II, Fructose, glucose, and sucrose: cake performance. *Cereal Chem., 55*(6): 945–952 (1978).
30. L. Hyvonen, R. Kurkela, P. Koivistoinen, and A. Raitilainen, Fructose-saccharin and xylitol-saccharin synergism, *J. Food Sci., 43:* 251 (1978).
31. L. Hyvonen, Research Report, Univ. of Helsinki, 1981.
32. L. Johnson, Unpublished Report, Hoffmann-La Roche, Nutley, N. J., 1982.
33. Aspartame to appear as an ingredient in new prepared foods, *Prepared Foods,* Aug. (1982).
34. B. Hunter, *The Sugar Trap,* Houghton-Mifflin, Boston, 1982.
35. E. A. Nikkila and J. K. Huttunen, Symposium on clinical and metabolic aspects of fructose, The Finnish Society for Endocrinology, Helsinki Finland, Jan. 20–22, 1972.

# 13

# High Fructose Corn Syrup

**John E. Long** *Archer Daniels Midland Company, Clinton, Iowa*

## INTRODUCTION

The commerical success of high fructose corn syrup (HFCS) has been remarkable (1,5). The first shipment in the United States of the modern-day product was made in 1968, and by 1988 shipments exceeded 11 billion pounds dry solids (d.s.). There are two major products: a 42% fructose syrup and a 55% fructose syrup. These sweeteners replace sucrose in food systems. In the remainder of the world, the 1989 production of HFCS was estimated at 3.8 billion pounds.

## MANUFACTURING

Since HFCS is a recent product, the manufacturing plants are new. They operate with continuous flow from starch to commercial fructose, and they are controlled by the most modern computer operations. This permits the industry to make the highest quality food ingredient.

### 42 HFCS

In the United States, HFCS is made from corn starch. The process to make the 42% fructose syrup is shown in Figure 1. The starch slurry is liquefied at high temperatures by $\alpha$-amylase to a dextrose equivalent (D.E.) of 10–20. In the second step glucoamylase saccharifies the starch to 94–96% dextrose. The saccharified liquor is refined by filtration and carbon and ion exchange treatment.

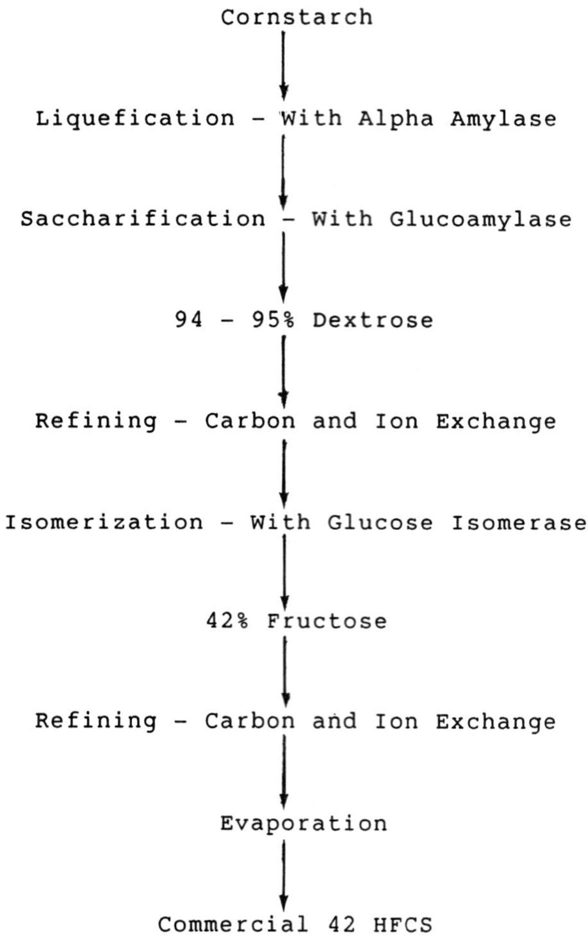


**Figure 1** Production of 42 HFCS.

**Table 1** Typical Composition of High Fructose Corn Syrup[a]

| | 42 HFCS | 55 HFCS | 80–95 HFCS |
|---|---|---|---|
| Fructose, % | 42 | 55 | 80–95 |
| Dextrose, % | 52 | 40 | 4–19 |
| Higher saccharides, % | 6 | 5 | 1–3 |
| Dry solids, % | 71.0 | 77.0 | 77.0 |

[a]Dry solids basis.

The refined dextrose hydrolyzate is then pumped through a column of immobilized glucose isomerase which converts about 45% of the dextrose to fructose. The outlet syrup at 42% fructose is again refined by carbon and ion exchange treatment and then evaporated to the present commercial 42% HFCS (Table 1).

## 55 HFCS

The process to make the 55% fructose syrup is shown in Figure 2. It is produced from the 42% syrup by a fractionation and blending process. The syrup is pumped into the fractionation unit where calcium sites retain the fructose while dextrose and the higher saccharides pass through. The dextrose and higher saccharide stream is returned to the 42% production channel. Water is pumped through the fractionation unit to release the fructose. The outlet stream generally is 80–95% fructose. It is refined and then blended with the 42 HFCS to make the 55% product (Table 1). A small amount of 80–95% HFCS is sold in commerce (Table 1).

The finished fructose syrup is shipped directly to customers in trucks or railcars. Another widely used procedure is to ship the syrup to the sweetener terminals, which are located throughout the country. From there it is trucked to the local food processor.

HFCS is a clear, colorless liquid. It tends to develop a light straw color after prolonged time and/or higher temperatures. However, the product remains usable for many months under the prescribed storage temperatures. Microbiologically the syrup is very stable (1).

The 42 HFCS may crystallize because of the higher levels of dextrose. Storage temperatures of 80–90°F will prevent this. The 55% product seldom crystallizes, and storage at 70°F should keep the product liquid. If either product does crystallize, the material can be reliquefied by warming the mass to resolubilize the dextrose.

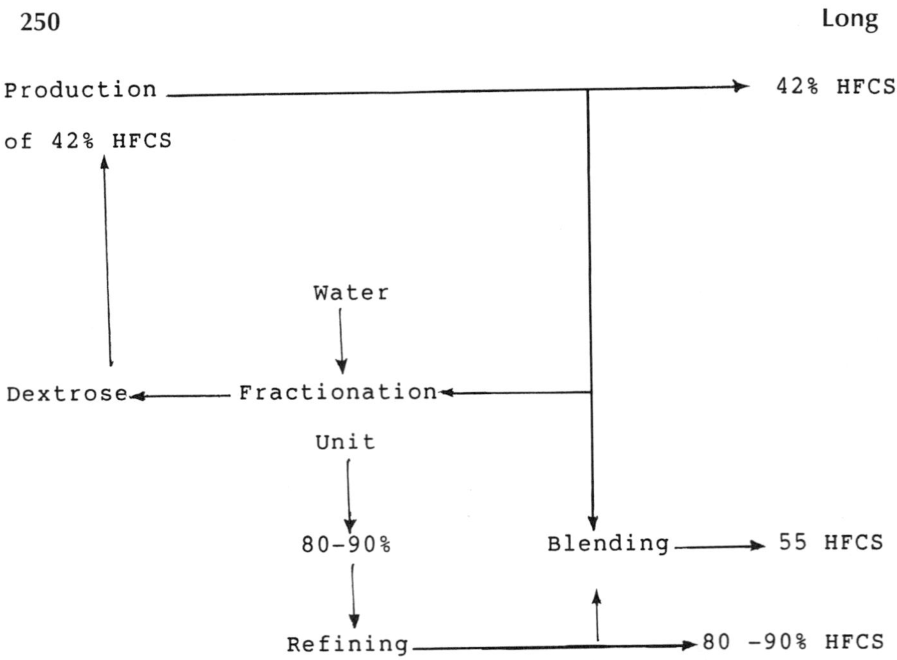


**Figure 2** Production of 55 HFCS

## PROPERTIES

HFCS's most important characteristic is sweetness (Table 2). In sweetness the 42 HFCS is about equal to sucrose, while the 55 HFCS is about equal to medium sucrose invert. The sweetness values vary with temperature, pH, and solids. In general the fructose syrups are sweetest in systems that are cold and at low pH. The 80–95% HFCS is considered to be sweeter than sucrose. Again, the value will relate to the food system.

**Table 2** Relative Sweetness of Nutritive Sweeteners

| Sweetener | Sweetness relative to sucrose |
|---|---|
| Sucrose | 1.0 |
| Crystalline fructose | 1.2–1.8 |
| 90% HFCS | 1.2–1.6 |
| 42% HFCS | 1.0 |
| 55 % HFCS | 1.0+ |
| Sucrose medium invert | 1.0+ |
| Dextrose | 0.7 |

**Table 3** Chemical Comparison Between Sucrose and HFCS

Sucrose inverts in acid foods or the body to dextrose and fructose.

100 g sucrose $\xrightarrow{\text{Total inversion}}$ 52.6 g dextrose + 52.6 g fructose

Commercially sucrose is usually marketed as sucrose or medium invert, which is a liquid and contains 50 parts sucrose, 25 parts dextrose, 25 parts fructose.
HFCS is similar to total sucrose invert.
HFCS has replaced sucrose, therefore, the per capita consumption of fructose remains the same.

Since HFCS replaces sucrose, we need to look at the chemistry of the two systems (Table 3). Sucrose is acid foods or in the body will invert to approximately equal amounts of dextrose and fructose. There is a 5.2% chemical gain. Sucrose is sold commercially as a dry crystalline material, as sucrose dissolved in water, or as medium invert (one-half of the sucrose is inverted).

Thus, the fructose corn syrups are similar to total sucrose invert. Since HFCS has replaced sucrose on a pound for pound basis, the per capita consumption of dextrose and fructose from these sweeteners remains the same. At times, one sees in the press references to higher consumption of fructose after HFCS came onto the scene. This is not true, as the consumption of fructose remains the same. HFCS has an energy value of approximately four calories per gram, d.s. This is the same as its major components—dextrose and fructose.

## ANALYTICAL AND QUALITY CONTROL

HFCS is the highest quality liquid sweetener on the U.S. market. This is because of refining to remove nonsweetener components and to extensive microbiological control. The predominant user of the HFCS is the soft drink industry. Therefore, the major specifications were developed with that trade (Table 4). The procedures for these tests are published by the Society of Soft Drink Technologists in April 1985 (2) and are periodically updated. This publication was the result of several years of effort between the Society, the Corn Refiners Association, the individual wet millers, and soft drink bottlers. Individual bottlers or wet millers may have specifications in addition to this list.

**Table 4** Quality Guidelines for High Fructose Corn Syrup

| 42 HFCS | | 55 HFCS |
|---|---|---|
| 71.0 ± 0.5% | Solids | 77.0 ± 0.5% |
| 0.05% Maximum | Ash (sulfated) | 0.05% Maximum |
| 42% Minimum | Fructose | 55% Minimum |
| 92% Minimum | Dextrose + fructose | 95% Minimum |
| 25 RBU, Maximum | Color | 25 RBU, Maximum |
| 2.0 ppm Maximum | Sediment | 1.0 ppm Maximum |
| 3.5 Minimum | pH | 3.5 Minimum |
| No obectionable flavor or odor | Flavor and ordor | No objectionable flavor or odor |
| 4.0 ml of 0.05 N sodium hydroxide maximum to pH 6.0 | Titratable acidity | 4.0 ml of 0.05 N sodium hydroxide maximum to pH 6.0 |
| 3 ppm, Maximum | $SO_2$ | 3 ppm, Maximum |
| 80 ppb, Maximum | Acetaldehyde | 80 ppb, Maximum |
| Maximum organisms/10 g sweetener solids | Microbiological | Maximum organisms/10 g sweetener solids |
| 200 Mesophilic bacteria | | 200 Mesophilic bacteria |
| 10 Yeast | | 10 Yeast |
| 10 Mold | | 10 Mold |

*Source:* The Society of Soft Drink Technologists.

## APPLICATIONS

The first shipment in the United States of the modern-day 42 HFCS was made in 1968, but industry volume figures are first available for 1972. The volumes for 1972–88 are shown in Table 5. In 1972 it was 246 million pounds d.s., and it grew to over 11.5 billion pounds in 1988. This is a dramatic increase for a food ingredient. The use has grown because of economics, quality of the product and ease of handling. This period of rapid growth has ended because the major markets have been saturated. Future growth will be due to increased consumption of present food items or to the development of new foods.

In Table 6 the world production outside the United States was estimated to be 3.8 billion pounds. By 1995 this is estimated to grow to 5.7 billion pounds.

The 42 HFCS was the only product on the market for the first few years. In 1977 the first large commercial 90% fructose plant came on stream, and this permitted the mass production of the 55 HFCS product.

**Table 5** Shipments of HFCS[a]

| Year | Pounds dry solids |
|---|---|
| 1972 | 246,347,280 |
| 1973 | 444,095,770 |
| 1974 | 592,118,700 |
| 1975 | 1,043,117,800 |
| 1976 | 1,505,309,300 |
| 1977 | 1,992,970,000 |
| 1978 | 2,454,975,600 |
| 1979 | 3,245,721,600 |
| 1980 | 4,356,906,000 |
| 1981 | 5,312,476,000 |
| 1982 | 6,216,176,000 |
| 1983 | 7,212,331,300 |
| 1984 | 9,915,003,200 |
| 1985 | 10,418,031,700 |
| 1986 | 10,500,000,000 |
| 1987 | 11,128,000,000 |
| 1988 | 11,500,000,000 |

[a]U.S. Shipments as estimated from USDA and CRA data.

The 55 HFCS was developed specifically for the soft drink market because it had similar sweetness to medium sucrose invert, the sweetener used by that industry.

The guidelines to the use of HFCS are shown in Table 7. In general, the syrup can replace sucrose or sucrose invert on a pound for pound dry solids basis. This may vary slightly in a specific good system. The HFCS cannot be used in dry food systems, e.g., dry mixes, nor where hygroscopicity is a problem, e.g., some confections.

The major uses in the United States are shown in Table 8. Beverages comprise over 70% of the volume and use primarily the 55% HFCS. In soft drinks alone the volume is over 7.0 billion pounds dry solids (3). HFCS has totally replaced sucrose in this industry. Cola drinks account for about 69% (3) of the soft drink volume. In 1980 55 HFCS first moved into a major cola at a 50% sweetener replacement level, and in 1984 all colas approved the syrup at the 100% level. In the late 1970s and early 1980s the noncola soft drinks switched to 100% HFCS. Some used 42 HFCS, but a majority used the 55 HFCS.

The nonbeverage industries primarily use 42 HFCS. In the baking trade it is used in yeast-raised goods and in cakes. In the canning area it

**Table 6** 1989–1995 Estimated World High Fructose Syrup Consumption (1000 Metric Tons Dry)

| | | 1989 | | 1995 | |
|---|---|---|---|---|---|
| | | 42% | 55% | 42% | 55% |
| *NORTH AMERICA* | | | | | |
| CANADA | | 50 | 56 | 56 | 62 |
| USA | | 1892 | 3507 | 2391 | 3947 |
| | SUBTOTAL | 1942 | 3563 | 2447 | 4009 |
| *SOUTH AMERICA* | | | | | |
| ARGENTINA | | 53 | 28 | 59 | 34 |
| URUGUAY | | — | 15 | — | 20 |
| | SUBTOTAL | 53 | 43 | 59 | 54 |
| *EUROPE-E.C.* | | | | | |
| BELGIUM | | 72.3 | — | 72.3 | — |
| FRANCE | | 22 | — | 22 | — |
| GREECE | | 13 | — | 13 | — |
| GERMANY | | 35.7 | — | 35.7 | — |
| ITALY | | 20.5 | — | 20.5 | — |
| NETHERLANDS | | 9.2 | — | 9.2 | — |
| PORTUGAL | | 10 | — | 10 | — |
| SPAIN | | 83 | — | 83 | — |
| U.K. | | 27.5 | — | 27.5 | |
| | SUBTOTAL | 293.2 | | 293.5 | |
| *SCANDINAVIA* | | | | | |
| FINLAND | | 8 | — | — | 16 |
| | SUBTOTAL | 8 | — | — | 16 |
| *EASTERN EUROPE* | | | | | |
| BULGARIA | | — | — | 30 | 30 |
| HUNGARY | | 50 | — | 70 | — |
| YUGOSLAVIA | | 70 | — | 100 | — |
| USSR | | — | — | 140 | — |
| | SUBTOTAL | 120 | — | 340 | 30 |
| *AFRICA* | | | | | |
| EGYPT | | — | — | 20 | 50 |
| | SUBTOTAL | — | — | 20 | 50 |
| *ASIA* | | | | | |
| CHINA | | 10 | — | 160 | 160 |
| INDIA | | 3 | — | 10 | — |
| INDONESIA | | 30 | 30 | 45 | 45 |
| JAPAN | | 300 | 435 | 340 | 490 |
| MALAYSIA | | 30 | — | 75 | — |
| PAKISTAN | | 40 | — | 80 | — |

Table 6 Continued

| | 1989 | | 1995 | |
|---|---|---|---|---|
| | 42% | 55% | 42% | 55% |
| PHILIPPINES | 4 | — | 5 | — |
| SINGAPORE | 3 | 3 | 5 | 5 |
| SOUTH KOREA | 20 | 130 | 30 | 190 |
| TAIWAN | 2 | 15 | 4 | 20 |
| THAILAND | 25 | — | 40 | — |
| | 467 | 613 | 794 | 910 |
| *OCEANIA* | | | | |
| AUSTRALIA | 10 | — | 12 | — |
| | 10 | | 12 | |
| WORLD TOTAL | 2,903 | 4,229 | 3,965 | 5,069 |
| GRAND TOTAL | 7,132 | | 9,034 | |

*Source:* McKeaney-Flavell Company, Oakland, CA.

Table 7 Guidelines for Use of HFCS

1. HFCS can replace sucrose or sucrose invert on a pound for pound dry solids basis.
2. There are two general exceptions:
   a. Where a dry sweetener is required, e.g., dry food mixes.
   b. Where hygroscopicity is a problem, e.g., some confections.

Table 8 HFCS Customer Profile, 1989

| Industry | Percent of market |
|---|---|
| Beverages | 75.2 |
| Baking | 7.7 |
| Canning | 8.5 |
| Dairy products | 4.3 |
| Other | 4.3 |

*Source:* McKeany-Flavell Company, Oakland, CA. Estimates for 1989 presented at Milling & Baking Seminar, June 5, 1989.

**Table 9** Market Division Between 42 and 55 HFCS

| Year | % 42 HFCS | % 55 HFCS |
|---|---|---|
| 1978 | 88 | 12 |
| 1979 | 78 | 22 |
| 1980 | 67 | 33 |
| 1981 | 57 | 43 |
| 1982 | 50 | 50 |
| 1983 | 45 | 55 |
| 1984 | 40 | 60 |
| 1985 | 36 | 64 |
| 1986 | 35 | 65 |
| 1987 | 35 | 65 |
| 1988 | 38 | 62 |

*Source:* Estimated from CRA Data and McKeany-Flavell, *Sweetener News,* Oakland, CA.

is used in canned fruits, while in dairy products the applications are in drinks and frozen desserts. Confections has been the one area where HFCS has had limited penetration. This is due to the hygroscopic nature of the syrup. Even crystalline fructose, unless protected, will rapidly pick up moisture and become gummy or in time go into solution.

The market division between 42 and 55 HFCS is shown in Table 9. Today, 55 HFCS comprises about 65% of the volume and is expected to stay in that range.

The consumption of HFCS and other corn sweeteners compared to sucrose is shown in Table 10. Since 1985, the consumption of corn sweetener has exceeded that of sucrose.

A short history of HFCS is shown in Table 11. In 1964 research for a commercial product was started and in 1988 over 11 billion pounds were sold. This is indeed a major innovation in the food industry.

## REGULATORY STATUS

HFCS conforms to the Standards of Identity for corn syrup, 21 CFR 168.120. 42% HFCS is listed as GRAS under 21 CFR 182.1866. The enzymes used to produce HFCS, glucoamylase, $\alpha$-amylase, and glucose isomerase are GRAS. They are listed in 21 CFR 184.1027 and 21 CFR 184.1372. Other listings may arise as approvals are requested for new enzyme sources.

**Table 10** U.S. per Capita Sweetener Consumption[a]

| Crop year | Refined sugar | HFCS | Glucose Syrups | Dextrose | Total corn sweeteners | Honey and edible syrups | Total caloric sweetener |
|---|---|---|---|---|---|---|---|
| 1959 | 96.4 | | | | 11.5 | 1.9 | 109.8 |
| 1960 | 97.6 | | | | 11.6 | 2.0 | 111.2 |
| 1961 | 97.8 | | | | 12.0 | 1.9 | 111.7 |
| 1962 | 97.9 | | | | 12.9 | 2.0 | 112.8 |
| 1963 | 97.3 | | | | 14.2 | 1.8 | 113.3 |
| 1964 | 96.8 | | | | 15.0 | 1.7 | 113.5 |
| 1965 | 97.0 | | | | 15.1 | 1.8 | 113.9 |
| 1966 | 97.3 | | | | 15.4 | 1.7 | 114.4 |
| 1967 | 98.5 | | | | 16.2 | 1.4 | 116.1 |
| 1968 | 99.2 | | | | 17.2 | 1.6 | 118.0 |
| 1969 | 101.0 | | | | 18.2 | 1.6 | 120.8 |
| 1970 | 101.8 | 0.7 | 14.0 | 4.6 | 19.3 | 1.5 | 122.6 |
| 1971 | 102.1 | 0.9 | 14.9 | 5.0 | 20.8 | 1.4 | 124.3 |
| 1972 | 102.3 | 1.3 | 15.4 | 4.4 | 21.1 | 1.5 | 124.9 |
| 1973 | 100.8 | 2.1 | 16.5 | 4.8 | 23.4 | 1.4 | 125.6 |
| 1974 | 95.7 | 3.0 | 17.2 | 4.9 | 25.1 | 1.1 | 121.9 |
| 1975 | 89.2 | 4.9 | 17.5 | 5.0 | 27.4 | 1.4 | 118.0 |
| 1976 | 93.4 | 6.9 | 17.5 | 5.0 | 29.4 | 1.3 | 124.1 |
| 1977 | 94.2 | 9.1 | 17.6 | 4.1 | 30.8 | 1.4 | 126.4 |
| 1978 | 91.5 | 11.2 | 17.8 | 3.8 | 33.8 | 1.5 | 125.8 |
| 1979 | 89.3 | 14.4 | 17.9 | 3.6 | 35.9 | 1.4 | 126.6 |
| 1980 | 83.6 | 18.0 | 17.6 | 3.5 | 39.1 | 1.2 | 123.9 |
| 1981 | 79.4 | 22.2 | 17.8 | 3.5 | 143.5 | 1.2 | 124.9 |
| 1982 | 73.6 | 26.7 | 18.0 | 3.5 | 48.2 | 1.3 | 127.8 |
| 1983 | 71.0 | 31. | 18.0 | 3.5 | 52.6 | 1.3 | 130.4 |
| 1984 | 67.6 | 37.3 | 18.0 | 3.5 | 58.8 | 1.4 | 128.7 |
| 1985 | 63.4 | 44.1 | 18.0 | 3.5 | 65.6 | 1.4 | 132.8 |
| 1986 | 60.8 | 46.0 | 18.0 | 3.5 | 67.5 | 1.4 | 135.5 |
| 1987 | 62.4 | 47.5 | 18.0 | 3.5 | 69.0 | 1.4 | 132.8 |
| 1988 | 66.4 | 48.5 | 18.0 | 3.6 | 70.1 | 1.4 | 133.9 |
| 1989[b] | 62.1 | 49.1 | 18.0 | 13.6 | 70.7 | 1.4 | 134.2 |

[a]Pounds, dry basis.
[b]Estimated.
*Source:* USDA, 1990 Corn Annual, CRA.

Table 11 Brief History of HFCS

| | |
|---|---|
| 1957 | Patent by Marshall and Kooi disclosed a microbial enzyme capable of isomerizing dextrose to fructose. |
| 1964 | Clinton Corn Processing Company started work to commercialize a process. |
| 1965 | Clinton Corn initiated work with Japanese Agency of Industrial Science and Technology to commercialize process. |
| 1967 | First commercial shipment HFCS by Clinton Corn was 15% fructose and was made by batch system using soluble glucose isomerase. |
| 1968 | First commercial shipment of 42% HFCS by Clinton Corn—batch process and with both soluble and immobilized enzyme. |
| 1972 | Clinton Corn brought on stream first continuous plant using immobilized enzyme. |
| May 1978 | First large-scale 90 HFCS plant by Archer Daniels Midland. Permitted volume production of 55 HFCS. |
| Jan. 1980 | HFCS approved as 50% of the sweetener in the drink Coca-Cola. |
| April 1983 | HFCS approved as 50% of the sweetener in the drink Pepsi-Cola. |
| Nov. 1984 | 100% level of HFCS approved for the drinks Coca-Cola and Pepsi-Cola. |

## REFERENCES

1. Seminar Proceedings, Products of the Corn Refining Industry in Food, 1978. Corn Refiners Association, Inc., 1001 Connecticut Avenue, N. W., Washington, DC 20036, pages 47–54.
2. Quality Guidelines and Analytical Procedures Bibliography for Bottlers High Fructose Corn Syrup, April 1985. Society of Soft Drink Technologists, 4419 41st Street, Brentwood, MD 20722.
3. Beverage World, March 1989. 7500 Old Oak Boulevard, Cleveland, OH 44130.
4. McKeany-Flavell Company, Oakland, California.
5. Corn Refiners Association Symposium on Food Products of the Wet Milling Industry, September 25, 1986. Corn Refiners Association Inc., 1001 Connecticut Avenue, N.W., Washington, DC 20036.
6. Sugar and Sweetener—Situation and Outlook Report, USDA. See reports from 1975–1989.

# 14

# Maltitol and Hydrogenated Starch Hydrolysate

**Alan H. Moskowitz** *Warner-Lambert Company, Morris Plains, New Jersey*

## INTRODUCTION

Maltitol, or hydrogenated maltose, was developed in Japan and has been in use since about 1964 in that country. Early grades of maltitol were low in purity and were only available in syrup form. Hydrogenated starch hydrolysate (HSH), first developed by the Swedish company Lyckeby in the 1960s, may be thought of as hydrogenated corn syrup, having similar functional properties but in a sugarless form. As maltitol is the main sweetening component of HSH and metabolism of both by mammalian systems is similar, these two sweeteners will be reviewed together.

## PRODUCTION

### Maltitol

As with other polyhydric alcohols (such as sorbitol, mannitol, and xylitol), maltitol is produced via the catalytic hydrogenation of a precursor, in this case, maltose.

A corn starch solution is hydrolyzed enzymatically into maltose syrup which, after purification and concentration, is hydrogenated via a nickel or other transition metal catalyst. Following additional purification steps to remove the starting materials and catalyst, the solution is concentrated to a syrup, then crystallized (1). Maltitol syrup, defined as that containing more than 50% maltitol, also contains low levels of sorbitol, maltotriitol, and higher hydrogenated oligosaccharides. Crystalline maltitol powders with maltitol contents as high as 99% are also available.

### Hydrogenated Starch Hydrolysate

Hydrogenated starch hydrolysate is produced from starch in the same manner as is maltitol (2,3). Here however, precise control of the starch enzymatic hydrolysis is essential to produce a product with a specific oligosaccharide profile. As discussed in a later section, the relative proportion of each of the hydrogenated species markedly affects the functionality and chemistry of the particular hydrogenated starch hydrolysate.

## PHYSICAL CHARACTERISTICS

### Maltitol

Maltitol ($\alpha(1\rightarrow4)$-glucosylsorbitol) is a crystalline polyhydric alcohol obtained by the hydrogenation of maltose, a disaccharide consisting of two glucose units linked via an $\alpha(1\rightarrow4)$ bond (Fig. 1).

Maltitol exhibits a negligible cooling effect in the mouth given its negative heat of solution (−23 kj/kg), which is much less than that of other carbohydrates: xylitol (−153 kj/kg), mannitol (−121 kj/kg), sorbitol (−111 kj/kg), isomalt (−39 kj/kg), dextrose (−104.6 kj/kg), and sucrose (−18 kj/kg).

Like the other polyols, maltitol has no reducing groups and will not undergo Maillard reactions to produce caramel-colored species in the presence of proteins. Crystalline maltitol of high purity has a sweetness of about 85–95% that of sucrose, making it sweeter than all other polyols except xylitol.

Currently, there are two categories of maltitol products (based upon on maltitol content) that are produced and marketed worldwide. Because these products are somewhat different, a discussion of each is warranted.

Towa Chemical Industry, Co., Ltd., an affiliate of Mitsubishi, Tokyo, Japan, produces and markets both crystalline maltitol and maltitol syrup under various brand names, including Amalty and Mabit. Towa licenses the patent for the process from Hayashibara Biochemical Laboratories, Okayama, Japan, which pioneered the process. Towa's license is by geographic region and includes Japan, the United States, and several other countries. Hayashibara also licenses the patents to ANIC and their subsidiary Melida, SpA, Milan, Italy, both marketing the product in

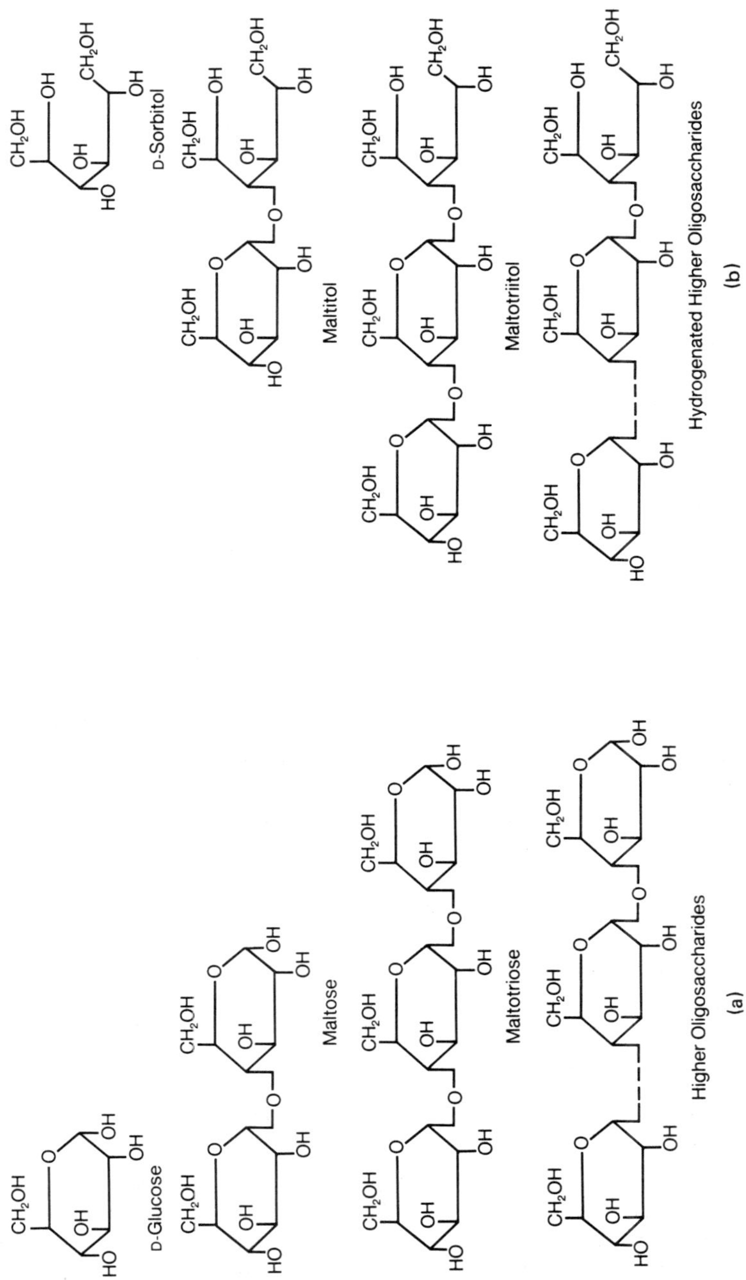


**Figure 1** (a) Chemical structures for monosaccharide, disaccharide, and polysaccharide polyol precursors. (b) Chemical structures for the polyol components of hydrogenated starch hydrolysate.

**Table 1** Specifications for Commercially Available Crystalline Maltitol Products

| | Roquette Maltisorb crystalline | Towa Amalty | Towa Amalty"MR |
|---|---|---|---|
| Maltitol (minimum %) | 99.0 | 88.5 | 93.5 |
| Sorbitol (max. %) | — | 3 | 1.5 |
| Maltotriitol (max. %) | — | 9 | 3.0 |
| Hydrogenated oligosac charides (max. %) | — | 3 | 2.5 |
| Reducing sugars (max. %, as dextrose) | 0.01 | 0.5 | 0.5 |
| Melting point (°C) | 149.4 | 130–142 | 140 |

Europe under the brand Malbit. Therefore the Towa and ANIC/Melida products, being produced from the same process, might be considered identical.

The specifications in Table 1 show the varying chemical compositions of several commercialized brands of maltitol, supplied by Towa Chemical Industry and Roquette Freres (Lestrem, France). While both Towa (4) and Roquette do have the technology for producing high purity crystalline maltitol, that type of product is not necessarily available in all countries. Patent constraints (5) often allow only one supplier in a given country.

For most applications high and low maltitol types (i.e., 99% and 88%) can be used interchangeably; exceptions include pan coating, where differences in crystallinity might be evident. One may expect that the high maltitol product would be a bit sweeter, as sorbitol and the other less sweet hydrogenated components are found in the lower maltitol product. However, there is probably little difference when comparing actual food products.

### Hydrogenated Starch Hydrolysate

Hydrogenated starch hydrolysate (HSH), also called hydrogenated glucose syrup, or, more recently, maltitol syrup (6)—for those products containing more than 50% maltitol—may be conveniently thought of as hydrogenated corn syrup. It is generally available as a 75% solids syrup composed of varying proportions of hydrogenated oligosaccharides of different chain lengths, characterized by their degree of polymerization (DP). These components consist then of sorbitol (DP-1), maltitol (DP-2), maltotriitol (DP-3), maltotetraitol (DP-4), etc., continuing above DP-20. The structures of these are shown in Figure 1, along with their precursors. Like maltitol, HSH does not have reducing groups, therefore minimizing Maillard browning reactions. As HSH is noncrystallizing, it is available only as a syrup.

In the United States that are two major suppliers of HSH, Roquette and Lonza, each with slightly different products (Lycasin and Hystar). A discussion of each brand is therefore necessary. Tables 2 and 3 show the polyol distributions of the most important commercially available products (2,3). Comparisons of viscosity, sweetness, and hygroscopicity are shown in Table 4.

*Lycasin®*

Roquette Corporation, Gurnee, Illinois (parent company Roquette Freres, Lestrem, France), manufactures and markets Lycasin in the United

**Table 2** Approximate Polyol Distributions for Lycasin

| Polyol | Percent composition |
|---|---|
| DP-1 (Sorbitol) | 6 |
| DP-2 (Maltitol) | 52 |
| DP-3 (Maltotriitol) | 18 |
| DP-4 | 1.5 |
| DP-5 | 2 |
| DP-6 | 2.7 |
| DP-7 through 10 | 7.8 |
| DP-11 through 20 | 9.2 |
| DP-20+ (max. 3% ) | 0.8 |

**Table 3** Approximate Polyol Distributions for Several Commercially Available Hystars

| | Percent Composition | | | | |
|---|---|---|---|---|---|
| | Hystars | | | | |
| Polyol | 3375 | 4075 | HM-65 | 5875 | 6075 |
| DP-1 (Sorbitol) | 14 | 13 | 15 | 7 | 14 |
| DP-2 (Maltitol) | 18 | 23 | 50 | 60 | 8 |
| DP-3 (Maltotriitol) | 10 | 12 | 12 | 11 | 10 |
| DP-4+ | 58 | 49 | 23 | 22 | 68 |

**Table 4** Properties of Several Hydrogenated Starch Hydrolysates

| | Lycasin | Hystars | | | | |
|---|---|---|---|---|---|---|
| | 80/55 | 3375 | 4075 | HM-75 | 5875 | 6075 |
| Viscosity (cp @ 25 °C) | 1900 | 7000 | 4000 | 2500 | 1500 | 10000 |
| Sweetness | high | med. | med. | high | high | low |
| Hygroscopicity | high | avg. | high | high | high | avg. |

States. There used to be Lycasin products with different specifications, however, the one now available in the United States and commonly used worldwide is Lycasin 80/55, often referred to as Lycasin. The Swedish company Lyckeby originated the concept of HSH in 1963 and developed several versions that were less cariogenic than sucrose by 1968. In that year Roquette Freres began a 10-year joint venture with Lyckeby to further develop the line. Roquette ultimately took over the patents in 1975, with the responsibility for developing future Lycasin products. Lycasin 80/55, with patents worldwide, is considerably improved with respect to noncrystallizing properties and noncariogenicity, compared to both the original "Swedish Lycasin" ( made from potato starch), and Lycasin 80/33, (made from corn starch), and, for that reason, is the only Lycasin type available today (2,7).

*Hystar*

In the United States, Lonza, Inc., Fairlawn, New Jersey (parent company Lonza Ltd., Basel, Switzerland), manufactures and markets several varieties of their Hystar brand of HSH with a range of functionalities.

## APPLICATIONS

### Maltitol

Maltitol has many attributes and properties that allow it to be used in a wide variety of food applications:

white crystalline powder
sweetness profile similar to sugar
nutritive, bulk sweetener
noncariogenic
non–insulin-dependent and can be used in diabetic diets
humectant

### *Confectionery Products*

There is considerable formulation and processing information available for producing maltitol-based confections and chewing gums (1,8,9). Confectionery types include chocolate, hard boiled candy, pressed tablets, caramels, jellies, gums, and pastilles.

High quality sugarless chocolates can be made by simply substituting maltitol for sucrose. Chocolate so made does not have the mouth-cooling property associated with sorbitol-based chocolate, and since the sweetness is greater than that made with sorbitol, the addition of a potent sweetener is not required.

Maltitol-based hard boiled candies can be manufactured with conventional "roping and forming" equipment, unlike sorbitol boiled candy, which generally must be seeded with crystalline sorbitol, deposited into plastic or metal molds, and allowed to cure (10). Resultant candy must be securely packaged in a moisture-proof wrapper to minimize crystallization. Here it is worth pointing out that there may be differences in stability with respect to "cold flow" (a distortion of shape over time), hygroscopicity, and crystallization, depending upon the purity of the maltitol used. Use of a higher purity maltitol would result in a boiled candy having lower hygroscopicity, for example.

### *Other Food Applications*

Dietetic foods for those on sugar-restricted diets can also be formulated with maltitol (1,9). These would include granola bars, jams with no added sucrose, ice cream analogs, cookies, and cakes. In most cases the use of a potent sweetener is not required as maltitol is nearly as sweet as sucrose when used at about a one-for-one replacement. In cases where sugarless products are to be formulated with lower caloric density, maltitol will work well with all potent sweeteners. A product sweetened with maltitol plus a potent sweetener will have a sweetness profile very similar to that of sucrose. For example, a table-top sweetener can be formulated with maltitol and aspartame.

## Hydrogenated Starch Hydrolysate

Hydrogenated starch hydrolysate has several attributes and properties that make it useful for many applications:

- reasonable level of sweetness
- generally noncariogenic
- noncrystallizing
- good humectant
- very low reducing sugar content minimizing browning during cooking

*Confectionery Products*

Many types of sugarless confectionery can be made by replacing sugar and corn syrup with HSH. Detailed information may be obtained from HSH suppliers and others, who have developed formulations for caramels, gummy bears, nougats, hard boiled candy, taffy, butterscotch, marshmallow, chewing gum, and pan coated candy such as jelly beans. Some general principles of sugarless candy making have been reviewed elsewhere (11). Because HSH is not available as a crystalline powder, it cannot easily be used as a sugar replacer in chocolates, or in pressed tablets where the presence of water would pose difficulties. It is generally not practical to cook the HSH to dryness and then grind the resulting mass into a powder for use in these applications.

In intermediate moisture confections such as gummy bears, jellies, taffys, nougats and caramels, direct substitution of HSH for corn syrup is a good starting point. To make the candy noncariogenic, additional ingredient substitutions are necessary. These would include removing any sugar-containing ingredients and replacing them with suitable alternatives. Examples would include substituting caseinates and milk fat for milk powder, and pectin, gelatin, and gum arabic blends for starch. One may find that higher levels of bodying agents (e.g., gelatin) are required for the sugarless version. Theoretically, there may exist a lower level of intermolecular hydrogen bonding than would be the case with sugar candy. Hydrogen bonding would be expected to aid in providing some degree of structure and firmness to the candy. Hydrogenated species have no carbonyl groups available and, therefore, the amount of hydrogen bonding occurring in sugarless confections might be lower.

Hard boiled candy made with HSH has some particular advantages, not the least of which is that it may be made with conventional candy-processing equipment, rather than by depositing. Processing costs are markedly lower for an HSH-based candy compared to that made with sorbitol, as the need for molding, curing, and demolding is eliminated. Regardless of the type of HSH used, it is critical that the final moisture content of the hard candy be very low (preferably 1% or less), and that the candy be packaged in a moisture-resistant container, such as a fin-sealed foil bag (12). In ideal circumstances the candy should be packaged while warm, so its tendency to absorb moisture will be reduced. This is because its water activity is raised, so there is less of a differential between candy's water activity and the relative humidity of the packaging area, thereby minimizing water vapor migration.

The reason for minimizing contact with the water is due to the nature of HSH candy, as compared with sucrose/corn syrup candy. In the latter case, moisture pickup on the surface of the candy leads to crystallization of sucrose (graining), which actually insulates against further moisture pickup. Because HSH cannot crystallize, moisture pickup at the surface causes a thin sticky layer of solubilized HSH (13). Extensive water uptake will cause the candy to partly dissolve.

Hard boiled candies made with HSH are also subject to "cold flow," due to a lack of crystal structure. This loss of shape tends to occur when the candy has an elevated moisture content even if protected from atmospheric water vapor. A product such as Hystar 6075, having a greater percentage of longer chain polyols to provide additional molecular structure, is more resistant to cold flow than, for example, Hystar 5875. The effectiveness of longer chain polymers in providing physical stability to HSH has been demonstrated with the use of polydextrose (14 ), CMC, and gum arabic (2). Maintaining dry conditions will ensure a long shelf life.

### *Other Applications*

Given its excellent humectancy, HSH can be used to partly replace sugar or corn syrup in baked goods such as cakes, cookies, and granola bars. HSH is also used in dentifrice, mouthwash, and medicinal syrups (e.g., cough syrup) as a noncariogenic sweetener and viscosity aid.

While reasonably sweet confections can be made with some of the higher maltitol content HSH syrups, products with sweetness profiles matching those of sugar-sweetened candies require the use of a potent sweetener. Aspartame or acesulfame-K can be used at levels ranging from 0.15% to 0.4%, depending upon both the sweetness of the HSH used and the level of acid in the candy (2,3). Better sweetness profiles generally are attainable when more than one sweetener is used; the sweeter HSH products therefore perform slightly better than the blander ones when used in conjunction with a potent sweetener.

## GENERAL COST AND ECONOMICS

Polyols, as a general rule, are more expensive than the common carbohydrate sweeteners sucrose and corn syrup, due to the additional processing they must undergo. Maltitol costs are in the \$1.00–\$1.25 per pound range. Although more expensive than crystalline sorbitol (typically around \$.70), maltitol's outstanding taste, sweetness, and functionality make it worth the added cost in many applications.

The cost of HSH is generally in the range of \$0.70–\$0.90 per pound of solids. This makes it somewhat more expensive than liquid sorbitol, the latter at about \$0.45–\$0.50 per pound of solids. Tanker wagon quantities would lessen the cost.

## METABOLIC ASPECTS

### General Metabolism

Mammalian metabolism of maltitol generally begins with the hydrolysis of maltitol into the monosaccharides glucose and sorbitol. While this hydrolysis may not be complete at high maltitol intakes, primarily glucose and sorbitol will be transported into the blood stream. Detailed reviews of the metabolism of sorbitol have appeared (7,15). Sorbitol is converted to fructose, which then is utilized via the glycolytic pathway.

Metabolically, hydrogenated starch hydrolysate is first enzymatically hydrolyzed into glucose, maltitol, and sorbitol (16,17). More than 96% of Lycasin 80/55 is broken down into sorbitol and glucose, the remainder being maltitol. Half the maltitol is excreted in the feces, the remainder in the urine (18). Metabolism of Lycasin and Hystar are thought to be nearly identical, as only the percentages and not the types of the component polyols differ. The rate of enzymatic hydrolysis might be a function of the component polyol's chain length.

### Cariogenicity

It is well known that dental caries are caused by the prolonged exposure of tooth enamel to acid. Dextran, the principal component of plaque, is produced by microorganisms such as *S. mutans* and is responsible for adherence of the organisms to the tooth. Metabolism of sugars and starches by these microorganisms in the mouth leads to the formation of acids which, if in sufficient quantity, will erode tooth enamel during prolonged contact. Polyols such as sorbitol, mannitol, maltitol, and xylitol are not metabolized by oral bacteria, and therefore no acid production occurs (19).

Laboratory assessment of cariogenic or acidogenic potential had been somewhat problematic. Quick preliminary screening of various materials for cariogenic potential had been accomplished with various bioassays, utilizing microorganisms (20) or laboratory animals (21). A rigorous test, now well accepted, is plaque pH telemetry in human subjects (22–24). Telemetry refers to the continuous electronic data-gathering process used in these experiments. In these tests, an indwelling pH electrode is secured in a denture appliance and implanted onto a tooth, replacing an existing appliance. Several days after sufficient plaque has built up on the appliance, the test subject is given a food and the pH is monitored as a function of time. If acid formation does not reach below pH 5.7 in 30 minutes, the food may be certified as safe for teeth ("zahnschonend" by the Swiss authorities). Demineralization of tooth enamel definitely occurs below pH 5.5; between 5.5 and 5.7 is a transition range where some demineralization may begin. The advantage of this in vivo experimental approach over other in vitro assay methods is that it takes into account salivary flow, the buffering capacity of saliva, and the removal of dissolved sugars from the mouth via swallowing. Recently, however, another simplified in vitro test has been developed that claims results comparable to that of the pH telemetry method (25). For a comprehensive discussion of cariogenicity in foods, the reader is referred to a special issue of the *Journal of Dental Research,* December 1986 (pp. 1473–1543), that is based upon the Scientific Consensus Conference on Methods for Assessment of the Cariogenic Potential of Foods that was held at the University of Texas Dental School, San Antonio, November 17–21, 1985. Among the many results of this conference was the promulgation of a set of criteria for assessment of cariogenic potential of sugarless foods.

There are extensive cariogenicity studies for both maltitol and hydrogenated starch hydrolysate. Several studies have shown that maltitol is noncariogenic (26–29), the plaque pH curve not dropping below pH 5.7 (26). Documentation of Lycasin and Hystar noncariogenicity has been carried out by Roquette and Lonza, respectively. Since cariogenic potential is a function of a specific HSH, it is proper to discuss each brand separately.

### *Lycasin*

Numerous studies of Lycasin have shown that it is noncariogenic and safe for teeth (23,30,31). One must be cautious in reviewing early studies not explicitly stating the use of Lycasin 80/55, as early versions of Lycasin, such as the 80/33, were more acidogenic than the 80/55, and were often just labeled as Lycasin. The Rundgren study (29) is the only study conducted after 1978 using an early version of Lycasin, while Imfeld's research (23,30) addresses only studies that utilize Lycasin 80/55 prior to 1978 (2). Comparing acidogenicity of several early Lycasin types with the 80/55 shows clearly that only the 80/55 is safe for teeth (31,32).

It has also been shown that most types of Lycasin are slightly fermentable, and that low levels of fermentation do not affect its noncariogenic status. Noncariogenicity has recently been demonstrated even in rats highly susceptible to caries (33). Fermentability increases with concentration of hydrogenated oligosaccharides with higher degrees of polymerization (31). Noncariogenicity reportedly has only been obtained if the test substance contains less than 3% of oligosaccharides with DP greater than 20 (7). The slight acidogenicity observed with Lycasin is caused by small amounts of fermentable sugars, such as maltose and maltotriose, that are hydrolyzed from the oligosaccharides by the action of salivary $\alpha$-amylase (34,35). The relatively low fermentability of Lycasin has been studied, and three mechanisms have been offered as explanation: low or no fermentation of maltitol and maltotriitol by dental plaque bacteria; maltotriitol inhibition of fermentation such that any cariogenic sugars released during hydrolysis are not acted upon; and reduction of the rate of amylase attack that produces small quantities of fermentable sugars (35).

### *Hystar*

A study by Jensen (36) assessed the acidogenicity of several of the Hystar products using an interproximal plaque pH telemetry study. The protocol in the first portion of the study involved placement of the appliance and a subsequent 3-day plaque accumulation period. After obtaining a stable pH, volunteers rinsed their mouths with the test solutions for 1 minute. Plaque pH was monitored for 30 minutes. Rinsing with distilled water was followed by a paraffin chew to raise the plaque pH to its resting state. Internal control rinses (50% sucrose solution, 50% sorbitol solution) were tested after evaluation of each experimental solution.

Results show that the overall mean pH minimum for sucrose was 4.06 ± 0.11 and for sorbitol was 6.48 ± 0.11. The pH minimum for Hystar 6075 was 6.39 ± 0.20; for Hystar HM-75, 6.23 ± 0.09; for Hystar 7570, 6.56 ± 0.25; for Hystar 8070, 6.37 ± 0.26; and for Hystar 7000, 6.42 ± 0.23. Comparison of experimental results with those obtained for sorbitol indicate that Hystar 6075, 7570, 7000, and 8070 were not significantly different. The mean minimum pH value for HM-75, 6.23, was found to be statistically significantly different from that of sorbitol, but still above the 5.7 "safe for teeth" value.

Recent studies have been conducted comparing the relative acidogenicity of Lycasin and Hystar. In one study (37) a microelectrode was used to probe 6 teeth in each of 20 subjects after rinsing with either control or test solutions. Plaque pH was monitored for 60 minutes. Results for mean minima are shown in Table 5.

Areas under the plaque pH curves were measured, with the observations that Lycasin and maltitol syrups were statistically equivalent to sorbitol, and Hystar 6075 and Hystar 3375 were statistically more acidogenic than sorbitol. Sucrose and the Hystars were statistically equivalent to each other. It is suggested that the level of the DP-20+ fraction is responsible for the increased acidogenicity of the Hystar 3375 and Hystar 6075. The DP-20+ fractions in Lycasin, Hystar 6075, Hystar 3375, and Maltisorb 75/75 are 1.0, 15.3, 11.8, 0.1 %, respectively (37). It should be noted that it is not possible to determine from this study

**Table 5** Plaque pH Minima After Rinsing with Test Solutions

| Product | pH minimum |
|---|---|
| Lycasin | 6.2 ± 0.4 |
| Hystar 6075 | 5.6 ± 0.4 |
| Hystar 3375 | 5.8 ± 0.4 |
| Maltisorb 75/75 (maltitol) | 6.4 ± 0.3 |
| Crystalline maltitol | 6.4 ± 0.4 |
| Sorbitol | 6.5 ± 0.4 |
| Sucrose 10% | 5.5 ± 0.5 |

*Source:* Adapted from Ref. 37.

whether either Hystar is safe for teeth, as the pH minimum even for sucrose borders on noncariogenicity. It may be that the method in this experiment is not sufficiently sensitive.

A study by a different group (38) used the indwelling plaque pH design to conclude that Lycasin and sorbitol promote less acidogenic challenge to in situ plaque than do Hystar 3375, Hystar 4075, and Hystar 6075. In that study panelists rinsed with a test solution for 2 minutes and plaque pH was monitored for 2 hours. pH minima, time under pH 5.5, and area under the pH curve below 5.5 were monitored. pH minima reported for sorbitol, Lycasin, Hystar 6075, Hystar 3375, Hystar 4075, and sucrose are 6.05 ± 0.24, 5.70 ± 0.27, 4.75 ± 0.29, 4.51 ± 0.16; 4.17 ± 0.07, and 4.24 ± 0.07, respectively. Exposure to Lycasin or sorbitol resulted in the plaque pH falling below pH 5.5 for significantly less time, and with smaller areas under pH 5.5, than did exposure to the Hystar products or sucrose. Interestingly, the data in this study do not support the idea that concentrations of the DP-20+ fraction exceeding 3% tend to promote acidogenicity (7). Composite plaque pH curves show that the pH drop is most for Hystar 4075, followed by Hystar 3375, and least for Hystar 6075; in inverse order expected by the concentrations of their DP-20+ fractions. Hystar products were not statistically different from sucrose for any parameter studied.

From this study it would appear that the Hystar products are higher in acidogenicity than the sorbitol, xylitol, mannitol, maltitol group but lower in acidogenicity than sucrose and other fermentable carbohydrates. Food manufacturers using any HSH would be wise in testing their final products to substantiate their acidogenic status before making any cariogenicity claims.

## Laxation

### *Maltitol*

As is the case for many polyols, particularly sorbitol and mannitol, excessive consumption of maltitol may have a laxative effect. The reason for this may be viewed as a direct consequence of the noncariogenicity of these materials. If these polyols are not readily fermented by oral microorganisms, it is because the microbes do not possess adequate quantities of the necessary enzymes, or their existing enzymes are not active on polyol substrates. Understandably, digestion by the human enzyme systems will be similarly impaired.

Hydrolysis of maltitol to glucose and sorbitol is significantly slower than is the hydrolysis of maltose (19). Therefore the intestinal brush border maltase enzymes may not have adequate time to complete hydrolysis prior to passage of the intestinal contents into the large intestine, where the delicate water balance is thereby upset. Sorbitol is absorbed slowly via passive diffusion, whereas glucose is readily absorbed by active transport. Not only do carbohydrates carry osmotically bound water into the large intestine, but microbiological fermentation of these polyols will create still a larger number of osmotically active components as well as excessive flatus. Excessive water absorption into the large intestine causes laxation.

Maltitol is less laxating compared to sorbitol; the maximum noneffective dose of maltitol (the dosage where no laxation or other discomfort occurs) is about twice that of sorbitol. This is consistent with the facts that disaccharides are less osmotically active, and also that half the maltitol molecule is glucose which, after hydrolysis, is readily absorbed (39,40). The U.S. Food and Drug Administration requires a food to be labeled with the warning "Excess consumption may have a laxative effect" if foreseeable daily consumption may result in ingestion of greater than 50 grams of sorbitol (21 CFR 184.1835 (e)). Towa Chemical Industries therefore asserts (39) that a warning statement may apply to maltitol only when greater than 100 g per day is likely to be ingested.

A study by Koizumi (40) concluded that the maximum noneffective laxative dose for both males and females was 0.3 g maltitol per kg body weight, and that the 50% single laxative dose, i.e., producing laxation in 50% of the subjects, was 0.8 g per kg body weight. Data cited by Towa (39) show that the estimated daily intake of maltitol is only about half the 50% single laxative dose, supporting their contention that significant laxation problems will not occur in most individuals.

### *Hydrogenated Starch Hydrolysate*

Lycasin. As sorbitol is present in Lycasin and is produced during its digestion, it is to be expected that some degree of laxation might occur. The mean laxation threshold in adult males is about 180 g per day; in females it is about 100 g per day. The laxation threshold in children is about 60 g per day (41). Symptoms of excess consumption are flatulence and diarrhea; however, after repeated daily consumption, the human digestive system appears to adapt, resulting in gradual disappearance of diarrhea and flatulence over 4–5 days (42–44).

The potential average daily consumption of Lycasin 80/55 by "eaters of Lycasin-containing products" is estimated to be 0.5–1.1 g per day, with 90th percentile consumption to be 1.1–2.6 g per day (18). These estimates are far below the level of consumption causing laxation.

Hystar. Several types of Hystars were tested to determine laxative responses in human volunteers (45). As expected, the Hystar composition containing the most sorbitol exhibited the most pronounced laxative effect. Hystars containing more of the DP-4+ fraction had significantly reduced degrees of laxation. Depending upon the Hystar tested, an acceptable range for consumption was determined to be between 50 and 100 g per day.

## Suitability in Diabetic Diets

### *Maltitol*

Because hydrolysis of maltitol is slower than that of sucrose in humans, it would appear that maltitol might have utility in diabetic diets. The goal in dietary management of this type is to minimize fluctuations in blood glucose by not overloading the insulin requirements at any one time. Indeed, many studies (40,41,46–50) have shown that ingestion of maltitol does not produce marked elevations in blood glucose levels in both diabetic and healthy individuals. It appears that any sorbitol produced from the hydrolysis of maltitol is absorbed very slowly and, to a certain extent, inhibits absorption of the glucose also released (51). Thorough reviews of the literature are available (39,51). Diabetics should consult their physician to determine suitability for their particular circumstances.

### *Hydrogenated Starch Hydrolysate*

Given the hydrolysis of HSH to glucose and sorbitol, its use in the diabetic diet is not supported. While it would appear that this is in conflict with the utility of maltitol in the diabetic diet, there is a significant difference. Mammalian digestive enzymes readily cleave glucose residues off the polyol chains, freeing up proportionally more glucose than sorbitol (2,19). There are, however, at least two studies that report lower blood glucose levels with HSH as compared to glucose, after oral administration (49,52). Here again, sorbitol may inhibit absorption of glucose.

## Caloric Content of Maltitol and HSH

With the high amounts of glucose hydrolyzed during digestion of HSH, it is believed that HSH is similar to other carbohydrates in caloric content; that is, 4 calories per gram. The caloric content of maltitol is still a matter of controversy. It has been claimed that maltitol's caloric value is reduced by about 50% (53), although many other studies point to full caloric utilization.

One of the earliest studies (54) showed that weanling rats placed on diets containing 13 or 26% maltitol for 9 weeks experienced reduced body weight gain, suggesting lower caloric utilization. Male mice given single oral doses of $^{14}C$-maltitol expired less $^{14}CO_2$, had less $^{14}C$ blood concentrations, and excreted more fecal $^{14}C$ than when given $^{14}C$ glucose, again suggesting lower caloric utilization (41). It has also been shown that maltitol can produce hyperglycemia in rats similar to that produced by an equivalent amount of sucrose or glucose but at a slower rate (55), suggesting slower or incomplete absorption. Supporting the conclusion of poor caloric utilization was a study using rats given an oral dose of $^{14}C$ maltitol (56). After 24 hours, about 60% of the radioactivity was localized in the cecum, the large intestine and the feces. Total respiratory $^{14}CO_2$ during a 24-hour period amounted to only 1.2% of the administered radioactivity, while 5% of the administered maltitol was excreted in the urine, indicating minimal transport across the intestinal lumen.

In contrast, in one of the first studies to suggest that maltitol was almost fully utilized in rats, dogs, and man (57), maltitol was shown to have a caloric utilization of about 90%. In that report it was suggested that hydrolysis to glucose and sorbitol occurs in the stomach, followed by rapid absorption of these components, and maltitol, in the intestine. Substantial recovery of $^{14}CO_2$, along with the presence of volatile fatty acids in the fecal samples, suggest that fermentation by the microflora of the large intestine and subsequent absorption and metabolism is the principal utilization pathway.

In another study (16), rats were given an oral dose of maltitol and were sacrificed after several hours. It was found that only a small quantity of nonhydrolyzed maltitol crossed the intestinal barrier, as the blood concentration was low. Urinary excretion of sorbitol or maltitol was only 1% of the total ingested. Increased doses of maltitol gave higher gastrointestinal concentrations of sorbitol and decreased concentrations of maltitol, suggesting that maltitol is hydrolyzed to its components. Other studies began to focus on the nutritional state of the subject animal in an attempt to understand the role of the microflora (58,59). Some work claims that the role of the microflora is minimal (59,60) although other studies (61,62) claim that certain microorganisms can utilize maltitol.

To further investigate the microflora effect, $^{14}C$-maltitol and $^{14}C$-sucrose were fed to dogs and rats (63). Rats oxidized both substrates to $^{14}CO_2$ (to a comparable extent) but the peak for $CO_2$ production was 1–2 hours after sucrose ingestion, compared with 4–6 hours after maltitol ingestion. This suggests that sucrose is absorbed and metabolized in the stomach and small intestine, while most of the maltitol is absorbed and metabolized in the large intestine. Dog data indicate similar absorption times for both substrates. Excreta from both species had little radioactivity, indicating extensive hydrolysis and metabolism.

Maranesi and co-workers (64,65) have reported slightly lower weight gains in young rats whose diets had one third of their carbohydrate content replaced with maltitol. The authors demonstrate a caloric yield of 3.64 kcal/g, based upon increases in body weights.

It appears the problem confounding the caloric determination is determining the extent of microflora metabolization of maltitol. If intestinal absorption of any polyol were indeed complete, the caloric content would be about 4 kcal/g. Qualitative indicators of the completeness of intestinal absorption are the absence of cecal enlargement in the rat and the absence of flatulence, cramping, etc. in man. Several mechanisms for diminution of physiologic caloric value have been discussed (19), such as:

1. Anaerobic conversion of carbohydrates into lower fatty acids dissipates heat, while only a minor part of the chemical energy is recovered in the fatty acids themselves.
2. Short-chain fatty acids may be utilized by the host animal or by the microflora for production of more biomass; the latter process contributes nothing towards the metabolic energy needs of the host, while the former enables additional calories to be derived from the original polyol substrate. Competition between these two processes will determine the maltitol's physiologically available caloric content.
3. Gases such as methane and hydrogen carry energy away from the host, the energy not being available.

Discussion of the caloric content of maltitol and other food ingredients was recently undertaken at the Symposium on Carbohydrate Digestion/Absorption and Energy Metabolism in Japan, held on July 24, 1987. While some investigators suggest an energy value of 1.2–2.1 kcal/g

(66), it was explained that these values are only preliminary estimates. In Japan the official method for determining caloric content is the Oxygen Digestion Method, established in 1978, when no consideration was given to any metabolic processes that may take place in the large intestine by the microflora (67). It is now known that microbiological considerations may be more important than previously thought. Dr. T. Oku (66) has suggested that the energy value of maltitol be computed by considering the value at each step of its metabolism: absorption in stomach and small intestine, absorption of volatile fatty acids in large intestine, and final excretion.

## CARCINOGENICITY AND OTHER TOXICITY

Numerous studies examining the acute, subacute, chronic, and subchronic toxicities, and carcinogenic, mutagenic, and teratogenic activities of maltitol and HSH have shown no evidence of any adverse effects (18,39,45,51). The Joint FAO/WHO Expert Committee on Food Additives (JECFA) first reviewed maltitol in conjunction with hydrogenated starch hydrolysate in 1980 when it concluded that the allocation of an "acceptable daily intake (ADI) need not be specified" (68). A second review in 1983 resulted in the allocation of a temporary ADI of 0–25 mg/kg body weight for hydrogenated glucose syrups containing 50–90% maltitol, mainly because the panel did not have, but requested, the results of chronic feeding studies (69). In 1985, following the acquisition of lifetime feeding studies, JECFA reevaluated the toxicology of maltitol and HSH and reverted to an ADI of "not specified" (51). As defined by JECFA, ADI "not specified" means that, on the basis of available scientific data, the total daily intake of a substance arising from its use at levels necessary to achieve the desired effect, does not, in the opinion of the Committee, represent a hazard to health. Therefore, establishment of an ADI in numerical terms was not necessary.

Another group of scientists, the Scientific Committee for Food of the European Economic Community (EEC), in its December 1984 report on sweeteners, found maltitol to be "acceptable." The committee found it inappropriate to establish an ADI for maltitol and other polyols.

## REGULATORY STATUS

### United States

#### *Maltitol*

Towa Chemical Industry Co., Ltd, filed a petition with the FDA on August 26, 1986, to affirm the Generally Recognized as Safe (GRAS) status of maltitol. The petition was accepted for filing on September 23, 1986, and was announced in the *Federal Register* on December 23, 1986. While the FDA is conducting their assessment of the petition, maltitol may be used in the United States, although potential customers should make their own assessment of the risk; that is, if the FDA rules adversely, products containing maltitol may need to be recalled from the marketplace. In this case the risk is minimal, as maltitol is a principal component in hydrogenated starch hydrolysate (hydrogenated glucose syrup) which has been used in the United States since around 1977, and also the subject of a longstanding, prior FDA GRAS petition.

The petition asks for usage of maltitol in a variety of applications, including flavoring agent, formulation aid, humectant, nutritive sweetener, processing aid, sequestrant, stabilizer and thickener, surface-finishing agent, and texturizer. The ingredient is used in food at levels not to exceed current good manufacturing practices: a maximum level of 99.5% in hard candy and cough drops, 99% in sugar substitutes, 85% in soft candy, 75% in chewing gum, 55% in nonstandardized jams and jellies, and 30% in cookies and sponge cake (39).

*Lycasin*

Roquette Corporation submitted a GRAS affirmation petition with the FDA on August 1, 1983, which later was accepted for filing December 16, 1983, and was announced in the *Federal Register* on February 27, 1984. The petition asks for usage of hydrogenated glucose syrup in candies and confections (18). Lycasin had been used as early as 1977 without the filing of the GRAS petition. Currently, Lycasin may be used, subject to the same risks as maltitol.

*Hystar*

Lonza Inc. filed a GRAS affirmation petition for Hystar 6075 in May 1985. The petition was accepted for filing, and subsequently was put into the same category for approval as Roquette's petition, that is, both petitions are now proceeding simultaneously in FDA's review process (70). The Hystar petition references most of its safety data from the Roquette petition, citing the similarity of the two hydrogenated starch hydrolysates. Lonza's petition asks for use of HSH as a humectant, flavoring agent and adjuvant, nutritive sweetener, processing aid, sequestrant, stabilizer, thickener, surface-finishing agent, and texturizer at levels not to exceed current good manufacturing practices in confectionery, candy, cookies, and other sweets (45).

It is likely that maltitol, Lycasin, and Hystar 6075 will be approved at the same time, as their chemical compositions and metabolisms are similar.

**Table 6** Worldwide Status of Lycasin and Maltitol

| COUNTRY | MALTITOL | LYCASIN |
|---|---|---|
| France | Approved | Approved |
| Switzerland | Approved | Approved |
| The Netherlands | — | Approved |
| Belgium | Approved | Approved |
| Denmark | Approved | Approved |
| Finland | Approved | Approved |
| Norway | Approved | Approved |
| Mexico | Approved | — |
| UK | Approved | Approved |
| Japan | Approved | Approved |
| Sweden | Approved | Approved |
| Germany | — | Petition filed |
| Austria | Approved | Approved |
| Spain | — | Petition filed |
| Australia | — | Approved |
| Canada | — | Petition filed |

## Worldwide

Table 6 shows the status of maltitol and HSH (i.e., Lycasin) in a number of countries. Where approval is shown, it signifies approval in at least one category of food even if restrictions such as preregistration may apply.

## REFERENCES

1. Towa Chemical Industry, technical literature.
2. Roquette Corporation, technical literature.
3. Lonza, Inc., technical literature.
4. M. Hirao, H. Hijiya, and T. Miyaka, Anhydrous crystals of maltitol and the whole crystalline hydrogenated starch hydrolysate mixture solid containing the crystals, and process for the production and use thereof, U.S. Patent 4, 789,559 (1988).
5. M. Mitsuhashi, M. Hirao, and K. Sugimoto, Low calorie sweetener mixture of maltitol and maltotriitol, U.S. Patent 3,705,039 (1972).
6. Proceedings of the Joint FAO/WHO Expert Committee on Food Additives, 33rd Session, Geneva, March 21–30, 1988, Technical Report Series 776 (1988).
7. P. J. Sicard, Hydrogenated glucose syrups, sorbitol, mannitol and xylitol, in *Nutritive Sweeteners* (G. G. Birch and K.J. Parker, eds.) Applied Science Publishers, New York and London, 1982, pp. 145–170.
8. Anonymous, Industrial applications of malbit, *Confectionery Production* (Dec.): 741–743 (1988).

9. I. Fabry, Malbit and its applications in the food industry, in *Developments in Sweeteners—3* (T. H. Grenby, ed.), Elsevier-Applied Science Publishers, London and New York, 1987, pp. 83–108.
10. K. J. Klacik, W. V. Vink, and P. R. Fronczkowski, Sorbitol containing hard candy, U.S. Patent 4,452,825 (1984).
11. M. Serpelloni, Sugarless confectionery—using sorbitol, mannitol and lycasin. *Confectionery Production* (June): 332–335. (1988).
12. P. J. Sicard, and P. Leroy, Mannitol, sorbitol and Lycasin: Properties and food applications, in *Developments in Sweeteners—2* (T. H. Grenby, K. J. Parker, and M.G. Lindley, eds.), Applied Science Publishers, New York and London, 1983, pp. 1–24.
13. D. A. Whitmore, Developments in the properties and applications of Lycasin and sorbitol. *Food Chemistry, 16:* 209 (1985).
14. H. Kastin, Hard candy composition, U.S. Patent 4,528,206 (1985).
15. B. K. Dwivedi, Food products for special dietary needs, in *Low Calorie and Special Dietary Foods* (B. K. Dwivedi, ed.), CRC Press, West Palm Beach, Florida, 1978, pp. 1–22.
16. F. Verwaerde and H. Dupas, Study of the in vivo digestion of Lycasin 80/55 in the rat, unpublished report from Roquette Freres, Lestrem, France. Submitted to WHO by Roquette Freres, 1982.
17. F. Verwaerde and H. Dupas, Study of the in vivo digestion of Lycasin 80/55 in the rat, Vol. III. A comparative study on fed and fasting rats, unpublished report from Roquette Freres, Lestrem, France. Submitted to WHO by Roquette Freres, 1984.
18. Roquette Corporation GRAS Affirmation Petition submitted to the U.S. Food and Drug Administration for hydrogenated glucose syrups, (1983).
19. S. C. Ziesenitz and G. Siebert, Polyols and other bulk sweeteners, in *Developments in Sweeteners—3* (T. H. Grenby, ed.), Elsevier-Applied Science Publishers, London and New York, 1987, pp. 109–149.
20. H. A. B. Linke, Sweeteners and dental health: The influence of sugar substitutes on oral microorganisms, in *Developments in Sweeteners—3* (T. H. Grenby, ed.), Elsevier-Applied Science Publishers, London and New York, 1987, pp. 151–188.
21. R. Havenaar, Dental advantages of some bulk sweeteners in laboratory animal trials, in *Developments in Sweeteners—3* (T. H. Grenby, ed.), Elsevier-Applied Science Publishers, London and New York, 1987, pp. 189–211.
22. T. Imfeld, Identification of low caries risk dietary components, *Monographs in Oral Science,* Vol. 11, (H.M. Myers, ed.), Karger, Basel, 1983, pp. 117.
23. T. Imfeld, and H.R. Muhlemann, Evaluation of sugar substitutes in preventing cariology, *J. Prev. Dent., 4:* 8 (1977).
24. T. Imfeld and H. R. Muhlemann, Cariogenicity and acidogenicity of food, confectionery and beverages, *Pharm. Ther. Dent., 3:* 53 (1978).
25. B. G. Bibby and J. Fu, Changes in plaque pH in vitro by sweeteners, *J. Dent. Res., 64* (9): 1130 (1985).
26. K. Matsuoka, On the possibility of maltitol and lactitol and SE-58 as non-cariogenic polysaccharides, *Nihon Univ. Dent. J., 49:* 334 (1975).

27. A. R. Firestone, R. Schmid, and H. R. Muhlemann, The effects of topical applications of sugar substitutes on the incidence of caries and bacterial agglomerate formation in rats, *Caries Res., 14:* 324 (1980).
28. T. Imfeld and F. Lutz, Malbit, ein zahnfreundlicher Zuckeraustauschstoff, *Swiss, Dent., 6:* 44 (1985).
29. J. Rundgren, T. Koulourides, and T. Encson, Contribution of maltitol and Lycasin to experimental enamel demineralization in the human mouth. *Caries Res., 14:* 67 (1980).
30. T. Imfeld, Evaluation of the cariogenicity of confectionery by intra-oral wire telemetry, *Helv. Odont. Acta., 21:* 437 (1977).
31. T. Imfeld and H. R. Muhlemann, Addendum to acid production from Swedish Lycasin (Candy Quality) and French Lycasin (80/55) in human dental plaques (G. Frostell, D. Birkhed), *Caries Res., 12:* 262 (1978).
32. G. Frostell and D. Birkhed, Acid production from Swedish Lycasin (Candy Quality) and French Lycasin (80/55) in human dental plaques (G. Frostell, D. Birkhed), *Caries Res., 12:* 256 (1978).
33. T. H. Grenby, Dental effects of Lycasin in the diet of laboratory rats, *Caries Res., 22:* 288 (1988).
34. D. Birkhed, S. Edwardsson, M.L. Ahldeh, and G. Frostell, Effects of 3 months frequent consumption of hydrogenated starch hydrolysate (Lycasin), maltitol, sorbitol and xylitol on human dental plaque, *Acta. Odontol. Scand., 37:* 103 (1979).
35. P. Wursch and B. Koellreutter, Maltitol and maltotriitol as inhibitors of acid production in human dental plaque, *Caries Res., 16:* 90 (1982).
36. M. E. Jensen, Human plaque acidogenicity studies with hydrogenated starch hydrolysates, Dows Institute for Dental Research College of Dentistry, The University of Iowa, Iowa City, Iowa. Unpublished study obtained through courtesy of Lonza, Inc.
37. D. C. Abelson, The effects of hydrogenated starch hydrolysate on plaque pH in vivo, *Clin. Prev. Dent., 11* (2): 21 (1989).
38. K. K. Park, B. R. Schemehorn, J. W. Bolton, and G. K. Stookey, Study on the acidogenic potential of hydrogenated starch hydrolysates, presented at the American Association for Dental Research, March 1989, San Francisco. Conducted at Indiana University School of Dentistry, Indianapolis, IN.
39. Towa Chemical Industry GRAS Affirmation Petition to the U.S. Food and Drug Administration for maltitol, (1986).
40. N. Koizumi, M. Fujii, R. Ninomiya, Y. Inoue, T. Kagawa, T. Tsukamoto, Studies on transient laxative effects of sorbitol and maltitol 1: Estimation of 50% effective dose and maximum non-effective dose. *Chemosphere, 12* (1):45 (1983).
41. Anonymous, Study of the in vivo digestion of Lycasin 80/55 in the rat, unpublished study performed by Roquette Research Laboratories.
42. P. J. Sicard and Y. LeBot, From Lycasin to crystalline maltitol: A new series of versatile sweeteners. Presented at the International Conference on Sweeteners—Carbohydrate and Low Calorie, September 22–25, 1988, Los Angeles, California. Sponsored by the American Chemical Society.

43. R. R. Abraham, M. Davis, J. Yudkin, and R. Williams, Controlled chemical trial of a new non-cariogenic sweetening agent, *J. Human Nutr., 35:* 165 (1981).
44. P. Leroy, Report on the tolerance test of Lycasin 80/55 administered in the form of sweets to adult human volunteers, unpublished report from Roquette Freres, submitted to WHO by Roquette Freres, 1982.
45. Lonza GRAS Affirmation Petition to the U.S. Food and Drug Administration for hydrogenated starch hydrolysate, (1985).
46. H. Atsuji, S. Asano, and S. Hayashi, A study on the metabolism of maltitol, a disaccharide alcohol, unpublished report from Keio University, Japan. Submitted to WHO by ANIC S.p.A. (1982).
47. K. Nishikawa, Variations in blood and urine sugar values with oral administration of maltitol, unpublished report from Nishikawa Hospital, Tadera, Japan. Submitted to WHO by ANIC S.p.A. (1982).
48. I. Takeuchi and M. Yamashita, Clinical test results of a disaccharide alcohol (maltitol), *Eiyo to Shokuryo (Nutrition and Food), 25:* 170 (1972).
49. T. Yamakubi, Malbit, a new sweetener and its applications for diabetics, *Proc. 14th Congress Jap. Soc. Diabet.,* April 3–4, 1971 (1971).
50. M. Kamoi, Y. Shimiza, M. Kawauchi, Y. Fuji, Y. Kikuchi, S. Mizukawa, H. Yoshioka, and M. Kibata, A study on the metabolism of maltitol (clinical study), *J. Jap. Diab. Soc., 18:* 451 (1975).
51. Toxicological evaluation of certain food additives and contaminants. Prepared by the 29th Meeting of the Joint FAO/WHO Expert Committee on Food Additives. June 3–12, 1985, Geneva, Switzerland. Published by Cambridge University Press on behalf of WHO, 1987, pp. 179–206.
52. Anonymous, Glucose tolerance and insulinemia after administration of Lycasin 80/55, sorbitol and sucrose, healthy individuals vs. diabetic patients, unpublished study performed by Dept. of Geriatrics, University of Uppsala, Sweden.
53. J. P. Felber, L. Tappy, D. Vouillamoz, J. P. Randin, E. Temler, and E. Jecquier, Comparative study of maltitol and sucrose by means of indirect calorimetry. Presented at the International Symposium of Sugar and Sugar Substitutes in Human Nutrition, Milan, 1984.
54. Y. Inoue, Effects of maltitol administration on the development of rats, unpublished Report from Nikken Chemicals Co., Omiya, Japan. Submitted to WHO by ANIC S.p.A., (1982).
55. J. Lederer, P. Delville, and E. Cevecoeur, Study of a new sugar replacement, maltitol from the Belgium Sugar Factory. *Sucr. Belge Sugar Ind. Abstr., 93:* 311 (1974).
56. T. Oku, Y. Inoue, and N. Hosoya, Absorption and Excretion of Maltitol-U-$C^{14}$ in Rat, *J. Jap. Soc. Food and Nutr., 24:* 399 (1971).
57. H. H. Rennhard and J. R. Bianchine, Metabolism and caloric utilization of orally administered maltitol-$^{14}C$ in rat, dog and man. *J. Agric. Food. Chem., 24*(2): 287 (1976).
58. T. Oku, S. H. Him, and N. Hosoya, Effect of maltose and diet containing

starch on maltitol hydrolysis in rat, *J. Jap. Soc. Food and Nutr., 34:* 145 (1981).

59. M. W. Kearsley, G. G. Birch, and R. H. P. Lian-Loh, The metabolic fate of hydrogenated glucose syrups, *Starche, 34:* 279 (1982).
60. C. Rosiers, F. Verwaerde, H. Dupas, and S. Bonquelet, New approach to the metabolism of hydrogenated starch hydrolysate: Hydrolysis by the maltase/glucoamylase complex of the rat intestinal mucosa, *Annals of Nutr. and Metab., 29:* 76 (1985).
61. T. Miksuoka, Utilization of maltitol by intestinal lactobacillus, unpublished report from the Institute of Physical and Scientific Research, Tokyo. Submitted to WHO by ANIC S.p.A. (1982).
62. T. Mitsuoka, Utilization of maltitol by intestinal flora, unpublished report from the Institute of Physical and Scientific Research, Tokyo. Submitted to WHO by ANIC S.p.A. (1982).
63. C. Mitoma, Metabolic disposition of maltitol in rats and dogs, unpublished report prepared by Stanford Research Institute for Towa Chemical Industry, Co., July 30, 1984.
64. M. Maranesi, P. Gentili, and G. Carenini, Nutritional studies on maltitol. Part 1: Acceptability, energetic yield, effects on growth, and blood biochemical parameters, *Acta. Vitamino. Enzymol., 6:* 3 (1984).
65. M. Maranesi, V. Barsanti, P. Gentili, and G. Carenini, Nutritional studies on maltitol Part 2: Effect on lipid metabolism and fatty acid composition of different rat tissues, *Acta. Vitamino. Enzymol., 6:* 11 (1984).
66. T. Oku, Unpublished studies presented at the Symposium on Carbohydrate Digestion/Absorption and Energy Metabolism, Tokyo, July 24, 1987.
67. Towa Chemical Industries Commentary on the Symposium on Carbohydrate Digestion/Absorption and Energy Metabolism, Tokyo, July 24, 1987. Obtained through Courtesy of Towa Chemical Industries.
68. Twenty-fourth Report of Joint FAO/WHO Expert Committee on Food Additives, Rome, 1980. WHO Technical Report Series, No. 653 (1980).
69. Twenty-seventh Report of Joint FAO/WHO Expert Committee on Food Additives, Geneva, 1983. WHO Technical Report Series, No. 696 (1983).
70. Anonymous, Lonza GRAS Affirmation Petition Linked with Lycasin, *Food Chemical News* (March 6): 6 (1989).

# 15

# Lactitol: A New Reduced-Calorie Sweetener

**John A. van Velthuijsen and Iwan H. Blankers** *CCA biochem b.v., Gorinchem, The Netherlands*

## HISTORY

Lactitol, under the trade name Lacty[R], is a bulk sweetener. It is a sweet-tasting sugar alcohol derived from lactose (milk sugar) by reduction of the glucose part of this disaccharide. It has not been found in nature and was described in the literature for the first time by Senderens (1) in 1920. The first useful preparation was made by Karrer and Büchi in 1937 (2). Lactitol is also named lactit, lactositol, and lactobiosit. The CAS registry number is 585-86-4. The molecular weight of anhydrous lactitol ($C_{12}H_{24}O_{11}$) is 344.32. The chemical structure of lactitol, 4-0-($\beta$-D-galactopyranosyl)-D-glucitol, is shown in Figure 1.

## PREPARATION

The principles of the preparation of lactitol are of general knowledge. The industrial process is a hydrogenation of a 30–40 percent lactose solution at about 100°C with a Raney nickel catalyst (3). The reaction is carried out in an autoclave under a hydrogen pressure of 40 bar or more. Upon sedimentation of the catalyst, the hydrogenated solution is filtered and purified by means of ion-exchange resins and activated carbon. The purified lactitol solution is then concentrated and crystallized. The monohydrate as well as the dihydrate can be prepared depending on the conditions of crystallization (4).

Hydrogenation under more severe conditions (130°C, 90 bar) results in partial epimerization to lactulose and partial hydrolization to galac-

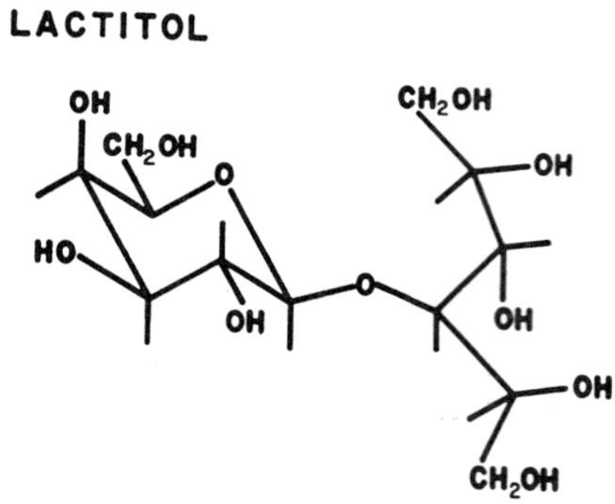


Figure 1 The chemical structure of lactitol.

tose and glucose and hydrogenation to the corresponding sugar alcohols lactitol, lactulitol, sorbitol, and galactitol (dulcitol) (5).

## TOXICOLOGY

All the required studies for a food additive have been carried out (CIVO-TNO, The Netherlands). There are no deleterious effects of feeding lactitol at levels up to 10% of the diet. The Joint FAO/WHO Expert Committee of Food Additives (JECFA) approved lactitol in April 1983. The Committee allocated an ADI (Acceptable Daily Intake) "not specified" to lactitol. (6).

The Scientific Committee on Food of the EC (SCF-EC) evaluated lactitol and considered it to be a safe product; they stated that "consumption of the order of 20 g per person per day of polyols is unlikely to cause undesirable laxative symptoms" (7). The same was stated for isomalt, maltitol, mannitol, sorbitol, and xylitol.

Safety studies in experimental animals included long-term feeding studies at high dietary levels for 2½ years in rats and for 2 years in mice. The safety data are not summarized here. Full data are described in CIVO-TNO (Zeist, The Netherlands) reports, which were not published. These reports have been evaluated by JECFA, SCF-EC, and the U.S. Food and Drug Administration.

## USE AND PURPOSE OF LACTITOL BASED ON BIOLOGICAL PROPERTIES

Lactitol is a sweet-tasting sugar alcohol with interesting nutritional, physiological, and pharmaceutical properties. It is not metabolized in animals and man as a carbohydrate since it is neither hydrolyzed nor absorbed in the small intestine. It is metabolized by bacteria in the large intestine where it is converted into biomass, organic acids, $CO_2$, and a small amount of $H_2$.

Lactitol consumption does not induce an increase of blood glucose or insulin levels, making it a desirable sugar substitute for diabetic patients. Furthermore, it has been demonstrated that its nutritional caloric utilization is half that of carbohydrates, with a metabolic energy value of maximum 2 kcal/g or maximum 8.4 kJ/g.

Lactitol is very slowly converted to organic acids (lactic acid) by tooth plaque bacteria, so that lactitol does not cause dental caries and satisfies the Swiss regulations as safe for teeth. These properties already suggest dietary applications, e.g., products for diabetic patients, non-cariogenic confections, and low-calorie foods.

## Biochemical Aspects

Karrer and Büchi (2) have studied the action of galactosidase-containing enzyme preparations on the splitting of lactitol into galactose and sorbitol. They found that lactitol is only hydrolyzed very slowly by these enzymes preparations. Later studies reported in a German patent (8) have confirmed that lactitol is only slowly split by enzymes at about a tenth the speed at which lactose is split. Based on both these in vitro studies, a reduced calorie value can be expected.

In patents of Hayashibara (9,10), it is claimed that lactitol has no caloric value because it is not digested or absorbed. The intestines of test rabbits not fed for 24 hours beforehand were closed at both ends and were injected with a 20% aqueous solution of lactitol or with an equimolar amount of a sucrose solution. After a lapse of several hours, the lactitol or sucrose left in the intestines was determined. While 85% of the sucrose intake had been lost due to absorption and digestion, all of the lactitol was still present.

## Metabolic Energy

At the Agricultural University of Wageningen in The Netherlands a study was carried out (11) determining the energy balances of eight volunteers on diets supplemented with either lactitol or sucrose. In this study volunteers were kept for 4 days in a respiratory room; this was done twice: once with 49 g of sugar a day in a diet; the other time, this sugar was replaced by lactitol. The dosage of 50 g lactitol monohydrate was ingested in 4–6 portions during the day. Intakes of metabolizable energy (ME) were corrected, within subjects, to energy equilibrium and equal metabolic body weight. Further correction of ME intake was made toward equal actometer activity. With regard to the value of ME to supply energy for maintaining the body in energy equilibrium, the energy contribution to the body of lactitol monohydrate was 60% less than for sucrose.

As the lactitol monohydrate contains 5% water and the results had an inaccuracy of SE 10%, the final conclusion is justified that the metabolic energy of lactitol, based on the dry substance, is at most 50% sucrose. Rounding off upwards, a caloric value of lactitol in man at 2 kcal/g seems fully justified.

### Lactitol and Diabetes

As lactitol does not have any significant effect on blood glucose or insulin levels, it is suitable for insulin-dependent diabetic patients (type I). It can also be consumed by noninsulin-dependent diabetic patients (type II). This group can limit its diabetic status by taking dietetic measures. Lactitol fits in their diets because of its reduced calorie value in comparison to sucrose and sorbitol, and its similar metabolism to dietary fiber.

### Tooth-Protective Properties

Sugars are a major factor in the pathogenesis of dental caries. Oral bacteria convert sugars into polysaccharides that are deposited on the teeth; these plaque sugars are then fermented into acids. The acid demineralizes the enamel and causes cavities.

#### *Dental Studies*

Three main types of experiments are used to evaluate the cariogenicity of foods and food ingredients:

Studies in vitro
Experiments in laboratory animals
Clinical trials and investigations in human subjects

#### *Studies In Vitro*

Among the earliest reports are those of Havenaar (12,13) on the utilization of lactitol by oral bacteria, with formation of acids. A number of plaque bacteria were found that could metabolize lactitol as a substrate, including *Streptococcus mutans*, which is known to possess cariogenic activity, and certain strains of *S. sanguis*, *Bifidobacterium*, and *Lactobacillus*. These first experiments did not establish the speed of fermentation, which because of the limited time that sugars and sweeteners remain in the mouth, is determinant of their cariogenicity. It was later shown that pH drop was slow, leading to the conclusion that lactitol could be fermented slowly by these microorganisms.

In addition to acid production, another important property of cariogenic bacteria is the capacity to synthesize extracellular polysaccharide from carbohydrate substrates, making dental plaque. No evidence for extracellular polysaccharide synthesis was found.

A more recent study on lactitol in vitro compared with five other bulk sweeteners, on incubation with mixed cultures of human dental plaque microorganisms, measuring acid development, insoluble polysaccharide synthesis, and the attack of the acid on enamel mineral, was done by Grenby (14). The six different sweeteners fell into three groups, with the highest acid generation, polysaccharide production, and enamel demineralization from glucose and sucrose, less from sorbitol and mannitol, and least of all from lactitol and xylitol.

### *Experiments in Laboratory Animals*

The cariogenicity of lactitol was also tested in rats (15). For this purpose, lactitol was incorporated in a powdered diet, consisting of 50% of a basal diet (SPP, Trouw & Co, Putten, The Netherlands), 25% wheat flour, and 25% of test substance. Lactitol was compared with sorbitol, xylitol, sucrose, and a control consisting of 50% wheat flour and 50% basal diet. The rats were program-fed. None of the animals suffered from diarrhea.

Especially in the lactitol group, no significant adverse effects on general health were observed. The results are shown in Table 1. Obviously the substitution of sucrose by lactitol reduces caries significantly.

**Table 1** Average of Carious Fissure Lesions and Weight Gains in Five Dietary Groups of 12 Rats

| Diet | Number of fissure lesions[a] | | Weight gain (g) |
|---|---|---|---|
| | B | C | |
| Sucrose | 5.75 | 2.58 | 82.0 |
| Sorbitol | 1.33* | 0.42* | 46.0** |
| Lactitol | 1.17* | 0.33* | 72.2* |
| Wheat flour | 0.67* | 0.25* | 80.8* |
| Xylitol | 0.42* | 0* | 48.6** |
| Pooled standard error | 0.56 | 0.43 | 5.4 |

[a]12 fissures at risk.
* = $p < 0.01$, significantly different from the sucrose group; ** = $p < 0.0001$ (Tukey's test).

The caries results for sorbitol and xylitol were in accord with other studies in which these polyols were used.

In the second experiment human food was fed to rats. Shortbread biscuits containing 16.6% sucrose or lactitol were incorporated at 66% in pulverized, blended diets fed to two groups of 21 rats, so that the final level of lactitol in the test diet was 11%. Again after a period of 8 weeks, caries attack showed highly significant reductions when replacing the sucrose in the biscuits with lactitol.

The most recent studies on the dental properties of lactitol in the rats tested it at lower levels in the diet and in the form of a finished human food product, rather than as a raw ingredient in a blended animal diet (16). At a level of 16% lactitol or 16% xylitol in the blended diet, the carious scores, lesion counts, and severity of the lesions were so close on the two polyol regimes that they were indistinguishable but significantly benificial compared to the 16% sucrose group.

### *Investigations in Man*

At the university of Zurich, a special method has been developed for the in vivo determination of the plaque pH in humans. After consumption

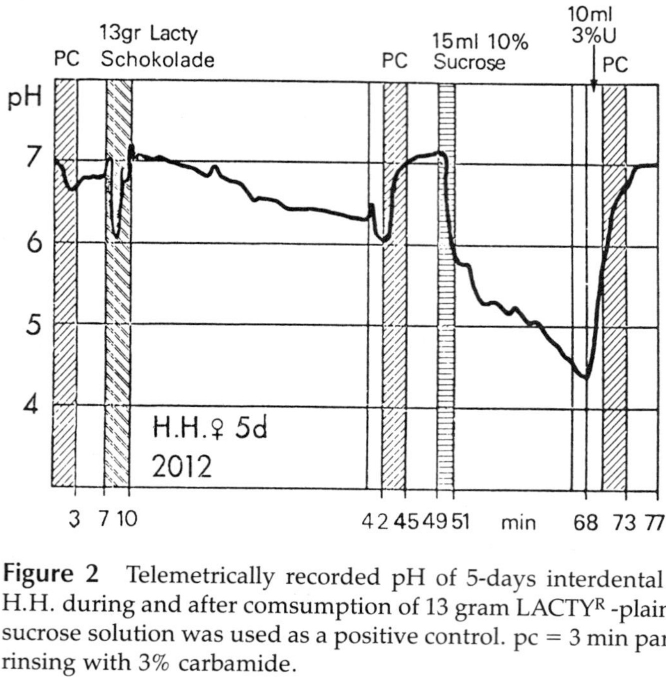


**Figure 2** Telemetrically recorded pH of 5-days interdental plaque in subject H.H. during and after comsumption of 13 gram LACTY[R] -plain chocolate. A 10% sucrose solution was used as a positive control. pc = 3 min paraffin chewing u = rinsing with 3% carbamide.

of chocolate or confections in which sucrose is replaced with lactitol, changes in plaque acidification are detected by the electrodes and transmitted electronically to a graph recorder.

The term "zahnschonend" (safe for teeth/friendly to teeth) is used officially in Switzerland when the pH of dental plaque does not drop below 5.7 during a 30-minute period. Professor Mühlemann demonstrated in chocolates that lactitol is "safe for teeth" (17,18).

In Figure 2 it is shown that during and after eating 13 g Lacty[R]-chocolate (plain lactitol-containing chocolate), the plaque pH is not affected, whereas with a 15 ml 10% sucrose solution, the plaque-pH is reduced to about 4.5, which indicates that sucrose is being fermented into acids by the oral bacteria.

## PHYSICAL PROPERTIES

Lactitol is a white crystalline powder available as a dihydrate or monohydrate. In the following sections the most important characteristics of lactitol will be described.

### Sweetness

Lactitol is less sweet than sucrose. Lactitol increases in relative sweetness when its concentration is increased (see Table 2). In most applications a sweetness of 0.4 times that of sucrose will be found. Lactitol has a clean sweet taste, without an aftertaste. In some products the lower sweetness will allow the flavor of other ingredients to develop better. In other products the sweetness will need to be increased. This is possible by using an intensive sweetener. To obtain a sweetness equivalent to sucrose, approx. 0.3% aspartame or acesulfame-K of lactitol, or approx. 0.15% saccharine needs to be added to lactitol.

**Table 2** Relative Sweetness of Lactitol Solutions Compared with Sucrose Solutions at 25°C

| Sucrose solution concentration (w/w) | Relative sweetness of lactitol (sucrose = 1.0) |
|---|---|
| 2% | 0.30 |
| 4% | 0.35 |
| 6% | 0.37 |
| 8% | 0.39 |
| 10% | 0.42 |

## Crystal Forms

Crystallization from aqueous solutions results in nonhygroscopic white monohydrate or dihydrate crystals. The dihydrate has a melting point of 76–78°C, and the monohydrate has a more complicated melting behavior: under the melting microscope, most of the crystals melt at 121–123°C, but part of the crystal bound water evaporates below 100°C, converting this monohydrate into the anhydrous form. The melting point determination of the monohydrate by differential scanning calorimetry shows the main endothermal peak in the range of 96–107°C (water evaporation).

## Hygroscopicity

Hygroscopicity is related to the amount of water absorbed by a product. Sucrose is not hygroscopic. Only under severe circumstances will sucrose absorb water. The same applies to lactitol. This can be seen in figure 3. When we look at sorbitol, we notice a high level of hygroscopicity. This means that, e.g., crisp bakery products will easily become soggy because they absorb water from the air.

## Water Activity

Water activity influences enzymatic activities, Maillard reaction, fat oxidation, microbial stability, and texture. These elements combined influence the shelf life. The influence of lactitol on water activity is very similar to that of sucrose (on a dry solids basis) since the molecular weight of lactitol is almost identical to that of sucrose. Sorbitol, being a much smaller molecule, has a much larger effect on the water activity.

## Solubility

There are many products for which solubility of the ingredients is very important, e.g., ice cream, hard boiled sweets. A very low solubility can make production processes more difficult (e.g., in hard boiled sweets, most of the water will have to be evaporated because more water will have to be added when the solubility is low). Lactitol's solubility at lower temperatures is less than that of sucrose, but still good enough not to cause inconveniences during processing (Fig. 4).

Some products are sensitive to solubility (e.g., ice cream can show sandiness, i.e., small crystals of undissolved sugars), but lactitol's solubility will not cause this sandiness.

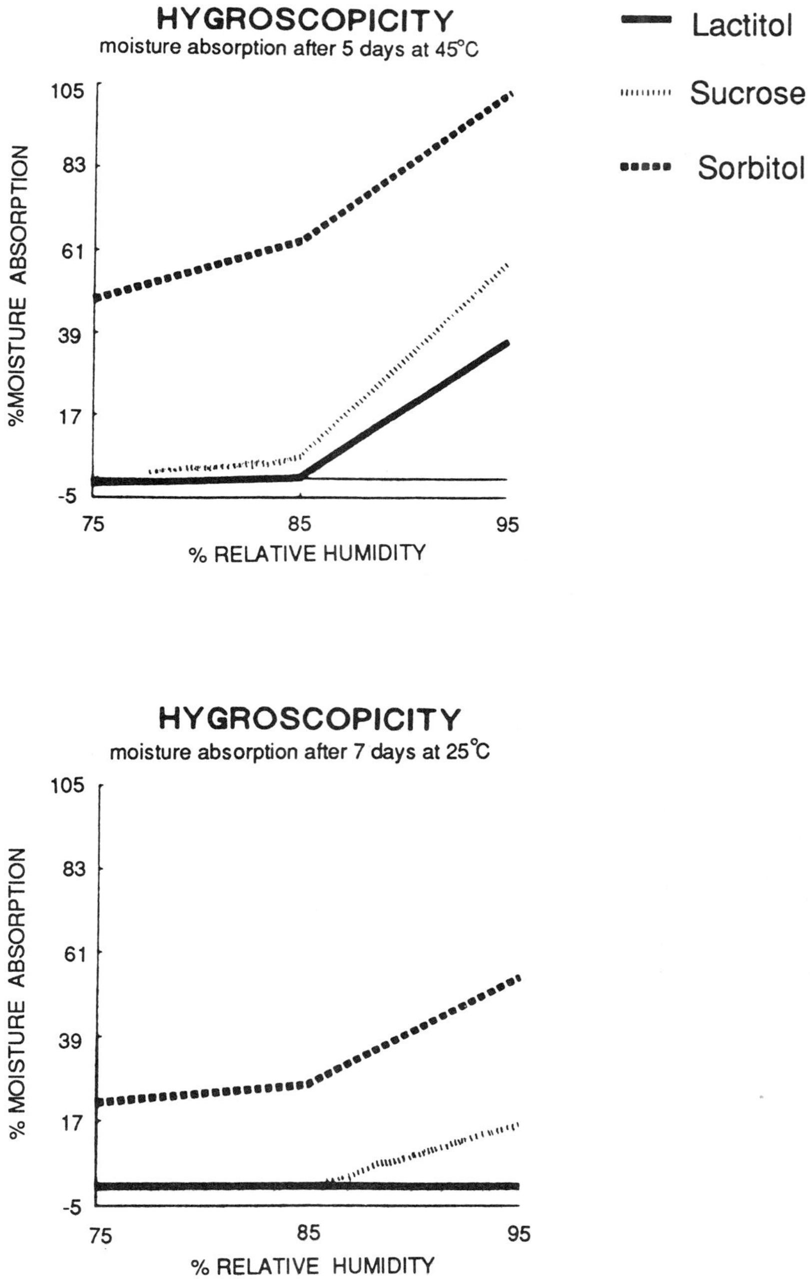


**Figure 3** Moisture absorption at different relative humidities.

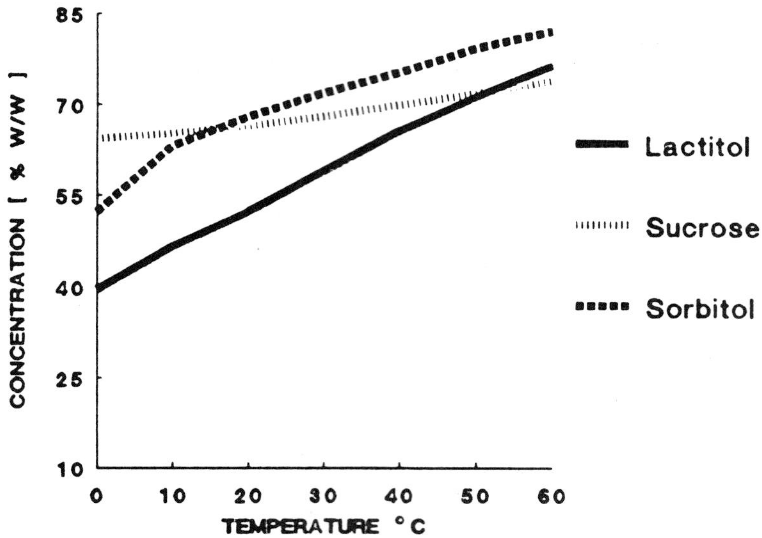


**Figure 4** Solubility of lactitol, sucrose, and sorbitol in water at different temperatures.

## Viscosity

The viscosity of lactitol in aqueous solution is slightly higher than that of sucrose (Fig. 5). In the case of hard boiled sweets, the viscosity of the lactitol-melt is more important. In this case a combination of lactitol and hydrogenated glucose syrup is used (compare sucrose and glucose syrup). For this combination a slightly lower viscosity is found than for its sucrose counterpart, but on cooling down, the same viscosity (or plasticity) will be reached, so lactitol will also be suitable for molded sweets.

## Heat of Solution

Lactitol has a cooling effect which is slightly stronger than that of sucrose but far less than that of sorbitol and xylitol (Table 3.).

## Freezing-Point–Depressing Effect

If sugars are replaced in ice cream, a replacer causing a similar freezing-point decrease is required. Lactitol is such a product, as shown in Figure

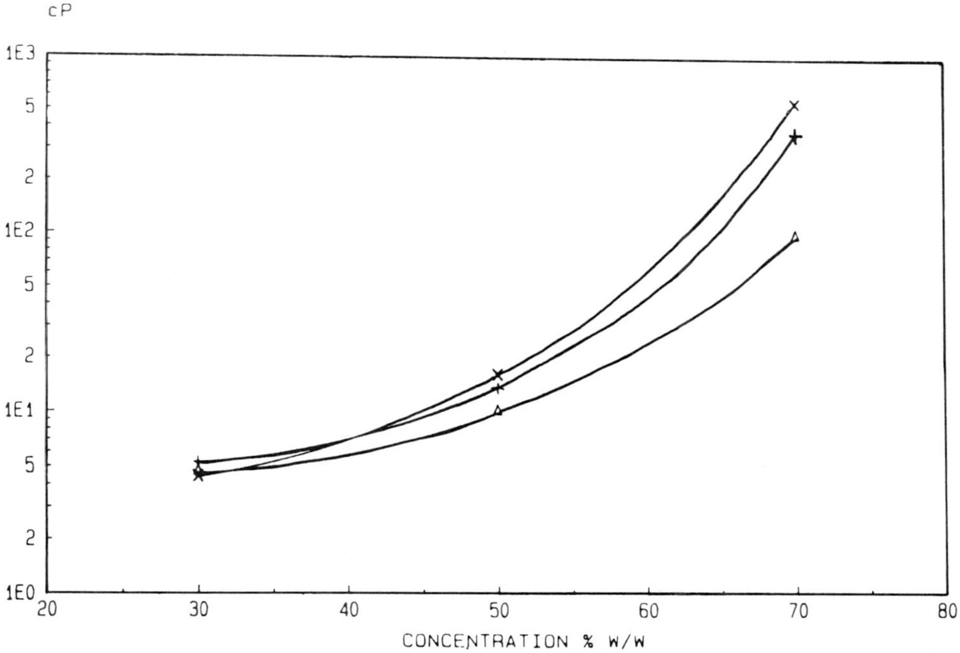


**Figure 5** Viscosity of lactitol [×], sucrose [−] and sorbitol [△] at 25°C.

**Table 3** Heats of Solution of Various Sweeteners

| Sweetener | cal/g |
|---|---|
| Sorbitol | −26.6 |
| Mannitol | −28.9 |
| Xylitol | −36.6 |
| Lactitol · $1H_2O$ | −12.4 |
| Lactitol · $2H_2O$ | −13.8 |
| Maltitol | −18.9 |
| Isomalt | −9.4 |
| Sugar | −4.3 |

6. This effect is closely related to the molecular weight since this is virtually the same for lactitol and sucrose, the effect is very similar.

## Chemical Properties

Lactitol is a polyol with nine OH-groups, which can be esterified with fatty acids resulting in emulsifiers (3), or which can react with propylene oxide (to produce polyurethanes). Due to the absence of a carbonyl group, lactitol is chemically more stable than related disaccharides like lactose. Its stability in the presence of alkali is much higher than that of lactose. The stability of lactitol in the presence of acid is similar to that of lactose. The absence of a carbonyl group also means that lactitol does not take part in nonenzymatic browning (Maillard) reactions.

Lactitol solutions have an excellent storage stability. A 10% lactitol solution in the pH range 3.0–7.5 at 60°C shows no decomposition after 1 month. After 2 months at pH 3.0, some decomposition (15%) is detected. At a higher pH this does not occur. With increasing temperature and especially with increasing acidity, hydrolytic decomposition of lactitol is observed. Sorbitol and galactose are the main decomposition products. At high pH lactitol is stable even at 105°C (Table 4).

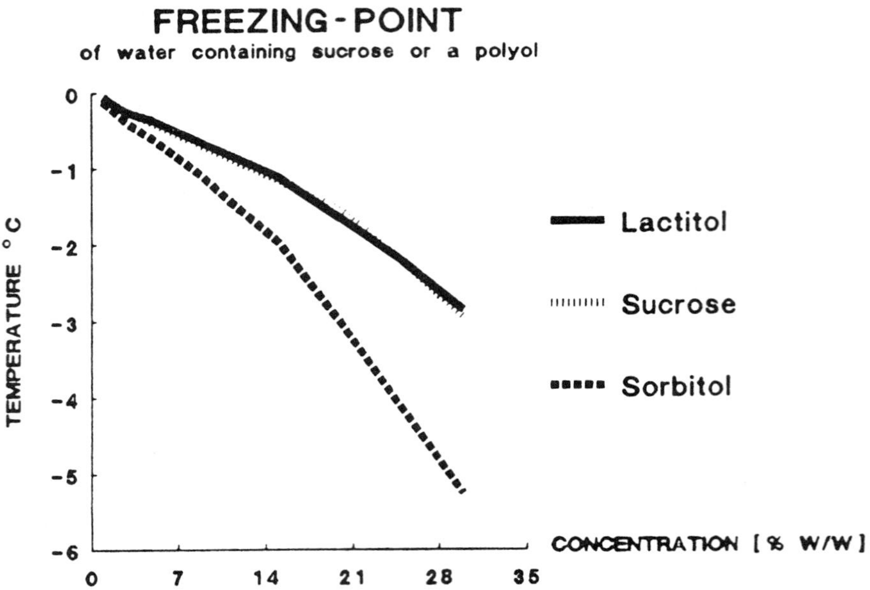


**Figure 6** Freezing point depressing effect.

**Table 4** Decomposition pf a 10% Lactitol Solution Under Extreme Conditions of pH and Temperature

| pH | % Weight recovery at 24 hr at 105°C | | | |
|---|---|---|---|---|
| | Lactitol | Galactose | Sorbitol | Not identified |
| 2.0 | 42.9 | 28.1 | 21.4 | 7.6 |
| 10.0 | 98.0 | — | — | 2.0 (acids) |
| 12.0 | 98.3 | — | — | 1.7 (acids) |

When heated to temperatures of 179–240°C, lactitol is partly converted into anhydrous derivatives (lactitan), sorbitol, and lower polyols.

## APPLICATIONS

Any application in which carbohydrates are used could be discussed, because lactitol could play a role in it. However, some general guidelines for specific application areas are offered.

Lactitol is used in products for its bulking properties. Therefore at most occasions a one-to-one replacement will be required. In some instances, where a solution is made, the replacement level may rise to one-to-one on a dry solids basis. In those cases where only the sweetness property of Lactitol is used, a replacement level up to 2.5 times may be required. In these circumstances a combination with an intense sweetener may be more appropriate.

### Bakery Products

In this field there are many products in which sugar can be replaced. For a large number of bakery products the crispiness of the product is very important. Most bulk sweeteners have the tendency to be hygroscopic, but if Lactitol products are handled like sucrose products, crispiness can be assured.

In the soft bakery products are, e.g., cakes, we find that lactitol (being nonhygroscopic) often benefits by a small addition of a humectant, e.g., sorbitol. This gives a smooth, moist feel without affecting the good structure and texture obtained by lactitol.

### Chewing Gum

In chewing gum the three main objectives for using a sugar replacer can be combined: low calories, tooth-friendliness, and suitability for diabetics. Besides these advantages, lactitol gives a more flexible structure, and there is no need to increase the gum base level as is required in sorbitol-containing chewing gum. Finally, because lactitol is not hygroscopic, it is also suitable to use as dusting powder or in hard panning.

### Chocolate

Lactitol can successfully be applied when producing a sugar-free chocolate. As discussed before, lactitol is not hygroscopic, however, it does contain crystal water. But when lactitol monohydrate is used, the crystal water will be bound sufficiently to use temperatures of up to about 60°C during the conching stage of plain chocolate.

### Confectionery

In confectionery there is a large range of products imaginable. Looking at sugar confectionery a combination of different sugars is often found. This is necessary to obtain the desired effect and results. In general lactitol replaces the sucrose part of recipes, whereas glucose syrup needs to be replaced by hydrogenated glucose syrup or maltitol syrup. These latter products are used as crystallization inhibitors. The right ratio differs from product to product; e.g., hard boiled sweets need a 70/30 ratio (lactitol/hydrogenated glucose syrup), chewies, fruit-gums, and pastilles a 40/60 ratio.

### Ice Cream

Lactitol shows a combination of physical properties (e.g., freezing-point–depressing effect, hygroscopicity, solubility) which are very suitable for ice cream. Not only will the use of lactitol reduce the caloric content of the ice cream, but it will also permit a low fat percentage, allowing even further reductions in calories.

### Preserves

The solution to low calorie preserves is often found in a low dry solids preserve. To combine this with lactitol will result in an even larger reduction in calories, while maintaining the good flavor and texture.

## REGULATORY STATUS FOR LACTITOL

| | |
|---|---|
| Austria | Allowed. |
| Belgium | Approval for ice cream, marzipan, and "speculoos." Other applications can be applied for and will most likely be granted. |

| | |
|---|---|
| Canada | Approval will be granted in 1991. |
| Denmark | Allowed, but for some applications maximum dosages are set. Chewing gum and pastilles have no maximum dosages set. |
| Finland | Application filed; requests by individual companies for specific applications are likely to be granted. |
| France | Approved in dietetic products. |
| Germany | Application filed. Approval by exception. |
| Israel | Allowed. |
| Italy | Application for approval has to be filed with help of industry. |
| Japan | Allowed. |
| Netherlands | Allowed. |
| Norway | Application filed; requests by individual companies for specific applications are likely to be granted. |
| Sweden | Application filed; requests by individual companies for specific applications are likely to be granted. |
| Switzerland | Allowed. |
| U.K. | Allowed. |
| U.S.A. | Application filed. |
| EC | Allowed in all EC countries after 1992. |

## CONCLUSION

Lactitol is a unique, commercially available bulk sweetener. Its physical properties guarantee optimal product performance during processing and storage of the food containing lactitol. Being so similar to sucrose, it is possible to develop new lactitol-based food products in a short period of time. It has also been proven to replace sorbitol powder successfully in several food products, whereby the overall quality increased remarkably.

## REFERENCES

1. J. B. Senderens, Catalytic hydrogenation of lactose. *Comptes Rendus, 170:* 47–50 (1920).
2. J. Karrer, Büchi. Reduktionsprodukte von Disacchariden: Maltit, Lactit, Cellobit, *Helvetica Chim. Acta, 20:* 86–90 (1937).
3. J. A. van Velthuijsen, Food additives derived from lactose: lactitol and lactitolpalmitate, *J. Agric. Food Chem., 27:* 680–686 (1979).
4. C. F. Wijnman, J. A. van Velthuijsen, and H. van den Berg, Lactitol monohydrate and a method for the production of crystalline lactitol, European Patent 39981 (1983).
5. T. Saijonmaa, M. Heikonen, M. Kreula, et al., Preparation and characteriza-

tion of milk sugar alcohol dihydrate (Abstract), Sixth European Chrystallograph Meeting, Barcelona, 1980, p. 26.
6. Joint FAO/WHO Expert Committee on Food Additives, IPCS Toxicological evaluation of certain food additives and contaminants: Lactitol, 27th report Geneva, WHO Food Additives Series 1983, pp. 82–94.
7. The Scientific Committee for Food, report on sweeteners, EC-Document III/1316/84/CS/EDUL/27 rev., 1984.
8. Maizena GmbH, Verwendung von Lactit als Zuckeraustauschstoff, German Patent 2,133,428 (1974).
9. K. Hayashibara, Improvements in and relating to the preparation of foodstuffs, U.K. Patent 1,253,300 (1971).
10. K. Hayashibara and K. Sugimoto, Containing lactitol as a sweetener, U.S. Patent 3,973,050 (1975).
11. A. J. H. Van Es, L. de Groot, and J. E. Vogt, Energy balances of eight volunteers fed on diets supplemented with either lactitol or saccharose, *Br. J. Nutr. 56:* 545–554 (1986).
12. R. Havenaar, Microbial investigations on the cariogenicity of the sugar substitute lactitol, unpublished report from the University of Utrecht (Netherlands), Department of Preventive Dentistry and Oral Microbiology, June 1976.
13. R. Havenaar, J. H. J. Huis in 't Veld, O. Backer Dirks, et al., Some bacteriological aspects of sugar substitutes, in *Health and sugar substitutes,* (Guggenheim B, ed), Proc. ERGOB Conf., Geneva, 1978. Basel, Karger, pp. 192–198.
14. T. H. Grenby and A. Phillips, Studies of the dental properties of lactitol compared with five other bulksweeteners in vitro, *Caries Res.* 1989; 23: 315–319.
15. J. S. van der Hoeven, Cariogenicity of lactitol in program-fed rats, *Caries Res. 20:* 441–443 (1986).
16. T. H. Grenby and A. Phillips, Dental and metabolic observations on lactitol in laboratory rats, *Brit. J. Nutr.* (1989), 61, 17–24.
17. H. R. Mühlemann, Gutachten über die zahnschonenden Eigenschaften von Lacty Schokolade und Lacty-Milch-schokolade der Firma CCA, Holland, unpublished report from Zahnärztliches Institut der Universität Zürich, July 1977.
18. T. H. Grenby and T. Desai, A trial of lactitol in sweets and its effects on human dental plaque, *Brit. Dent. J., 164:* 383–387 (1988).

# 16

# Isomaltulose

**William E. Irwin** *Palatinit Süßungsmittel GmbH, Elkhart, Indiana*

**Peter J. Sträter** *Palatinit Süßungsmittel GmbH, Mannheim, Germany*

## INTRODUCTION

Isomaltulose is a reducing disaccharide. Its systematic name is 6-*O*-$\alpha$-D-glucopyranosyl-D-fructofuranose (CAS Reg. No. 13718-94-0). Isomaltulose is a natural constituent of honey (1), occurring at levels up to 1%, and has also been found in sugar cane extract (2).

Isomaltulose was discovered in 1952 in a sucrose-containing substrate of *Leuconostoc mesenteroides* (3). Süddeutsche Zucker AG in 1957 first reported their interest in isomaltulose (4). This isomer of sucrose is produced commercially by Palatinit® Süßungsmittel GmbH using a sucrose–glucosylfructose–mutase enzyme. Isomaltulose (Palatinose®) is beginning to be used as a food ingredient and is the basic raw material for the production of isomalt (Palatinit®).

The chemical formula of isomaltulose is $C_{12}H_{22}O_{11} \cdot H_2O$. The molecular weight is 360.32. The chemical structure of isomaltulose has been confirmed by Loss (6) and is shown in Figure 1. The crystalline structure of isomaltulose has been studied by Dreissig and Lugar (7).

## PRODUCTION

The transglucosidation of sucrose to produce isomaltulose is accomplished using *Protaminobacter rubrum* (ATCC No. 8457) (5). In addition to this strain, other microorganisms are capable of carrying out this transformation, e.g., *Serratia plymuthica* (8) and *Erwinia rhapontici* (9).

The immobilized cells of the bacterium are used to pack a reactor column. A concentrated sucrose solution is pumped through the column, where the sucrose is transformed into isomaltulose.

The clarified solution from the fermenter is concentrated by evaporation and is then crystallized in two stages. The first crop is obtained by

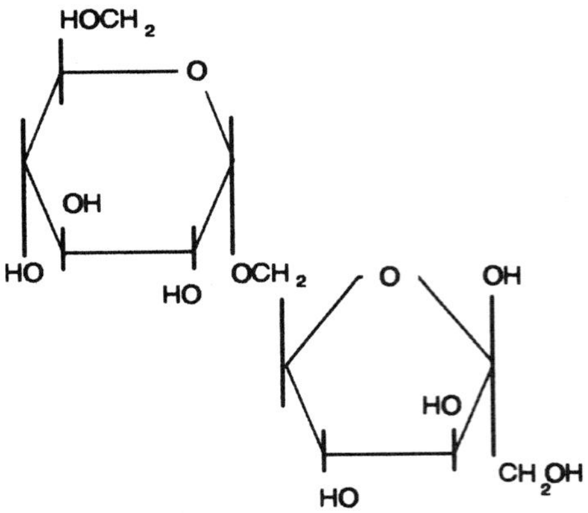


**Figure 1** Chemical structure of isomaltulose.

**Table 1** Food-Grade Isomaltulose: Typical Analysis

| Test | Results | Method |
|---|---|---|
| Appearance | White crystalline substance | FCC |
| Melting Range | 120–126°C | FCC |
| Water | 5.0–6.5% | FCC |
| Specific rotation $[\alpha]^{20}$ | + 97–104° (C = 2, water) | FCC |
| Ash | 0.02–0.10% (dry basis) | AOAC |
| Color of solution | max. 80 ICUMSA units | ICUMSA |
| Isomaltulose | NLT 98.0% | HPLC or GC |

FCC = Food Chemicals Codex.
ICUMSA = International Commission for Uniform Methods of Sugar Analysis.
AOAC = Associaton of Official Analytical Chemists.

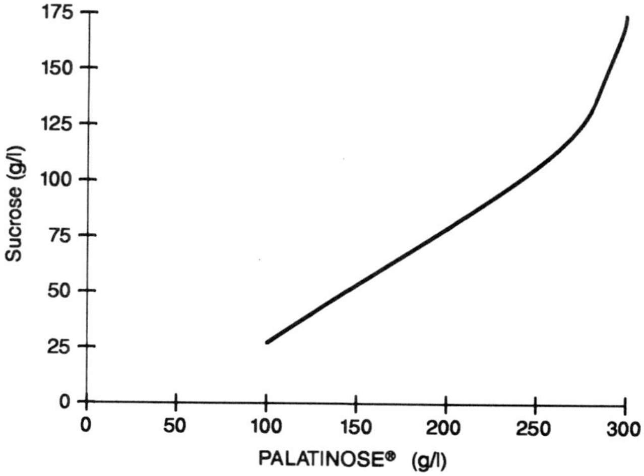


**Figure 2** Isosweet aqueous solutions of Palatinose(R) and sucrose at 20°C.

evaporation and the second by cooling. The crystals are dissolved in demineralized water and recrystallized. Isomaltulose crystallizes as a monohydrate.

The typical analysis of food-grade isomaltulose is given in Table 1.

## SENSORY PROPERTIES

The sweetness quality of isomaltulose is similar to sucrose. It is perceived quickly, it is refreshing and heavy without any aftertaste. Its sweetening power in comparison to a 10% sucrose solution is 0.48, but there is an increase in sweetening power as the concentration is increased, as seen in Figure 2 (10).

Isomaltulose is reported to mask the off-flavors of some intense sweeteners (11).

## PHYSICAL AND CHEMICAL PROPERTIES

### Physical Properties

The appearance of isomaltulose, a white crystalline material, is similar to that of sucrose. Its melting point (123–124°C) is lower than that of sucrose (190°C). It is practically nonhygroscopic. It is easily ground, an important property in many applications.

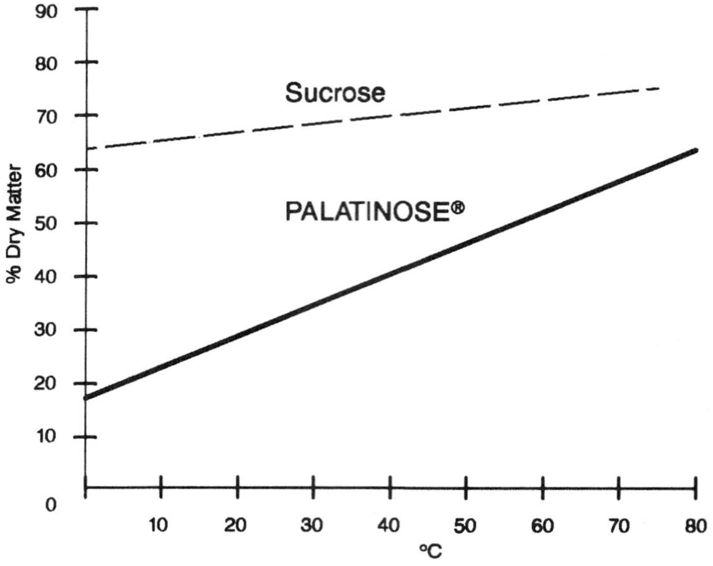


Figure 3 Solubility of Palatinose(R) and sucrose in water.

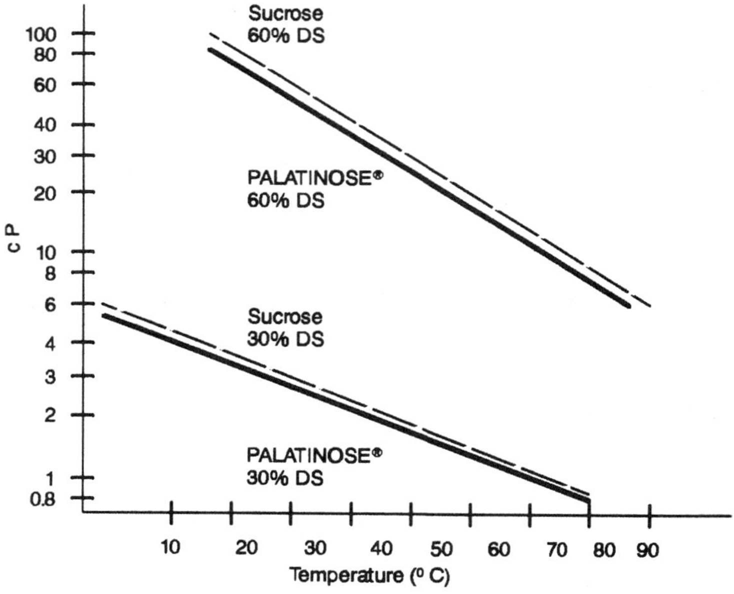


Figure 4 Viscosity of Palatinose(R) and sucrose solutions.

The solubility of isomaltulose, as seen in Figure 3, while lower than that of sucrose, is adequate for most applications. The viscosity of isomaltulose solutions shown in Figure 4 is very similar to that of sucrose solutions.

### Chemical Properties

Isomaltulose is very acid stable when compared with sucrose. At pH 2.0 with HC1, a 20% isomaltulose is stable for over 60 minutes when heated at 100°C, whereas a 20% sucrose solution is completely hydrolyzed under these conditions. Isomaltulose is a reducing sugar and as such undergoes Maillard reactions. The microbial stability of products prepared with isomaltulose is very good.

## METABOLISM

Isomaltulose has been shown to be hydrolyzed to glucose and fructose by intestinal $\alpha$-glucosidases (12). The relative cleavage rate of isomaltulose by homogenates of jejunal mucosa is 44 compared to a sucrose rate of 100. Relative cleavage rates of 30 and 100 are found with rat tissue (13).

Experiments with rats have shown that the metabolism of isomaltulose is independent of the size of the dose and sex of the animals, and compares in all respects to that of sucrose. It was concluded in the study by Macdonald and Daniel (14) that the $V_{max}$ for the enzymatic hydrolysis of isomaltulose is about half that for sucrose.

In a study with pigs, one group of animals was provided with a reentrant fistula in the end of the ileum, which made possible the examination of the enzymatic gastrointestinal digestion process in the small intestine separately from the fermentative digestive processes in the large bowel. A second group of nonfistulated animals were used to study digestibility over the whole gastrointestinal tract. Study diets included either 20% sucrose or isomaltulose. Analysis of the ileal chyme from the fistulated pigs showed that there was no passage of isomaltulose (<0.3%) at the terminal end of the ileum. Both sugars were almost completely digested and absorbed in the small intestine; fermentation in the large intestine was not a significant factor (15).

### Cariogenicity

Isomaltulose has been shown to be less cariogenic than sucrose. Ciardi has reported that isomaltulose does not support growth by oral streptococci and, in addition, has the added advantage of inhibiting bacterial growth and/or glucan synthesis (16). A similar conclusion was reached by Gehring (17), who reported the final pH values of nutrient media containing different sugar substitutes after incubation with 100 *Streptococcus mutans* strains. These results have been confirmed in vitro by pH-telemetric studies according to Muhlmann with pure isomaltulose and isomaltulose based chewing gum (18,19).

It has been reported that when isomaltulose was added to reaction mixtures which contained various concentrations of sucrose, a significant reduction in the yield of insoluble glucan was noted. With the addition of 4% isomaltulose, insoluble glucan synthesis in a 1% sucrose solution was reduced from 100 to 4%. The addition of 0.125% isomaltulose reduced it to approximately 50% (20). The competition of isomaltulose in glucosyl reception is believed to cause the reduction in the amount of insoluble glucan produced from sucrose (21).

### Effect on Blood Glucose and Insulin

Either isomaltulose or sucrose, in doses ranging from 0.25 to 1.0 g/kg of body weight, was given to healthy adult males. The insulin and fructose concentrations in the serum measured after isomaltulose ingestion were approximately half of those measured after sucrose ingestion. At all doses the rate of rise of blood glucose, insulin, and fructose levels and the peak values were lower for isomaltulose than for sucrose. When the area under the glucose response curve was calculated, there was no significant difference between sucrose and isomaltulose, except at the highest dose, whereas the area under the insulin response curve was lower in the case of isomaltulose at all dose levels. As a result of the

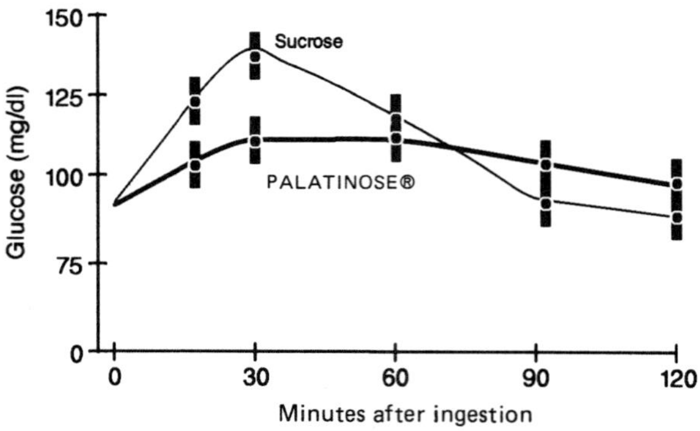


**Figure 5** Plasma glucose levels after oral administration of 50 g sucrose or Palatinose(R) in man.

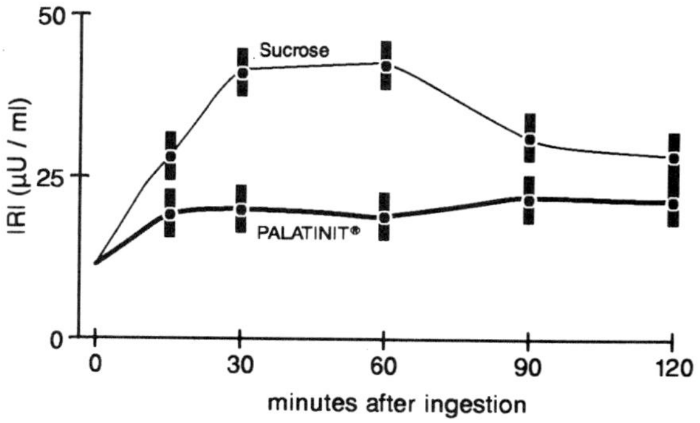


**Figure 6.** Plasma insulin levels after oral administration of 50 g sucrose or Palatinose[(R)] in man.

slower cleavage rate and therefore slower absorption of the monosaccharides from isomaltulose, no rise in any intestinal discomfort was observed. This is compatible with the view that isomaltulose is completely absorbed, although more slowly than is sucrose (14). These results are similar to those reported by Kawai et al. (22) as seen in Figures 5 and 6.

### Tolerance

No evidence of intestinal discomfort has been observed in human studies with isomaltulose (14,22).

### Toxicological Evaluations

The monomers and chemical bonds comprising the chemical structure of isomaltulose are the same as those found in sucrose or starch. It is not surprising that only a limited number of toxicological studies have been conducted. No evidence of toxicity was noted in a 26-week toxicological study conducted with rats (23).

## APPLICATIONS AND PRODUCT DEVELOPMENT

In the manufacture of most foods, isomaltulose can be substituted for sucrose without any changes in the traditional manufacturing processes. Product development studies have been successful in many applications (24). The following applications are representative of these studies.

| | |
|---|---|
| Anticaking powder | Frostings |
| Baked goods | Hard and soft candy |

| | |
|---|---|
| Canned fruits | Puddings |
| Chewing gum | Sports drinks |
| Chocolate products | Toothpaste |
| Confections | |

The solubility of isomaltulose is adequate for most applications. An exception is in the manufacture of traditional jams and jellies, where the lower solubility of isomaltulose does not allow the production of high solids products. Isomaltulose is suitable for use in the manufacture of the newer "light" fruit spreads.

## REGULATORY STATUS

In Japan isomaltulose (Palatinose$^{(R)}$) is considered a food and is marketed as such. Regulatory status has not yet been established in other countries.

## REFERENCES

1. R. Siddiqua and B. Furhala, Isolation and characterization of oligiosaccharides from honey, *J. of Apicultural Res.*, *6* (3): 139 (1967).
2. I. Takazoe, New trends on sweeteners in Japan, *International Dental J.*, *35:* 58 (1985).
3. F. H. Stodala, H. J. Hoepsell, und E. S. Sharpe, Ein neues von *Leuconostoc mesenteroides* gebildetes Disaccharid, *J. Am. Chem. Soc.*, *74:* 3202 (1952).
4. R. Weidenhagen und S. Lorenz, Palatinose (6-α-glucopyranosido-fructofuranose), ein neues bakterielles Umwandlungsprodukt der Saccharose, *Zeitschr. Zuckerindustrie*, *7:* 533 (1957).
5. H. Schiweck, Palatinit—Herstellung, technologische Eigenschaften und Analytik Palatinit haltiger Lebensmittel, *Alimenta*, *19:* 5 (1980).
6. F. Loss, Über einige Derivate der Palatinose (Isomaltulose), zugleich ein Beitrag zur Frage der Konstitution, *Zeitschr. Zuckerindustrie*, *19:* 323 (1969).
7. V. W. Dreissig und P. Lugar, Die Strukturbestimmung der Isomaltulose, $C_{12}H_{22}O_{11}H_2O$, *Acta Cryst. ect. B*, *29:* 514 (1973).
8. S. Schmidt-Berg-Lorenz und W. Mauch, Ein weiterer Isomaltulose bildender Bakterienstamm, *Zeitschr. Zuckerindustrie*, *14:* 625 (1964).
9. P. S. J. Cheetham, C. E. Imber, and J. Isherwood, The formation of isomaltulose by immobilized *Erwinia rhapontici*, *Nature*, *299:* 628 (1982).
10. K. Paulus, Abschlussbericht, sensorische Untersuchungen zur Süßkraft von Palatinose, private communication, May 28, 1985.
11. Anonymous, Low cariogenic sugar with a bright future; Palatinose, *Food Engineering*, *57* (5): (1985).
12. A Dahlquist, S. Auricchio, G. Semenza, and A. Prader, Human intestinal disaccharidases and hereditary disaccharide intolerance. The hydrolysis of sucrose, isomaltulose (Palatinose), and a 1,6-α-oligiosaccharide (isomalto-oligosaccharide) preparation, *J. Clin. Invest.*, *42:* 556 (1963).

13. U. Grupp and G. Siebert, Metabolism of hydrogenated Palatinose, an equimolar mixture of $\alpha$-D-glucopyranosido-1,6-sorbitol and $\alpha$-D-glucopyranosido-1,6-mannitol, Res. Exp. Med. (*Berl.*), *173:* 261 (1978).
14. I. Macdonald and J. W. Daniel, The bioavailability of isomaltulose in man and rat, *Nutr. Rep. Intern., 28:* 1083 (1983).
15. I. E. R. van Weerden, L. J. Huisman, and P. Leeuwen, *Digestion processes of Palatinose*$^{(R)}$*and saccharose in the small and large intestine of the pig,* ILOB-report 520, 20 June, 1983.
16. J. Ciardi, W. H. Bowen, G. Rolla, and K. Nagorski, Effect of sugar substitutes on bacterial growth, acid production and glucan synthesis, *J. Dent. Res., 62:* 182 (1983).
17. Gehring, Über sie Säurebildung kariesätiologisch wichtige Streptokokken aus Zuckern und Zuckeralkoholen unter besonderer Berücksichtigung von Isomaltit und Isomaltose, *Ztschr. für Ernährungswiss., Suppl. 15:* 16 (1973).
18. H. R. Mühlmann, private communication, Zahnärztliches Institut der Universität Zürich, 1983.
19. T. Imfeld, private communication, Zahnärztliches Institut der Universität Zürich, 1983.
20. I. Takazoe, G. Frostell, K. Ohta, V. Topitsoglou, and N. Saski, Palatinose—a sucrose substitute, *Swed. Dent. J. 9:* 81 (1985).
21. Y. Nakajima, M. Takaoka, K. Ohta, and I. Takazoe, Inhibitory activity of Palatinose on the synthesis of insoluble glucan from sucrose by *Streptococcus mutans, Seito gijutsu kenkyukaishi, 34:* 58 (1985).
22. K. Kawai, Y. Okuda, and K. Yamashita, Changes in blood glucose and insulin after an oral Palatinose administration in normal subjects, *Endocrinol. Japan, 32* (6): 933 (1985).
23. K. Yamaguchi, S. Yoshimura, H. Inada, E. Matsui, T. Ohtaki, and H. Ono, 26-week Palatinose toxicity study with rats through oral administration, *Pharmacometrics 31* (5): 1015 (1986).
24. P. J. Sträter, Palatinose, *Zuckerindustrie, 112* (10): 900 (1987).

# 17

# Isomalt

**Peter J. Sträter** *Palatinit Süßungsmittel GmbH, Mannheim, Germany*

**William E. Irwin** *Palatinit Süßungsmittel GmbH, Elkhart, Indiana*

## INTRODUCTION

### Properties of an Ideal Sugar Substitute

When alternate sweeteners replace sucrose and/or glucose as ingredients in a food or beverage, they ideally should have certain properties: first, the new sweetener should confer body and texture and a sweet flavor without any aftertaste; second, they should perform essentially the same functions as sucrose or glucose in the food or beverage; and third, they should, when compared to sucrose, undergo different physiological and nutritional metabolism in the gastrointestinal tract and biochemical changes in the mouth.

The new ideal sweetener should be stable chemically and biologically for an indefinite period of time and should provide the same properties and undergo the same processes as sucrose. This is critical for the industrial production of foods and beverages because the raw material must be transported, warehoused, and distributed easily inside the factory; ground, dissolved, heated, sterilized, or melted without appreciably changing its molecular structure; and maintained under specific conditions in a solidified or melted state and recrystallized if required.

A further desired characteristic would be that existing processing equipment could be used for all of these operations without requiring

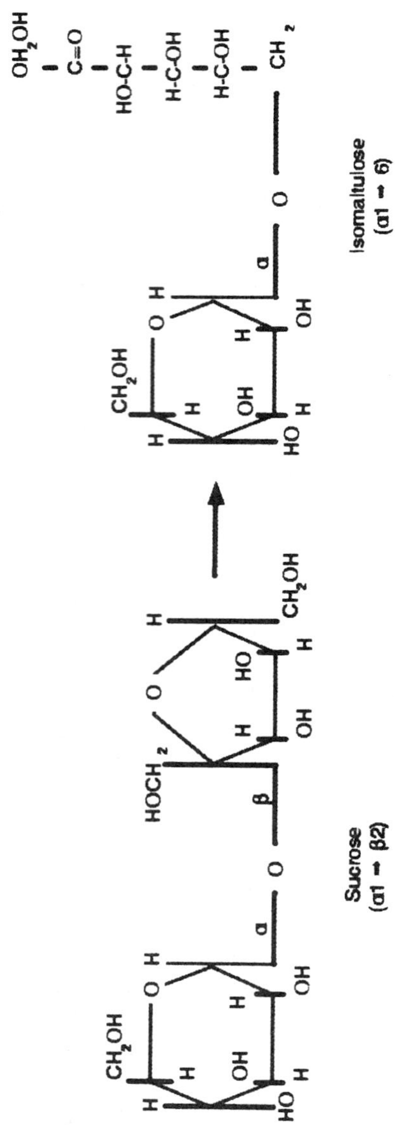


Figure 1 The enzymatic conversion of sucrose to isomaltulose.

major changes. In addition the finished products should have practically the same taste and appearance as those of traditional products and should be readily accepted by the consumer. They should have an extended shelf life as well.

Isomalt is a sweet, low calorie bulking agent with properties and characteristics similar to sucrose. It is odorless, crystalline, and nonhygroscopic (1,2). It is a nonreducing sugar and is optically active. However, unlike sucrose, it is extremely stable with respect to chemical and enzymatic hydrolysis, and it cannot be fermented by a large number of yeasts and other microorganisms found in nature. The molecular structure of isomalt has been established by Gau et al. (3).

Isomalt (Palatinit®) is manufactured and marketed by Palatinit® Süßungsmittel GmbH, a wholly owned subsidiary of Südzucker AG.

## PRODUCTION

Isomalt is produced in two steps:

1. The enzymatic transglucosidation of $\alpha$-D-glucopyranosyl (1 → 2)-ß-D-fructofuranoside (sucrose, saccharose) into $\alpha$-D-glucopyranosyl (1 → 6) fructose (isomaltulose, Palatinose®. See Chem. Abstracts no. 13718-94-0). By changing the disaccharide bond from (1 → 2) to (1 → 6), as seen in Figure 1, a reducing disaccharide, isomaltulose, is obtained from a nonreducing disaccharide, sucrose. Isomaltulose is obtained in crystalline form and crystallizes with one mole of water of crystallization.
2. The hydrogenation of isomaltulose in a neutral aqueous solution (pH 6–8) by hydrogen with Raney nickel as a catalyst, as seen in Figure 2, yields an equimolar mixture of $\alpha$-D-glucopyranosyl-1,6-

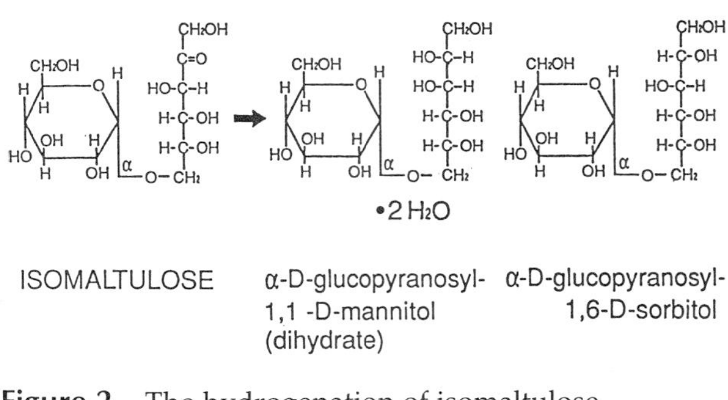


**Figure 2** The hydrogenation of isomaltulose.

D-sorbitol (GPS. See Chem. Abstracts no. 534-73-6) and α-D-glucopyranosyl-1,1-D-mannitol (GPM. See Chem. Abstracts no. 20942-99-8). In aqueous systems, GPS forms anhydrous crystals and GPM has two molecules of water of crystallization. As a result, isomalt contains about 5% bound water. Additional details regarding the production process have been reported by Schiweck (4). The current specifications for Palatinit® are shown in Table 1.

**Table 1** Palatinit®—Current Specifications

| Test | Specification | Method |
|---|---|---|
| Description | White, odorless, sweet, crystalline substance | |
| Water (free and hydrated) | max 7.0 % | FCC |
| GPS content (based on total solids) | 43–57% | HPLC or GC |
| GPM (based on total solids) | 43–57% | HPLC or GC |
| D-Mannitol (based on total solids) | max 0.5% | HPLC or GC |
| D-Sorbitol (based on total solids) | max 0.5% | HPLC or GC |
| Reducing substances (calc. as isomaltulose) | max 0.8% | ICUMSA |
| Melting range | 145–150°C | FCC |
| Color of solution | max 75 ICUMSA-Units | ICUMSA |
| Sulfated ash | max 0.05% | AOAC |
| Arsenic | max 0.2 ppm | FCC |
| Total heavy metals (as lead) | max 10 ppm | FCC |
| Nickel | max 2.0 ppm | AOAC |
| Solubility | soluble in water Insoluble in ethanol | |

FCC = Food Chemicals Codex.
ICUMSA = International Commission for Uniform Methods of Sugar Analysis.
AOAC = Association of Official Analytical Chemists.

## SENSORY PROPERTIES

### Sweetness and Taste

The sweetening power of isomalt varies with the concentration, temperature, and form of the product. Paulus and Fricker (5) have studied the sweetening power of isomalt solutions at low concentrations. Figure 3 plots the sweetening power of Palatinit® solutions versus concentration, in comparison with sucrose isosweetened solutions. Figure 3 indicates that the relative sweetness of Palatinit® increases from 0.45 to 0.64 as the concentration in solution increases from 100 to 500 g/liter.

Isomalt has a pure sweet taste similar to sucrose without any aftertaste. Isomalt reinforces flavor transfer in foods. Synergistic effects occur when isomalt is combined with other sugar alcohols, e.g., xylitol, sorbitol, mannitol, hydrogenated glucose syrup, etc. When 10% of these sugar alcohols are added to isomalt, a sweetening power comparable to that of sucrose is obtained. Similar synergistic effects are observed when isomalt is used in combination with high intensity sweeteners such as

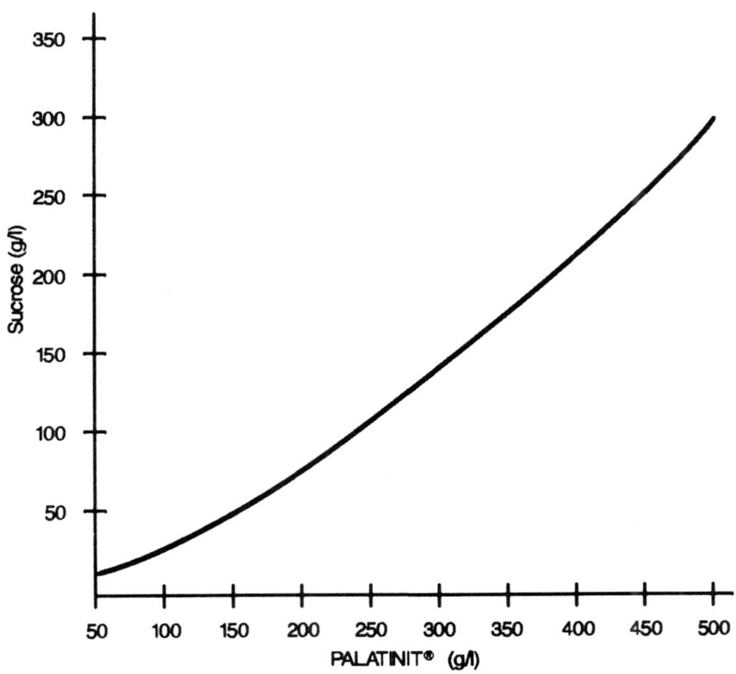


**Figure 3** Isosweet aqueous solutions of Palatinit® and sucrose.

**Table 2** Heat of Solution of Various Sugars

| | kJ/mol | kJ/kg |
|---|---|---|
| Sucrose[a] | −6.21 | −18.16 |
| Isomaltulose | −7.80 | −21.58 |
| Sorbitol[b] | −20.20 | −110.99 |
| Mannitol[b] | −22.00 | −120.88 |
| Xylitol[a] | −23.27 | −153.07 |
| GPM | −21.82 | −57.42 |
| GPS | −7.37 | −21.42 |

[a]From Ref. 7.
[b]From Ref. 8.

acesulfame-K, aspartame, cyclamate or saccharin. Additionally isomalt tends to mask the bitter metallic aftertaste of some intense sweeteners.

### Cooling Effect

Several sugar alcohols, used as sugar substitutes, have a very high negative heat of solution (6–8). A greater or lesser degree of a mouth-cooling effect is experienced when these products dissolve, in the mouth, from a crystalline or solid state. This mouth-cooling effect is desirable for peppermint or menthol products but is considered undesirable for many other products, such as baked goods and chocolates. Isomalt has a very low negative heat of solution (Table 2), comparable to that of sucrose, and therefore does not produce a cooling effect.

## PHYSICAL AND CHEMICAL PROPERTIES

### Physical Properties

Isomalt is an equimolecular mixture of two stereoisomers, $\alpha$-D-glucopyranosyl-(1 → 6)-sorbitol (GPS) and $\alpha$-D-glucopyranosyl-(1 → 1)-mannitol (GPM). These stereoisomers are in crystalline form. Isomalt is homogeneous and nonhygroscopic. Therefore, it can be transported, stored, and distributed without any special precautions.

The sintering point of isomalt is between 145 and 150°C. It can be very easily ground to obtain 100 $\mu$m granules. This property is important for the manufacture of chocolates and chewing gum. Larger agglomerates can easily be formed from finely ground isomalt. This product is suitable for use in instant products and can be molded into tablets.

If isomalt is heated above its melting point until the water content is reduced to below 2%, a liquid of low viscosity results, which will set to a transparent melt if cooled rapidly. This melt does not recrystallize, even after being stored for more than 2 years. The surface of the melt does not become sticky, because it is not hygroscopic, an ideal property in hard candies.

Figure 4 plots the boiling temperature of Palatinit® and sucrose masses as a function of water content. It can be seen that above 86% solids, a higher temperature is required with isomalt-melted masses to obtain a given solids content than is needed with sucrose masses. This is probably due to the presence of the relatively stable bond with the water of crystallization in the GPM crystal. To reduce the moisture content of a melted isomalt mass to less than 2%, a temperature of about 160°C is necessary. At this temperature, there is no change in the molecular structure of isomalt. There will be only a slight change in the color of the melt, even if the melt is held at this temperature for a relatively long time.

Isomalt is considered nonhygroscopic. The absorption isotherms seen in Figure 5 show that at 25°C isomalt absorbs very little water below

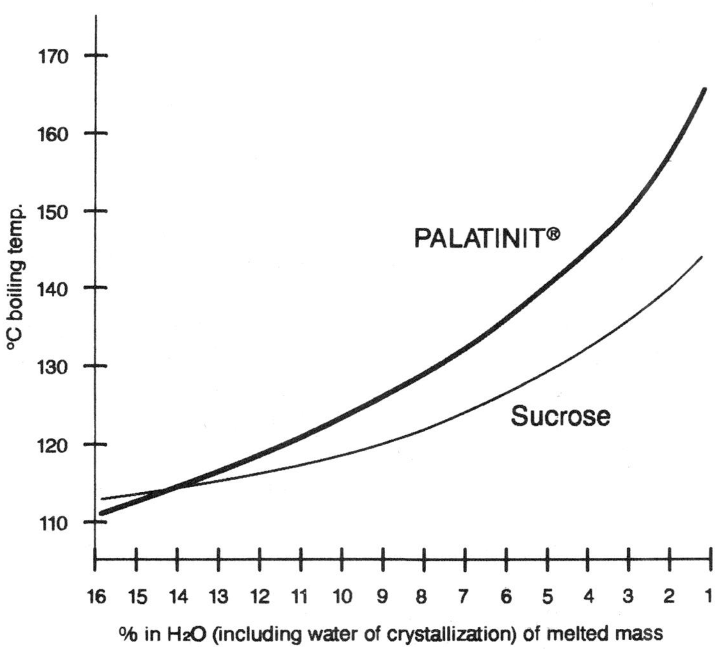


**Figure 4** Boiling temperatures of melted Palatinit® and sucrose masses as a function of water content.

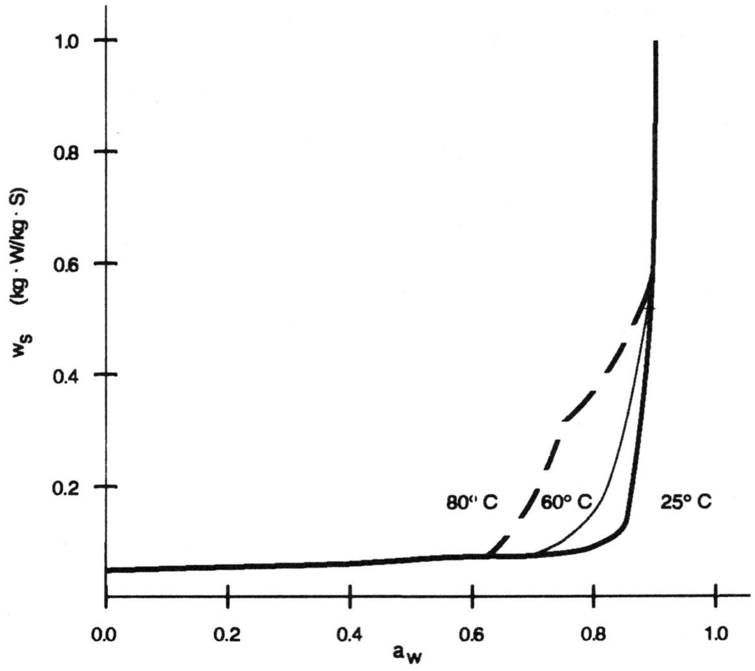


**Figure 5** Absorption isotherms of isomalt at various temperatures.

85% relative humidity. At temperatures of 60 and 80°C, the relative humidity must be 75 and 65%, respectively, for significant water absorption. Isomalt may, therefore, be warehoused for extended periods of time under normal conditions.

If a melt of isomalt with a relatively high water content (> 10%) is cooled or if saturated aqueous solutions of isomalt at a high temperature are stirred or agitated, then GPM, because of its lower solubility, crystallizes out first, forming fine crystals. The higher solubility of GPS increases the percentage of solids in the aqueous phase; therefore, isomalt can be used as a melt for the manufacture of soft caramels, chewing gums, and soft candies. For this application, due to the water content (< 10%), the addition of a crystallization inhibitor such as hydrogenated glucose syrups may be desirable. The prompt crystallization of isomalt simplifies the coating of chewing gum, hard boiled and chewable candies. In this instance, one of the applications might be to enhance the shelf life of products that are generally hygroscopic.

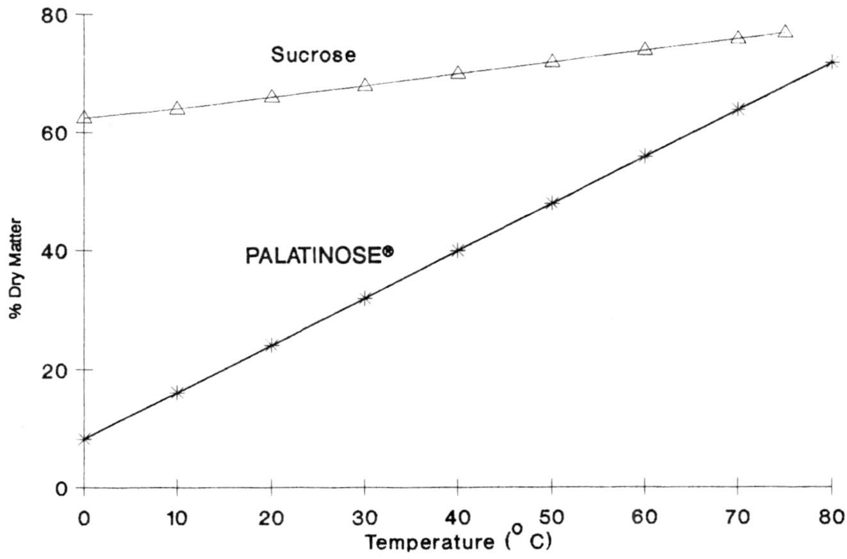


**Figure 6** Solubility of Palatinit® in water.

The solubility of Palatinit® in water is shown in Figure 6. It is evident that the solubility of isomalt in water is adequate for most applications. In the manufacture of jellies and jams, the solids content should not exceed 26% in the product in order to avoid the recrystallization of isomalt. The solubility of Palatinit® in water and ethanolic solutions (20–60°C) is shown in Figure 7. As the percentage of alcohol increases, the solubility of isomalt decreases almost linearly. Isomalt crystallizes from hot supersaturated water and ethanolic solutions on the container walls in a thin layer of fine crystals, which prevents the passage of solvents and flavors. Due to this property, liquor-flavored centers (crusted liqueurs) can be produced from supersaturated isomalt solutions.

The viscosity of aqueous solutions of Palatinit® as shown in Figure 8 is comparable to that of sucrose solutions.

## Chemical Properties

GPS and GPM are very stable chemically and, therefore, so is isomalt. If the crystalline substance is heated above the melting point or if the aqueous solutions are heated above the boiling point, no changes in the molecular structure are observed in either sugar alcohol. Also, GPS and

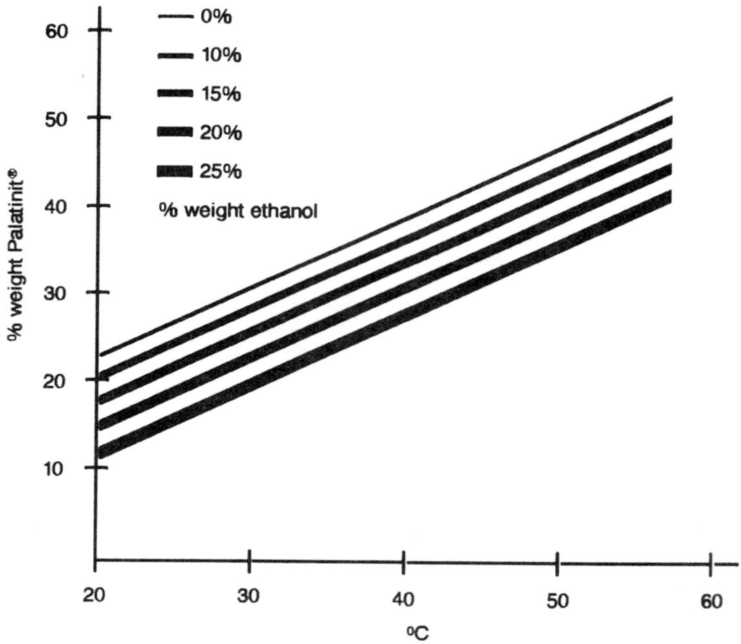


**Figure 7** Solubility of Palatinit® in ethanol solutions.

GPM do not react with other ingredients in food such as amino acids that can cause a Maillard reaction. Reducing agents such as $SO_2$ are not needed to inhibit the browning of isomalt-sweetened foods.

The energy required to hydrolyze a disaccharide, either by acid hydrolysis or enzymatic hydrolysis, is directly proportional to the energy of the bond linking the two saccharides. The energy required to hydrolyze two monosaccharides linked by their glycosidic hydroxyl groups, as in sucrose, is very low compared to that required to hydrolyze (1 → 6) linkages, such as found in isomaltulose, gentiobiose, and isomalt.

In the production of hard boiled candy, acid is added during the cooking phase. Therefore hydrolysis of isomalt was studied after the addition of acid to the melts (15 g of citric acid/kg of melted mass). Under normal conditions there is no hydrolysis. Even when the melt is held at a temperature of 150°C for 30 min after acid has been added, the percentage of mannitol and sorbitol is less than 1%. Isomalt is very stable under these conditions.

Sucrose dissolved in a 0.01 N solution of hydrochloric acid at 100°C is completely hydrolyzed in less than 30 min, whereas GPS is completely

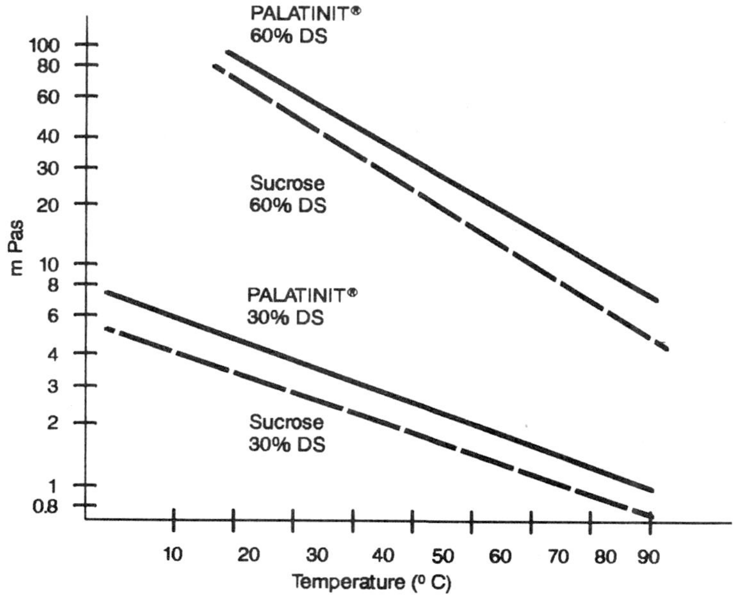


**Figure 8** Viscosity of Palatinit® and sucrose solutions.

hydrolyzed in a 1.0 N solution of hydrochloric acid at 100°C only after 2 hr. When GPS is placed in 0.1 N solution of hydrochloric acid at 100°C, only two thirds is hydrolyzed after 10 hr (Fig. 9). Under the severest conditions that can be encountered during food production (pH 2; heating at 100°C for more than 1 hr), less than 10% of GPS will be hydrolyzed. The disaccharide bond of GPM is even more stable than that of GPS. GPM dissolved in 1.0 N solution of hydrochloric acid at 100°C requires more than 3 hr for complete hydrolysis to occur.

GPS and GPM are cleaved by α-D-glucosidases; however, the disaccharide bond of GPM and GPS has been shown to be very resistant to enzymatic hydrolysis. Therefore the rate of hydrolysis is slow. These disaccharide alcohols must be cleaved to monosaccharides (glucose, sorbitol, and mannitol) before microbiological activity will occur. It is not surprising, therefore, that isomalt is not a substrate for most microorganisms (yeast, mold, and bacteria) associated with food and beverages. The cleavage rate of GPS and GPM by yeast α-glucosidases is approximately $10^3$ less than that of maltose (9). Under normal conditions, the enzymatic cleavage of GPS and GPM can be considered negli-

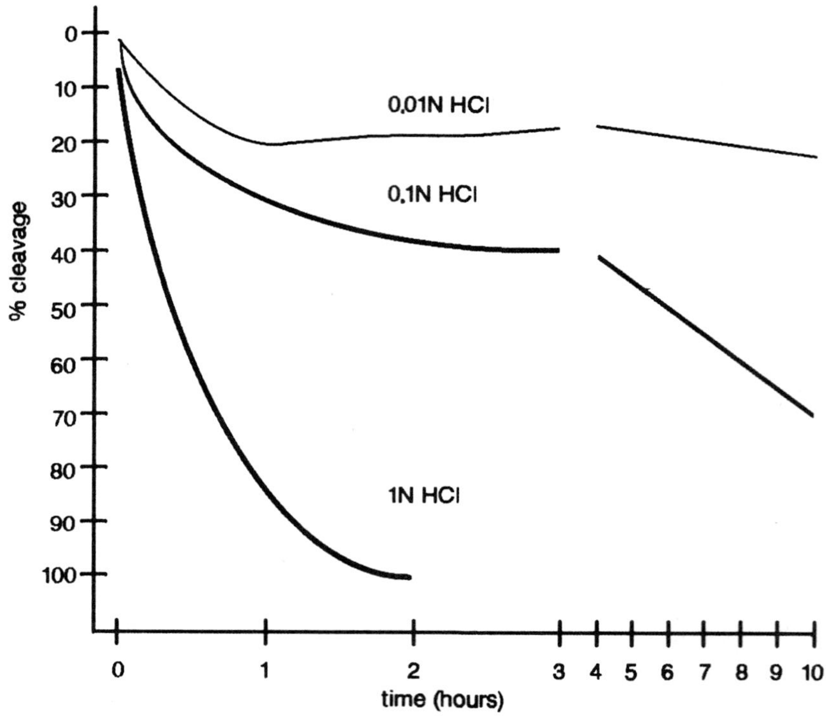


**Figure 9** Hydrolysis of GPS over time at different acid concentrations.

gible. Products manufactured with isomalt exhibit high microbiological stability (10).

To assess the stability of carbohydrates in alkaline medium, the changes in solution color and acid (lactic) formation were studied. In Figure 10, curves are plotted showing the increase in color, measured at 420 mm, versus time of a 10% Palatinit® solution mixed with 0.01 N, 0.1 N, and 1.0 N solutions of sodium hydroxide, at 100°C, as compared with color change of a 10% sucrose solution. The curves show that isomalt solutions change color only during the first hours. After that a reduction in the extinction coefficient is observed. On the other hand, the increase in the color of sucrose solutions is almost exponential. In addition, the magnitude of the color changes in the isomalt solutions is significantly less (in absolute values) than in the sucrose solutions.

As illustrated in Figure 11, the chromophores formed with Palatinit® in an alkaline medium appear to be similar to those formed with sucrose.

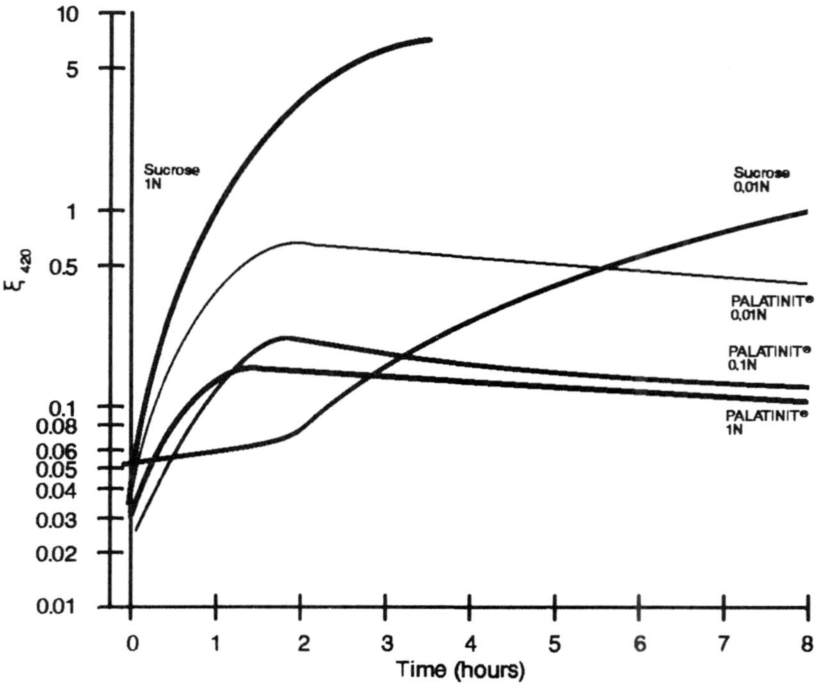


**Figure 10** Discoloration of a 10% Palatinit® solution at 100°C in NaOH solutions, measured at 420 nm, with a 10% sucrose solution as the control.

Absorption curves obtained with solutions of isomalt or sucrose, in an alkaline medium, plotted against wave length, from 200 to 500 nm, are shown in Figure 11. Maximum absorption was observed at 266 nm for both isomalt and sucrose. Von Prey and Andres have obtained a similar value for this maximum for sucrose in an alkaline medium (11). The differences observed in the extinction coefficients indicate that isomalt is more stable in alkaline solutions than sucrose.

The amounts of lactic acid found after holding isomalt or sucrose alkaline solutions for 7½ hours at 100°C also indicate that there is a difference in alkaline resistance between isomalt and sucrose. The percentage of D- and L-lactic acid in the isomalt solutions is less than 0.01% for all NaOH concentrations. The percentage of D- and L-lactic acid in sucrose solutions treated with a 1 N solution of sodium hydroxide is, after 4 hr, greater than 0.1%. These values are 10 times higher than those obtained with corresponding isomalt solutions (Table 3).

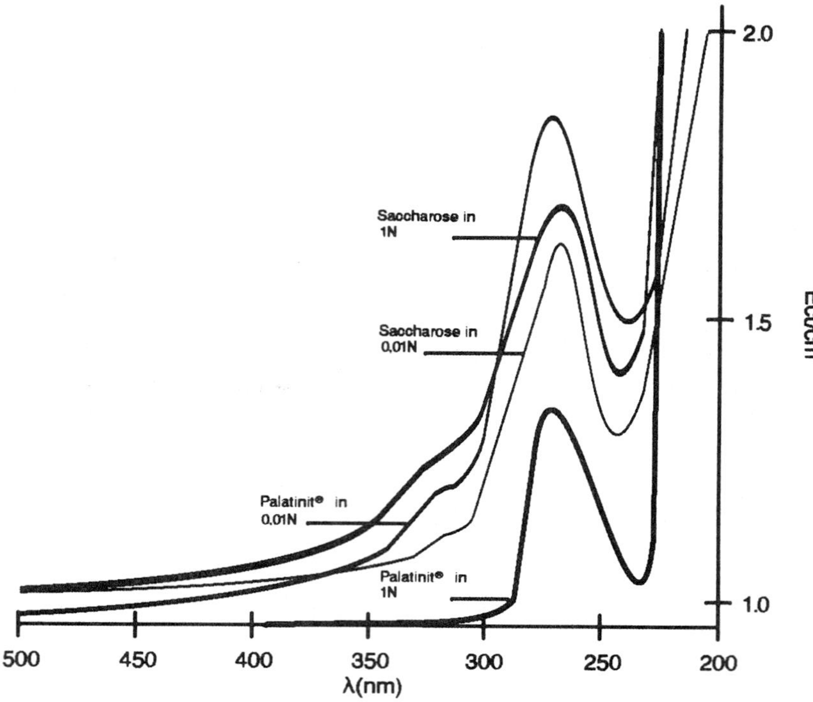


**Figure 11** Absorption curves of Palatinit® and sucrose solutions after 7 hr at 100°C.

**Table 3** The Formation of Lactic Acid by the Cleavage of 10% Palatinit® or Sucrose Solutions after 7.5 hr Incubation with Various NaOH Solutions at 100°C

| | | Lactic acid (%) | |
|---|---|---|---|
| | NaOH (N) | D form | L form |
| Palatinit® | 1.00 | 0.008 | 0.009 |
| Palatinit® | 0.10 | 0.005 | 0.006 |
| Palatinit® | 0.01 | 0.006 | 0.006 |
| Sucrose | 1.00[a] | 0.120 | 0.180 |
| Sucrose | 0.01 | 0.005 | 0.005 |

[a]Experiment interrupted after 4 hr.

## METABOLISM

A property of primary importance in the metabolism of isomalt is the slow cleavage rate of its disaccharide bond in comparison to other water-soluble carbohydrates (12). The cleavage rate is low with most chemical and biochemical agents over a wide range of pH and temperatures. This low cleavage rate is significant in the human digestive tract, which must break down the higher saccharides to monosaccharides before they can be absorbed in the small intestine and subsequently metabolized. The relative activity of maltase/glucoamylase and sucrase/isomaltase enzymes, isolated from human small intestine mucosa, on a isomalt substrate is less than 4% of that on a sucrose and/or maltose substrate (13). Acid production in the mouth occurs only after the bacteria present cleave ingested carbohydrates. In a number of experiments, as will be discussed, it has been shown that in comparison to sucrose, isomalt has the following properties after oral ingestion:

Only about 50% of isomalt ingested is converted into energy.
Isomalt is noncariogenic.
Isomalt is suitable for use by diabetics.

### Cariogenicity

The indigenous bacteria in the mouth require the presence of fermentable sugars to produce lactic acid, which is the principal cause of tooth decay. If fermentable sugars are not present in the mouth, the bacteria does not produce this acid. Isomalt cannot be used as a substrate by these bacteria, and as a result it does not lead to acid formation in the mouth.

The experiments of Karle and Gehring (14) with normal and xerostomized (surgical removal of salivary glands) rats show (Fig. 12) the low cariogenicity of isomalt in comparison with several other carbohydrates. These results were confirmed in a number of in vitro and in vivo experiments with animals and man (15,16) and in pH-telemetric studies, according to Muhlemann and Imfeld (17). The conclusions of these studies are that isomalt is not a substrate for plaque formation by microorganisms in the mouth, acid formation is minimal, and it does not influence oral microflora.

### Effect on Serum Glucose and Serum Insulin

The effects of isomalt were studied in a number of clinical trials. The results of earlier studies by Grupp and Siebert (9) with healthy volunteers and by Drost et al. (18) with diabetic patients have been confirmed

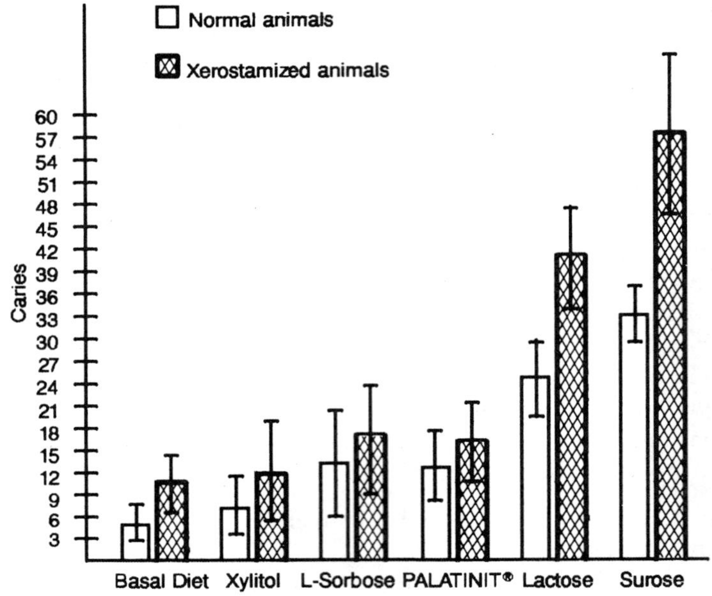


**Figure 12** Caries incidence in rats after programmed feeding of different carbohydrates. (Fre. Ref. 14.)

by other investigators. The changes in serum insulin and blood glucose reported by Thiebaud et al. (19) are shown in Figures 13 and 14. The conclusion is that ingestion of isomalt causes insignificant changes in serum insulin and blood glucose levels in man.

## Caloric Value

For carbohydrates that are not completely metabolized and absorbed in the small intestine, there are no standardized methods today for assessing the energy released and therefore their nutritive value. In order to evaluate the energy value of isomalt, a wide variety of in vitro and in vivo studies have been conducted at various research institutions. These studies have been reviewed by Sträter and Sentko (20) and Berschauer and Spengler (21).

These studies have shown that as a result of its slow cleavage rate, in comparison to maltose, only a small portion of ingested isomalt is absorbed, as glucose, mannitol, and sorbitol, in the small intestine (9). Additionally, inhibitory effects on the hydrolysis of maltose and sucrose were observed (12). The unabsorbed isomalt is transported to the large

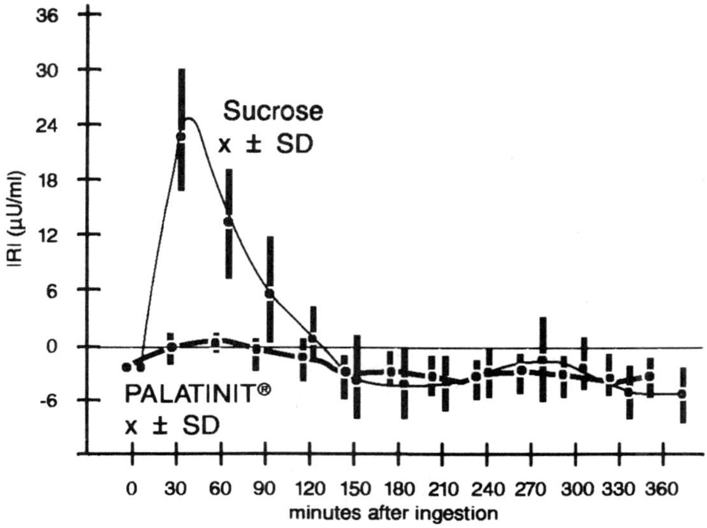


**Figure 13** Serum insulin levels (difference compared with initial value, $n = 10$) after oral administration of 30 g sucrose or 31.6 g Palatinit® in man.

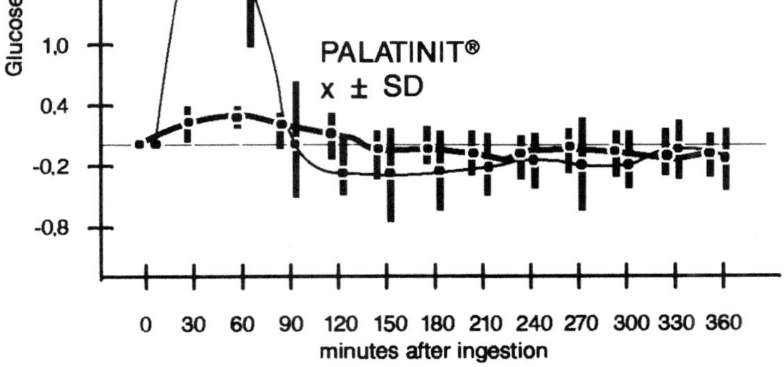


**Figure 14.** Blood glucose levels (difference compared with initial value, $n = 10$) after oral administration of 30 g sucrose or 31.6 g Palatinit® in man.

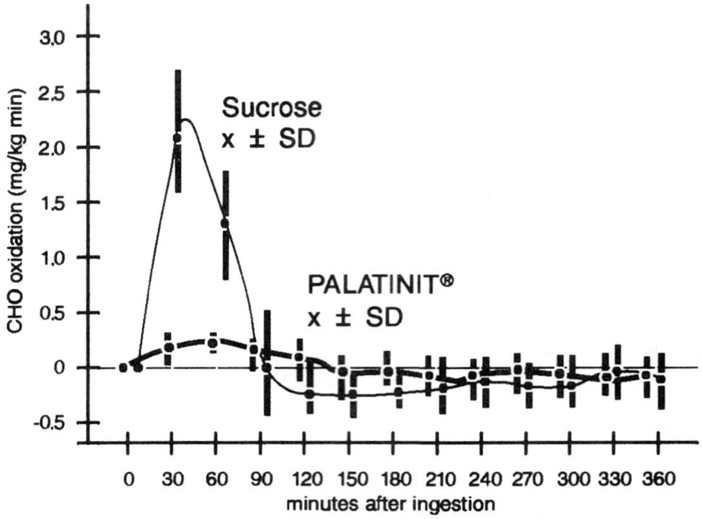


**Figure 15** Carbohydrate oxidation in man versus time after ingestion of 30 g of Palatinit® or sucrose.

intestine, where it is fermented by the microflora present (22), resulting in the production of carbon dioxide, hydrogen, methane, and volatile fatty acids. An unknown portion of the free fatty acids are absorbed in the colon and used for growth and maintenance requirements.

In the case of isomalt, a wide variety of studies have shown that its apparent digestibility is significantly lower than other carbohydrates, as evidenced by a greater percentage of biomass in the feces. Carbohydrate oxidation is almost 60% lower for Palatinit® than for sucrose after the administration of 30 g in man (19), as seen in Figure 15.

It has been concluded by Berschauer and Spengler (21) that from the calculation of the energy utilization of isomalt based on comparative carcass experiments with rats and pigs as well as on respiration experiments with man and pigs, that for food labeling purposes an energy value of 2 kcal/g for isomalt seems to be justified. This value has been adopted by various government regulatory agencies that regulate the caloric value of specific food substances in countries such as Switzerland, France, Austria, Luxembourg, and Australia.

### Gastrointestinal Tolerance

The tolerance of isomalt has been closely monitored in many studies. In addition, a number of trials designed to study tolerance in particular have been conducted (23–26). These studies have shown that:

1. The relative tolerance of isomalt and other sugar alcohols depends, in large measure, upon the form in which the product is ingested. The osmotic load after intake of isomalt in a liquid food is higher than after the intake of the same quantity in a solid form.
2. Chronic ingestion of isomalt, when adapted to the dose, does not cause either increased diarrhea, flatulence, or metabolic disturbance in experimental animals.
3. Isomalt is at least as well tolerated by nonadapted humans as comparable doses of other well-known sugar alcohols. The osmotic load after the intake of a disaccharide alcohol, like isomalt, is one half of that after the intake of a monosaccharide alcohol. The tolerance of substances like sugar alcohols and dietary fiber increases significantly after a few days of adaptation.
4. A study with nonadapted children and adults showed no differences in intestinal manifestations with once-only doses of up to 35 g between the isomalt and sucrose groups (25). In another study with healthy adults, daily doses up to 48 g per day were well tolerated (26).

### Toxicological Evaluations

Isomalt has undergone the full range of generally accepted chronic and embryotoxicological tests. The results of the tests conducted have been summarized in a WHO report prepared by the Joint FAO/WHO Expert Committee on Food Additives (JECFA) (27). This report concludes by assigning isomalt (Palatinit®) an ADI "not specified," the safest rating assigned to any evaluated food substance. The most significant of these studies have recently been published in detail (28–31).

## APPLICATIONS AND PRODUCT DEVELOPMENT

The physical and chemical properties of isomalt that are important in the chocolate, confectionery, baking, and dairy industries have been studied in product development laboratories and reviewed by Irwin (32). Experimental products have been produced in our own laboratories, in independent laboratories, and in the laboratories of interested companies.

In all tests conducted to date, the traditional sweeteners could be totally replaced using isomalt. In some cases, the flavoring of some of these products was modified or intense sweeteners were utilized to compensate for the lower sweetness due to isomalt. These tests have shown that isomalt can be used by most food companies, with only minor changes in their traditional production methods. In comparison with other alternate bulk sweeteners, isomalt offers significant advantages in a number of production processes.

The properties of isomalt which are of particular interest to the product development scientist include the following:

Pure, sweet taste
No aftertaste
Synergism with intense sweeteners
Low caloric value
Low hygroscopicity
Excellent stability
Use of standard processes and equipment
Suitability for diabetics
Noncariogenicity

Product development studies have been successful in the following representative food groups.

Baked goods (33–35)
Breakfast cereals
Chewing gum
Chocolate products (36,37,38)
Confections (39,40)
Hard (44) and soft candy
Fillings
Frosting, icings
Frozen dairy desserts
Fruit and water ices
Jams, jelly, and fruit
Puddings, gelatin desserts
Sauces
Snack foods (42)
Spreads
Table-top sweeteners

## REGULATORY STATUS

Regulatory approval for the use of isomalt as a food ingredient or food additive has been obtained in most European countries. In the Far East, approval for use has been obtained or a petition has been filed in a number of countries. The use of isomalt has been approved in Australia, and approval is pending in New Zealand. A GRAS (Generally Recognized as Safe) affirmation petition was submitted to the U.S. Food and Drug Administration on September 17, 1986. This petition, GRASP 6G0321, was accepted for filing by this agency on October 11, 1990.

## CONSUMER ACCEPTANCE

The consumer acceptance of ingredients like isomalt is based mainly on three factors: (a) taste, (b) mouthfeel, and (c) the appeal of its product claims. Isomalt has no off-flavors and is less sweet than sucrose. The relative sweetness lies between 45 and 65% of that of sucrose, depending upon its concentration in the product. This leaves the sweetness level of the final product totally under the control of the formulator.

When isomalt is used in combination with intense sweeteners such as saccharin, cyclamate, aspartame, and acesulfame-K, the level of sweetness in the final products, such as baked goods, candy, or chocolate, is easily adjusted. Isomalt tends to mask the bitter aftertaste of some of the intense sweeteners, so the consumer is usually not able to differentiate, on the basis of sweetness, between isomalt and sucrose-sweetened products.

The mouth feel of isomalt-containing products is the same as those that are based upon sucrose. In solid products in which isomalt is used in its crystalline form (e.g., chocolate), the one molecule of water of crystallization may influence the smoothness of the product. With appropriate processing conditions, a mouthfeel that is comparable to sucrose-sweetened products can be obtained. In solid products in which isomalt is used as a solution (e.g., chewable candies), isomalt may recrystallize, whereas glucose/sucrose systems do not. This tendency can be minimized by combining isomalt with a cyrstallization inhibitor such as a high molecular weight sweetener (e.g., hydrogenated glucose syrup). Combination with glucose syrup would give the same effect, but at the possible cost of some product claims.

The consumer appeal of product claims that may be made for isomalt-containing products, such as reduced calories, noncariogenicity, and suitability for diabetics, is well known. Isomalt-containing hard boiled candies have been tested by consumers against their sucrose- or glucose-sweetened analogs. The test results have shown clearly that isomalt candies, with the above product claims, have good overall acceptance and are preferred by a large percentage of the population over their sucrose or glucose counterparts.

## REFERENCES

1. P. J. Sträter, Palatinit (isomalt), an energy-reduced bulk sweetener derived from saccharose, in *Low-Calorie Products* (G. G. Birch and M. G. Lindley, eds.), Elsevier Applied Science, London, 1988, p. 63.
2. P. J. Sträter, Palatinit—technological and processing characteristics, in *Alter-*

native Sweeteners, (L. O. Nabors and R. C. Gelardi, eds.), Marcel Dekker, New York, 1986 p. 217.

3. W. Gau, J. Kurz L. Müller, E. Fischer, G. Steinle, U. Grupp, and G. Siebert. Analytische Charakterisierung von Palatinit. *A. Lebensmittel-Untersuchung und -Forschung. 168:* 125 (1979).
4. H. Schiweck, Palatinit—Herstellung, technologische Eigenschaften und Analytik Palatinit haltiger Lebensmittel, *Alimenta, 19:* 5 (1980).
5. K. Paulus and A. Fricker, Zucker-Ersatzstoffe: Anforderungen und Eigenschaften am Beispiel von Palatinit, *Z. Lebensmitteltechnologie und Verfahrenstechnik, 31:* 128 (1980).
6. F. Voirol, Xylitol: Food applications of a noncariogenic sugar substitute, in *Health and Sugar Substitutes,* Verlag S. Karger, Basel, 1979, p. 130.
7. J. N. Counsel, *Xylitol,* Applied Science Publishers, London, 1978, p. 7.
8. F. R. Benson, Alcohols, Polyhydric (Sugar), in *Kirk-Othmer Encyclopedia of Chemical Technology,* 3rd ed., Vol. 1, 1978, p. 754.
9. U. Grupp and G. Siebert, Metabolism of hydrogenated Palatinose, an equimolar mixture of $\alpha$-D-glucopyranosido-1,6 sorbitol and $\alpha$-D-glucopyranosido-1,6 mannitol, *Res. Exp. Med. (Berl.), 173:* 261 (1978).
10. C. C. Emeis and S. Windisch, Palatinosevergärung durch Hefen, *Z. Zuckerindustrie, 10:* 248 (1960).
11. V. Prey and H. Andres, Nebenreaktionen der Saftreinigung, *Z. Zuckerindustrie, 21:* 268 (1971).
12. U. Nilsson and M. J. Jägerstad, Hydrolysis of lactitol, maltitol and Palatinit by human intestinal biopsies, *British J. Nutrition, 58:* 199 (1987).
13. F. Heinz, *Enzymatische Spaltung von Zuckeraustauschstoffen durch isolierte Enzyme und Enzymkomplex der Dünndarmmukosa,* Forschungsvorhaben Nr. 6539, Medizinische Hochschule Hannover, (1987).
14. E. J. Karle and F. Gehring, Kariogenitätsuntersuchungen von Zuckeraustauschstoffen an xerostomierten Ratten, *Dtsch. Zahnärtzl. Z., 34:* 551 (1979).
15. J. Ciardi, W. H. Bowen, G. Rolla, and K. Nagorski, Effect of sugar substitutes on bacterial growth, acid production and glucan synthesis, *J. Dental Res., 62:* 182 (1983).
16. F. Gehring and E. J. Karle, Der Saccharoseaustauschstoff Palatinit unter besonderer Berücksichtigung mikrobiologischer und kariesprophylaktischer Aspekte, *Z. Ernährungswiss. 20:* 96. (1983).
17. T. N. Imfeld, *Identification of Low Caries Risk Dietary Components,* Karger, Basel, 1983, p. 117.
18. H. Drost, M. Spengler, P. Gierlich, and K. Jahnke, Palatinit ein neuer Zuckeraustauschstoff: Untersuchungen bei Diabetikern vom Erwachsentyp, *Aktuelle Endokrinologie,* Vol. 2, Georg Thieme Verlag, Stuttgart, New York, 1980, p. 171.
19. D. Thiebaud, E. Jacot, H. Schmitz, M. Spengler, and J. P. Felber, Comparative study of isomalt and sucrose by means of continuous indirect calorimetry, *Metabolism, 33* (9): 808 (1984).
20. P. J. Sträter and A. Sentko, "Energy utilization of isomalt," presented at

Sweeteners: Carbohydrate and Low Calorie International Conference, 196th meeting of American Chemical Society, Los Angeles, CA, Sept. 22–25, 1988.

21. F. Berschauer und M. Spengler, Energetische Nutzung von Palatinit, *Dtsch. Zahnärztl. Z., 42:* 145 (1987).
22. P. Guinand, *Kohlenhydrat-Abbau durch Beteiligung von Darmbakterien,* Diplomarbeit von Februar 1980, Institut für Biologische Chemie und Ernährungswissenschaft, Universität Hohenheim, Stuttgart, 1980.
23. M. Spengler, J. Sommer, and H. Schmitz, *Vergleich der Palatinit-Tolerance gegenüber der Sorbit-Toleranz gesunder Erwachsener nach einmaliger oraler Gabe in aufsteigender Dosierung,* Bayer AG, Pharma Bericht Nr. 8457, 1979.
24. M. Spengler, H. Schmitz, and C. Biermann, *Vergleich der Palatinit-Tolerance gegenüber der Sorbit-Toleranz gesunder Erwachsener nach 14-tägiger oraler Gabe,* Bayer AG, Pharm Bericht Nr. 8449, 1979.
25. D. M. Paige, T. M. Bayless, and L. R. Davis, *The Evaluation of Palatinit Digestibility—Revised Report.* Johns Hopkins University, Baltimore, MD, May 1986.
26. M. Spengler, J. C. Somogyi, E. Pletcher, and K. Boehme, Tolerability, acceptance and energetic conversion of isomalt (Palatinit) in comparison with sucrose, *Akt. Ernährung, 12:* 210 (1987).
27. Anonymous, Isomalt, in *Toxcological evaluation of certain food additives and contaminants,* WHO Food Additive Series: 20, Cambridge University Press, Cambridge, UK, 1987, p. 207.
28. D. H. Waalkens-Berendsen, H. B. W. M. Koeter, G. Schlüter, and M. Renhof, Developmental toxicity of isomalt in rats, *Food Chem. Toxicology, 27* (10): 631 (1989).
29. D. H. Waalkens-Berendsen, H. B. W. M. Koeter, and M. W. van Marwijk, Embryotoxicity/teratogenicity of isomalt in rats and rabbits, *Food Chem. Toxicology, 28* (1): 1 (1990).
30. D. H. Waalkens-Berendsen, H. B. W. M. Koeter, and E. J. Sinkeldam, Multigeneration reproduction study with isomalt in rats, *Food Chem. Toxicology, 28,* (1); 11 (1990).
31. A. E. Smits-Van Prooije, A. P. De Groot, H. C. Dreef-Van Der Meulen, and E. J. Sinkeldam, Chronic toxicity and carcinogenicity of isomalt in rats and mice, *Food Chem. Toxicology, 28* (4): 243 (1990).
32. W. E. Irwin, "Palatinit—technological properties. An overview," presented at Sweeteners: Carbohydrate and Low Calorie International Conference, 196th meeting of American Chemical Society, Los Angeles, CA, Sept. 22–25, 1988.
33. W. Seibel, G. Brack, und F. Bretschneider, Backtechnische Wirkung des Zuckeraustauschstoffes Palatinit (isomalt); Feinteige mit Hefe, *Getreide, Mehl und Brot, 40* (8): 239 (1986).
34. G. Brack, W. Seibel, und F. Bretschneider, Backtechnische Wirkung des Zuckeraustauschstoffes Palatinit (isomalt); Feinteige ohne Hefe, *Getreide, Mehl und Brot, 40* (9): 269 (1986).
35. G. Brack, W. Seibel, und F. Bretschneider, Backtechnische Wirkung des

Zuckeraustauschstoffes Palatinit (isomalt); Massen mit Aufschlag, *Getreide, Mehl und Brot, 40* (10): 302 (1986).

36. H. Bollinger and T. Keme, How to make sucrose-free chocolate: The use of the reduced calorie sugar substitute Palatinit, *Zucker-und Süsswaren Wirtschaft, 41* (1): (1988).
37. W. E. Irwin, "The use of Palatinit in the production of reduced calorie chocolate ingredients", presented at 197th meeting of American Chemical Society, Dallas, TX, April 9–14, 1989.
38. W. E. Irwin, Sugar substitute in chocolate, *Manufacturing Confect., 70* (5): 150 (1990).
39. W. E. Irwin, Reduced calorie bulk ingredients: isomalt, *Manufacturing Confect., 70* (11): 55 (1990).
40. H. Bollinger, Palatinit (isomalt)—ein kalorienreduzierter Zuckeraustauschstoff. *Gordian* (May): 92 (1987).
41. H. Bollinger und A. Lieser, The continuous production of sugar-free, stamped boiled sweets using the sugar substitute, Palatinit, *Süsswaren* (June): (1988).
42. H. Bollinger und H. Steinhage, Herstellung von kalorienreduzierten Getreideextrudaten, *Zucker-und Süßwaren Wirtschaft, 42* (2): 48 (1989).

# 18

# Sorbitol and Mannitol

**Basant K. Dwivedi** *NuGen Foods, Ltd., Paterson, New Jersey*

## INTRODUCTION

The polyalcohols sorbitol and mannitol are widely distributed in nature. Sorbitol, which was first isolated in 1872 by Joseph Boussingault, a French chemist, from the berries of the mountain ash tree, is found in a variety of fruits and marine vegetation (1) (Table 1). In the mammalian system, sorbitol is present in the fetal blood and in the seminal vesicles and plasma, where it is undoubtedly a normal metabolite (2). Mannitol is found in sweet exudate from the ash tree (*Fraxinus ornus*). The exudate is called manna, thus the mannitol. Mannitol is widely distributed in the plant kingdom. Seaweeds and mushrooms (3) are particularly rich in mannitol. The relationship of these polyalcohols to the corresponding monosaccharides is shown in Figure 1.

Being specialty sweeteners, the current world production and usage of these polyalcohols is relatively small, as compared to sucrose and glucose (Table 2).

## COMMERCIAL PRODUCTION

Although sorbitol and mannitol occur in significant quantities in a number of fruits and berries, it is not commercially economical to extract them from these sources. Thus, these polyalcohols are produced primarily by hydrogenation of corresponding reducing sugars.

**Table 1** Natural Occurrence of Sorbitol

| Fruit juice | Sorbitol (g/liter) |
|---|---|
| Apples | 2.6–9.2 |
| Pears | 11.0–26.4 |
| Cherries | 14.7–21.3 |
| Sour cherries | 13.1–29.8 |
| Plums | 1.8–13.5 |
| Peaches | 0.7–1.2 |
| Black currents | Trace |
| Red currents | Trace |
| Raspberries | Trace |
| Elderberries | 0.17–0.28 |

*Source:* Ref. 1.

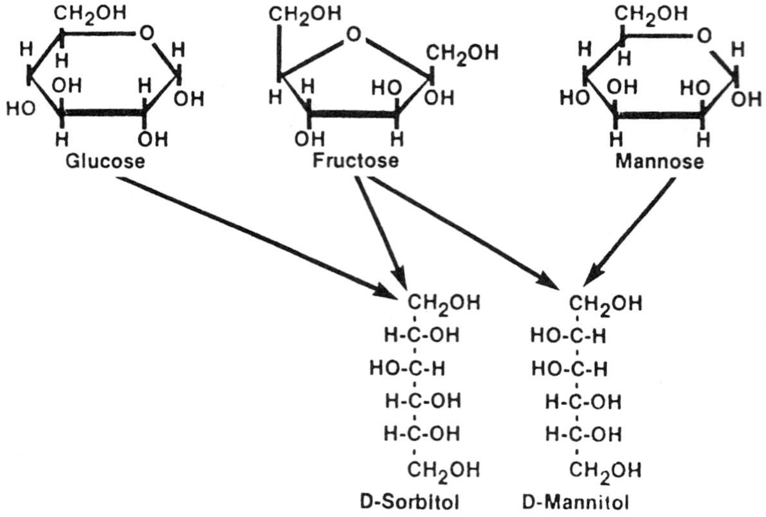


**Figure 1** Sorbitol, mannitol and corresponding monosaccharides.

## Sorbitol

Glucose and starch are preferred raw materials for the production of sorbitol. If a material other than glucose is used, it is first hydrolyzed to glucose. Glucose is then hydrogenated at 120–160°C at hydrogen pressure of 70–140 atm in the presence of a supported nickel or Raney nickel

**Table 2** Annual World Production and Current Price of Nutritive Sweeteners

| Sweetener | 1987 World production in metric tons | Current selling price relative to sucrose (U.S.) |
|---|---|---|
| Sucrose | $0.7–0.8 \times 10^8$ | 1.0 |
| Corn sweeteners | $0.2 \times 10^8$ | 0.7 |
| Liquid sorbitol (70% w/w)[a] | 360,000 | 1.0 |
| Crystalline sorbitol[a] | 55,000 | 2.3 |
| Crystalline mannitol | 10,000 | 4.0 |
| Hydrogented starch Hydrolysates (75% dry wt.) | 3,000 | 2.0 |
| Crystalline xylitol | 2,000 | 10.0 |

[a]The actual production of sorbitol is higher because of its captive use for vitamin C synthesis.

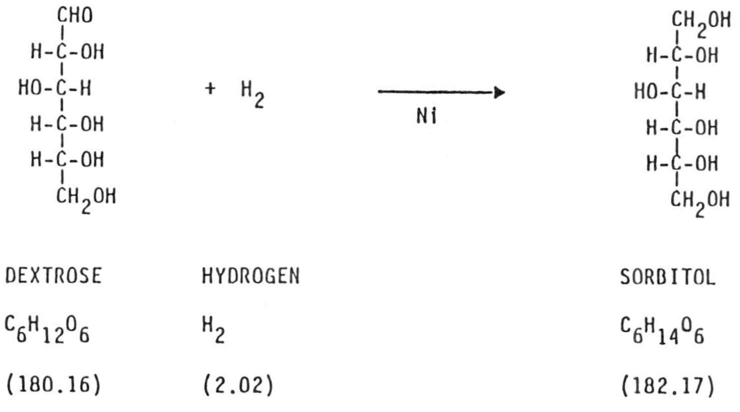


**Figure 2** Sorbitol by hydrogenation.

catalyst (Figs. 2,3). Manufacturers that use sucrose as a raw material often produce sorbitol (75%) and mannitol (25%) simultaneously by the hydrogenation of invert sugar.

## Mannitol

Seaweeds may be used as raw materials for the production of mannitol. For example, the upper part of the seaweed *Laminaria digitara* contains

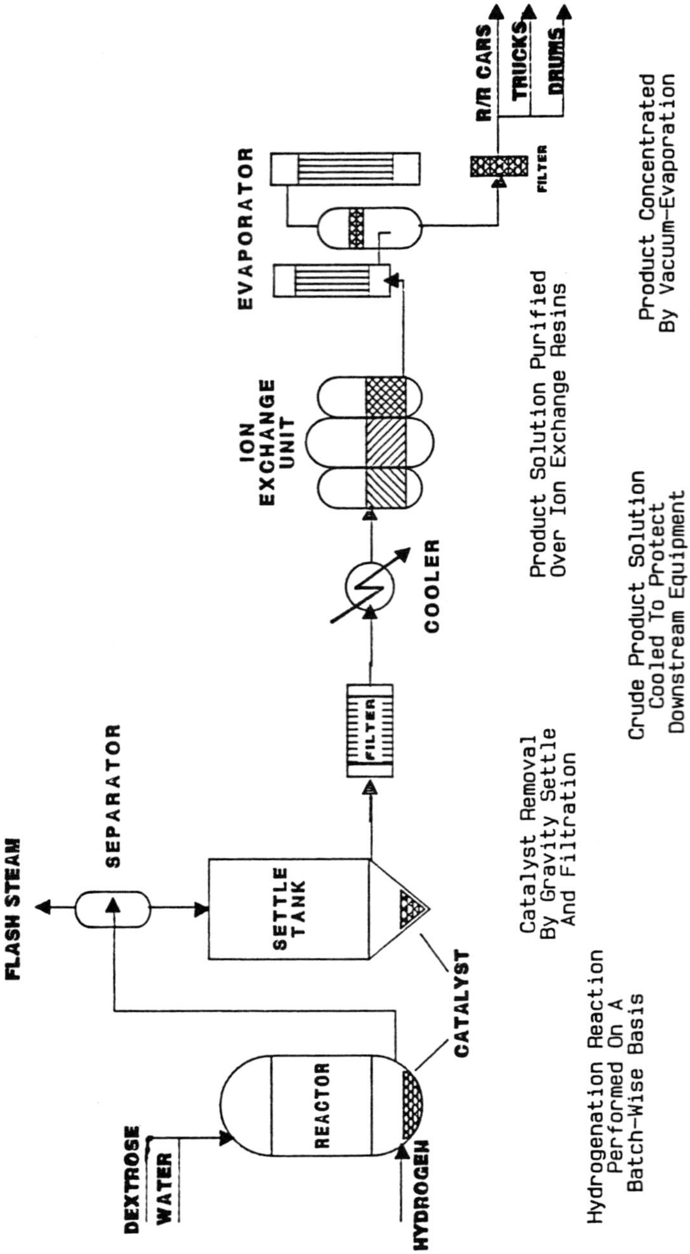


Figure 3 Sorbitol plant process flow diagram.

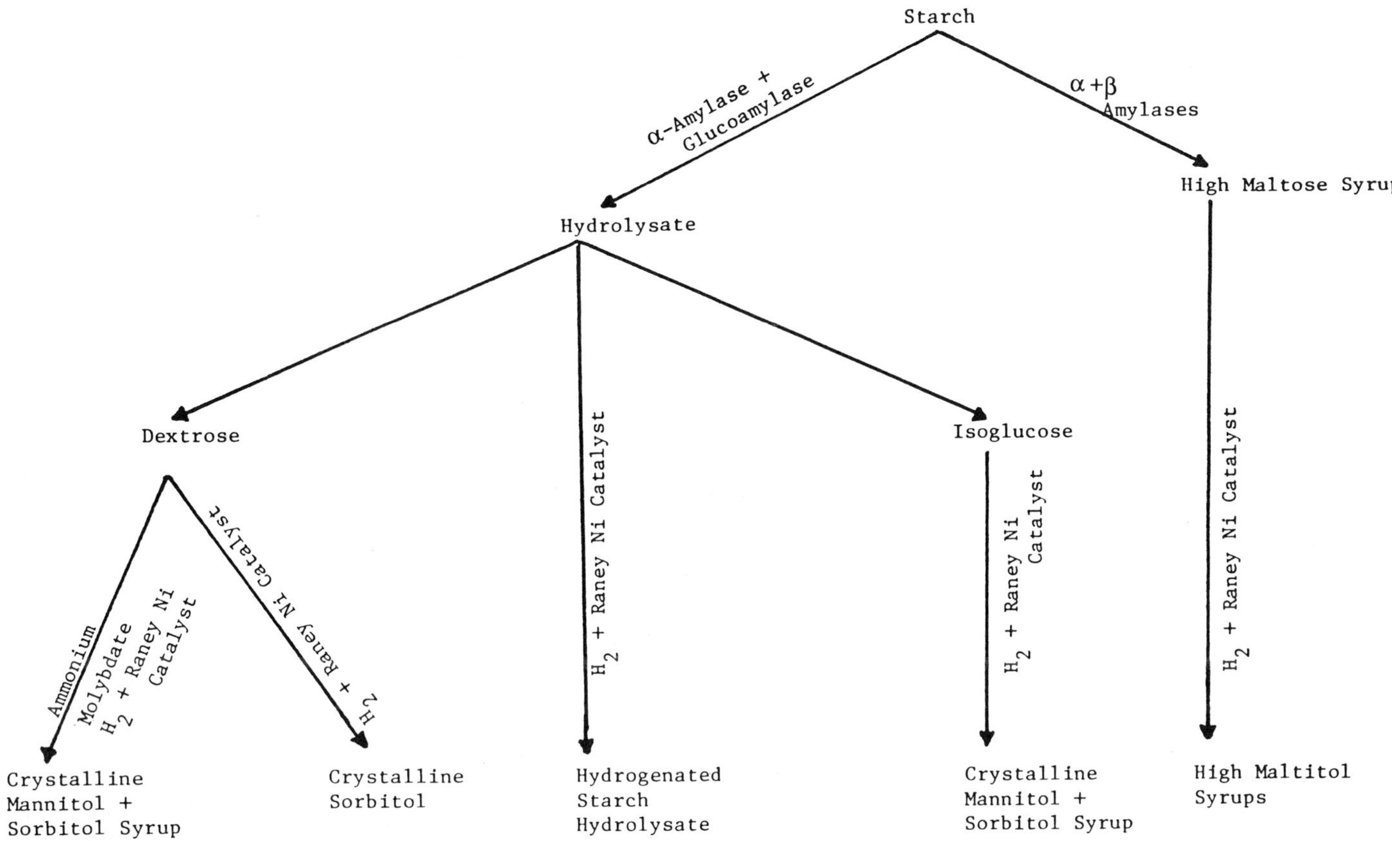


**Figure 4** Polyalcohols from starch. (From Ref.4)

up to 10% mannitol. Thus, it is possible to couple the production of alginates from seaweeds and get mannitol as a byproduct (4).

Currently, invert sugar and dextrose are preferentially used as raw materials for the production of mannitol. As discussed earlier, hydrogenation of invert sugar produces a mixture of sorbitol and mannitol. Mannitol, because of its limited solubility in water, readily crystallizes from the solution. The mother liquor, which contains predominately sorbitol and some mannitol, is concentrated to 70% solids and is used in various applications. If the production of mannitol is the primary objective, its yield can be improved by carrying the hydrogenation under alkaline conditions.

Recently, starch has emerged as a preferred material for the production of various polyalcohols (Fig. 4).

## PHYSICOCHEMICAL PROPERTIES

Various physicochemical properties of sorbitol and mannitol are listed in Table 3. Polyalcohols in general have a considerably higher endothermic heat of solution, as compared to sugars, and thus produce a cooling sensation in the mouth.

Sorbitol and mannitol, because of the absence of free carbonyl groups, do not undergo Maillard reactions and are more heat stable than corresponding mono- and disaccharides. The polyalcohols are also more resistant to microbial degradation than the sugars. Sorbitol is also very hygroscopic and requires care during processing. Mannitol is only moderately hygroscopic.

The relative sweetness of various polyalcohols is given in Table 4.

**Table 3** Physicochemical Properties of Sorbitol and Mannitol

| | Sorbitol | Mannitol |
|---|---|---|
| Chemical formula | $C_6H_{14}O_6$ | $C_6H_{14}O_6$ |
| Molecular weight | 182.17 | 182.17 |
| Melting point (⁰C) | 96–97 | 165–168 |
| Heat of solution at 25°C (cal/g) | −28 | −28.9 |
| Solubility in water at 25° (g/100 g water) | 235 | 22 |
| Caloric value (cal/g) | 3 | 2 |

*Source:* Ref. 4.

Table 4 Relative Sweetness of Various Polyalcohols

| Polyalcohol | Sweetness relative to sucrose (= 100) |
|---|---|
| Sorbitol | 60 |
| Mannitol | 50 |
| Maltitol | 90 |
| Hydrogenated starch hydrolysates | |
| Lycasin[a]80/33a | 40 |
| Lycasin[a]80/55a | 55 |

[a]Lycasin is a registered trademark of Roquette Freres S.A., Lilli, France.

## ABSORPTION AND METABOLISM

Sorbitol is well utilized in the body. Orally administered mannitol is only moderately utilized. The main metabolic reactions of absorbed polyalcohols involve their oxidation to the corresponding ketose/aldose, followed by phosphorylation to the polyol-l-phosphate, i.e., D-glucose⇆sorbitol⇌D-fructose. The direct phosphorylation of polyalcohols is known to occur in microorganisms only (2).

### Sorbitol

Within the last few years, several excellent reviews on sorbitol metabolism have appeared (4–10).

Intestinal absorption of sorbitol occurs slowly by means of passive diffusion. The molecular weight of sorbitol, 182, is close to the limit at which free diffusion is still possible. Generally, no more than 10–20 g of sorbitol can be taken up from the intestinal lumen per hour. If an excess dose is given, it often leads to diarrhea (8).

In the mammalian system, sorbitol is oxidized to fructose by the enzyme sorbitol dehydrogenase, which is then metabolized by the fructose-l-phosphate pathway (Fig. 5). The first reaction of the pathway involves the conversion of fructose to fructose-l-phosphate by fructokinase. In contrast to the corresponding enzyme of glucose metabolism, glucokinase, the activity of fructokinase is not dependent on or regulated by insulin. Fructose-l-phosphate is split by liver aldolase B into dehydroxyacetone phosphate and glyceraldehyde. The dehydroxyacetone phosphate is mixed into the triose-phosphate pool and metabolized further either to pyruvate or to glucose and glycogen, depending on the metabolic state of the organism, i.e., need for glycolysis or gluco-

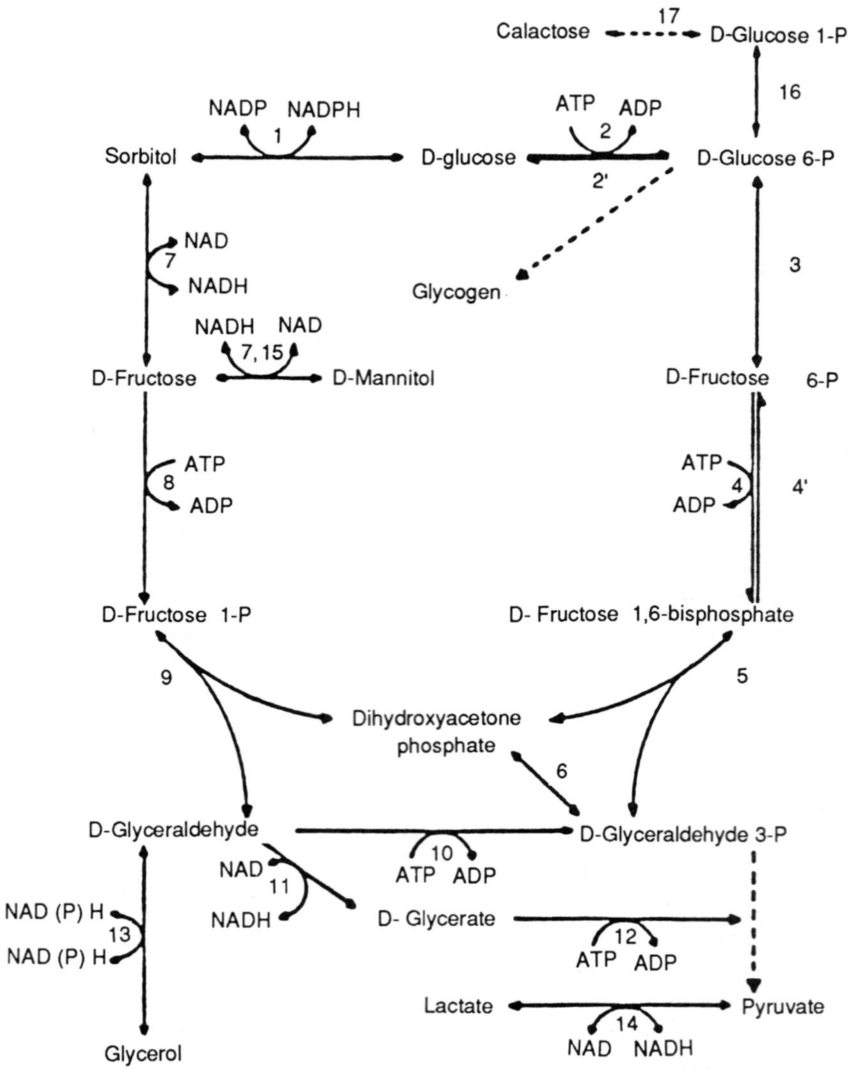


**Figure 5** Schematic diagram of hepatic metabolism of mannitol, sorbitol, galactose, and related carbohydrates.

neogenesis. For example, in the fasting state, during recovery after vigorous exercise, or during cold exposure when glycogen reserves of the body are depleted, or in untreated or poorly controlled diabetes, gluconeogenesis proceeds at a higher rate and metabolic products of sorbitol are primarily converted to glucose and glycogen. Sorbitol may also be converted to glucose in the presence of aldose reductase enzyme and metabolized further.

### D-Mannitol

Both mannitol and sorbitol share a common metabolic route—via fructose. After passive diffusion through the intestinal lumen, D-mannitol is oxidized to D-fructose by the action of mannitol dehydrogenase (4):

$$\text{D-Mannitol} + \text{NAD}^+ \rightleftharpoons \text{D-fructose} + \text{NADH}$$

Some early studies on mannitol suggested that mannitol is not absorbed (11,12) or metabolized (13). However, studies by Iber and Nasrallah (14) clearly demonstrated that mannitol is absorbed and partially metabolized in humans. In the study, (U-$^{14}$C)mannitol was administered orally in doses of 28–100 g. Over 65% mannitol was absorbed, of which 17.5% was recovered in the urine, suggesting that approximately 50% of the ingested mannitol was utilized in the body. Based on these studies, the caloric value of mannitol is only 2 cal/g. Most mannitol calorically available is secondary to microbial fermentation.

## APPLICATIONS

### Dietary Management of Diabetes

Insulin, secreted by the beta cells of pancreas, stimulates and controls the following metabolic reactions:

Oxidation of glucose for energy←Blood glucose→Storage of glycogen
↓
Synthesis of fatty acids
↓
Storage of fat in adipose tissue

Insulin controls the passage of glucose from the blood through the hepatic cells. In diabetics, because of the lack of insulin, this mechanism fails, resulting in an abnormal increase in blood glucose levels. High blood glucose levels have been shown to be associated with many tissue toxicities, including those of the eye, the kidney, and the large and small blood vessels. Excess blood glucose also leads to the accumulation of sorbitol in many cells, with increased osmotic pressure. The localized accumulation of the fluid is thought to contribute to the development of diabetic complications, such as cataracts and polyneuritis.

The blood glucose level in diabetics is controlled by insulin injections (in severe diabetes), and by excluding the quickly absorbable carbohydrates such as D-glucose, invert sugar, and disaccharides from the diet.

Sorbitol has been used as a sweetening agent for diabetics since it was reported in the late 1920s that moderate amounts of sorbitol taken by normal or diabetic subjects caused only a small and insignificant rise in the blood sugar concentration. Adcock and Gray (15) confirmed these observations by feeding uniformly labeled D-($U^{14}C$)sorbitol and D-($U^{14}C$)glucose to normal and diabetic subjects. Data reported in Table 5 show that in normal and mildly diabetic subjects, over 80% of ingested sorbitol is quickly metabilized to carbon dioxide. In severe type I diabetes, the recovery of labeled carbon dioxide amounted to only 65%.

In a study by Akgun and Ertel (16), 35 g each of sucrose, sorbitol, and fructose were fed to normal and type II diabetic subjects as part of a 400-calorie breakfast. In both normal and diabetic subjects, the average plasma glucose level was highest after the sucrose-containing breakfast (44.0 mg/dl for normal subjects; 78.0 mg/dl for type II diabetics), and lowest after the sorbitol-containing breakfast (9.3 mg/dl for normal subjects; 32.2 mg/dl for type II diabetics). Fructose produced an intermediate rise in plasma glucose (Fig. 6).

Some health professionals have discouraged the use of sorbitol by diabetics on grounds that sorbitol is capable of being converted to glucose, and eventually requires insulin for its metabolism. Forster and Mehnert disagree (8). Their research shows that although glucose can be produced during sorbitol metabolism, its production is delayed to such an extent that hyperglycemia in diabetics after ingestion of sorbitol is insignificant, as compared to other carbohydrates, including starch-containing foods.

Another concern, that ingested sorbitol may be associated with diabetes complications brought on by the intracellular accumulation of sorbitol, has been shown to be without foundation (17). The intracellular accumulation of sorbitol is directly related to hyperglycemia produced due to the lack of insulin. Because sorbitol ingestion results in lower blood sugar levels, sorbitol is expected to reduce the severity of diabetes complications.

Sorbitol is also suitable in the diets of diabetic children according to a study carried out by Steinke et al. (18). Sorbitol was added to the diets of diabetic children at a summer camp for periods varing from 8 to 48 days. Up to 40 g of sorbitol daily, given in three daily doses during meals, was tolerated without diarrhea and without evidence of increased insulin requirements.

**Table 5** Percent Radioactivity Recovered in Urine, Feces, and Breath After Ingestion of 35 g Labeled Glucose and Sorbitol

| | Urine (0–24 hr) | | Feces (0–30 days) | | Breath (0–24 hr) | | Total recovered | |
|---|---|---|---|---|---|---|---|---|
| Subject | Glucose | Sorbitol | Glucose | Sorbitol | Glucose | Sorbitol | Glucose | Sorbitol |
| Normal | | 1.5 | | 13.1 | | 80 | | 95 |
| Mildly diabetic | 5.4 | 2.7 | 6.9 | 0.9 | 76 | 87 | 88 | 90 |
| Severely diabetic | 40.9 | 9.0 | 2.7 | 4.9 | 44 | 65 | 88 | 90 |

*Source:* Ref. 15.

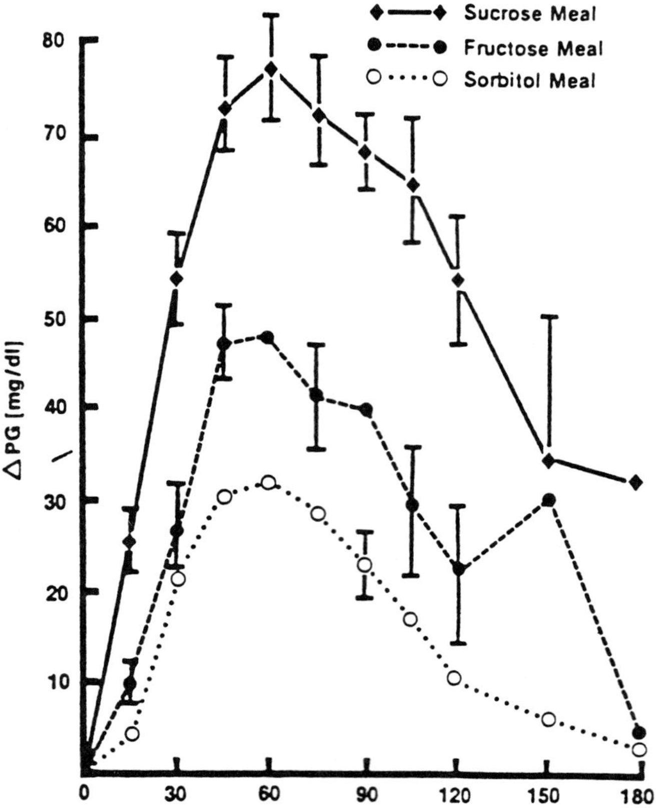


**Figure 6** Changes in plasma glucose and insulin concentration after sucrose, sorbitol, or fructose meal in diabetic subjects. (From Ref.16)

Based on available data, Mehnert and Forster (8) concluded that by replacing insulin-requiring carbohydrates with sorbitol, it is possible to improve carbohydrate tolerance to the extent that in borderline cases, the need for oral hypoglycemics will no longer be necessary. In the same way, stabilization of the metabolism by administration of sorbitol, even in patients requiring insulin, is to be expected. Mannitol, because of its very low laxative threshold, is less desirable in a diabetic diet.

## Prevention of Dental Caries

Prerequisites for dental caries development include a pH drop below 5.5, fermentable carbohydrate, and plaque formation. Sorbitol, mannitol, and other polyalcohols are accepted as being less cariogenic than sucrose (19,20). Studies have shown that sugars and polyalcohols behave differently with respect to their degradation products, their influence on the physical properties and the chemical composition of plaque and saliva, and their fitness for microbial metabolism; hence, their suitability as substrate for the oral microflora (21,22). Hassell (23) studied acid production after the ingestion of sucrose, sorbitol, mannitol, and xylitol. Whereas sucrose ingestion resulted in immediate acid production and a drop in pH, sorbitol, mannitol, and xylitol ingestion produced a much smaller drop in pH, which also stayed above the critical value of 5.5. Several studies have supported these findings (21–24).

### Other Applications

Besides its application in special dietary foods, sorbitol is also used as a:

Humectant
Sequestrant for certain multivalent metals
Stabilizer
Viscosity control agent

In addition, sorbitol is used in the production of polyethers for the manufacture of urethane foams and as a raw material for the synthesis of resins which are utilized as surface coatings and printing-ink vehicles.

Sorbitol, with six hydroxy groups, can also be esterified with a variety of organic acids. Esterification with long-chain fatty acids to form internal ethers, i.e., sorbitans, is by far the most important reaction. Partial fatty acid esters of sorbitan are extensively used in surface-active applications, such as emulsification, wetting, detergency, and solubilization. The sorbitan fatty acid esters can be further modified by the attachment of polyoxyethylene chains. These ethoxylated sorbitan fatty acid esters are also used as surface active agents.

Mannitol, because of its nonhygroscopic nature, is used as a dusting powder and an anticaking agent, besides its special dietary foods application.

## REGULATORY STATUS

The regulatory status of sorbitol and mannitol in various countries is given in Table 6. Sorbitol is permitted in 22 countries and mannitol in 16 countries for applications in foods and beverages, (25). Current U.S. food and drug laws covering these sweeteners are discussed below.

**Table 6** Regulatory Status of Sorbitol and Mannitol

| Sweetener | Legislation | Country |
|---|---|---|
| Sorbitol | Listed as permitted sweetener | Japan, South Africa, Sweden, United States |
| | Permitted except where standard prohibits use: not listed as sweetener | United Kingdom, Federal Republic of Germany |
| | Listed as sweetener: permitted in certain foodstuffs only | Belgium, Denmark, Australia, Finland, Greece, Norway, Spain, Switzerland |
| | Listed as food additive only | France, Italy, Netherlands, Brazil, Canada |
| | Not described | Austria |
| Mannitol | Listed as permitted sweetener | South Africa, United States |
| | Permitted except where standard prohibits use: not listed as sweetener | United Kingdom, Federal Republic of Germany |
| | Listed as sweetener: permitted in certain food stuffs only | Belgium, Denmark, Australia, Greece, Spain, Sweden, Switzerland |
| | Listed as food additive only | France, Canada, Japan |
| | Not described | Italy, Netherlands, Austria, Finland, Norway |

*Source:* Refs. 28, 29

## Sorbitol

Sorbitol has been affirmed as Generally Recognized as Safe (GRAS) (26) and may be used in food at levels not to exceed Good Manufacturing Practices. Current GMPs permit the use of sorbitol up to the following maximum levels (21 CFR 184.1835):

Hard candy and cough drops, 99%
Chewing gum, 75%
Soft candy, 98%
Jams and jellies (nonstandardized), 30%
Baked goods and baking mixes, 30%
Frozen dairy desserts and mixes, 17%
Other foods, 12%

The Food and Drug regulations also require the following label statement for foods whose reasonably foreseeable consumption may result in the daily ingestion of 50 g of sorbitol: "Excess consumption may have a laxative effect."

### Mannitol

In the United States, mannitol has interim food additive status (21 CFR 180.25) (27) and may be used at levels not to exceed the following:

Pressed mints, 98%
Hard candies, 5%
Cough drops, 5%
Chewing gum, 31%
Soft candies, 40%
Confections and frostings, 8%
Nonstandardized jams and jellies, 15%
Others, 2.5%

The regulation also stipulates that the label of food whose reasonably foreseeable consumption may result in a daily ingestion of 20 g of mannitol must carry the statement: "Excess consumption may have a laxative effect."

## REFERENCES

1. K. C. F. Heaton, F. D. Robinson, and M. Lewin, Joint Meeting Midland Branch IFST and British Soc. Flavorist, Sutton Bonington, March 11, 1980.
2. O. Touster, in *Physiological Effects of Food Carbohydrates* (A. Jeanes and J. Hodges, eds.), ACS Symp. Series 15, American Chemical Society, Washington, D.C., 1975, p. 123.
3. G. K. Parrish, R. B. Beelman, and L. R. Kneebone, *Hortscience, 11*(1): 32 (1976).
4. P. J. Sicard, in *Nutritive Sweeteners* (G. Birch and K. G. Parker, eds.), Applied Science Publishers, London/New Jersey, 1982, p. 145.
5. B. K. Dwivedi, in *Carbohydrates and Health* (L. R. Hood, E. K. Wardrip, and G. Bollenback, eds.), AVI Publishing Co., Westport, Conn., 1977, p. 27.
6. B. K. Dwivedi, in *Low Calorie and Special Dietary Foods* (B. K. Dwivedi, ed.), CRC Press, West Palm Beach, Fla., 1978, p. 1.
7. J. D. Brunzell, *Diabetes Care, 1*(4): 223 (1978).
8. H. Forster, and H. Mehnert, *Akt Ernahrung, 5:* 245 (1979).
9. J. M. Olefsky, and P. Crapo, *Diabetes Care, 3*(2): 390 (1980).
10. B. K. Dwivedi, *Food Reviews Int'l., 2*(2): 171 (1986).
11. J. S. Fordtran, F. C. Rector Jr., F. M. Ewton, N. Soter, and J. Kinney, *Clin. Invest., 44:* 1935 (1965).
12. H. Kamada, T. Abei, S. M. Nasrallah, and F. L. Iber, *Am. J. Physiol., 214:* 1090 (1968).

13. F. P. Stuart, E. Torres, R. Fletcher, D. Crocker, and F. D. Moore, *Ann. Surg., 172:* 190 (1970).
14. F. L. Iber, and S. M. Nasrallah, *A. J. Med. Sci., 258:*80 (1969).
15. L. H. Adcock, and C. H. Gary, *Biochem. J., 65:* 554 (1957).
16. S. Akgun, and N. H. Ertel, *Diabetes Care, 3*(5): 582 (1980).
17. K. H. Gabbay, *N. Engl. J. Med., 288*(16): 831 (1973).
18. J. Steinke, F. C. Wood, L. Donenege, A. Marble, and A. E. Arnold, *Diabetes, 10:* 218 (1961).
19. O. Larje, and R. H. Larson, *Arch. Oral Biol., 15:* 805 (1970).
20. D. Birkhed, and S. Edwardsson, *Health and Sugar Substitutes,* Proc. ERGOB Conf., Geneva, 1978, p. 211.
21. F. Bramstedt, and K. Trautner, *Dtsch. Zahnarztl Z., 26:* 1135 (1971)
22. S. Edwardsson, D. Birkhed, and B. Mejar, Acta Odon, Scand., 35: 257 (1977).
23. T. M. Hassell, *Dtsch. Zahnarztl. Z., 26:* 1145 (1971).
24. H. Yamamoto, S. Kawanishi, Y. Mckia, T. Matsukubo, and T. Takaesu, *Dental. J. Res., 66:* 1557 (1987).
25. G. T. Molitor, Calorie Control Council 1987 Annual Meeting, Scottsdale, Ariz.
26. *Fed. Reg., 39:* 34180 (1974).
27. *Fed. Reg., 39:* 34178 (1974).
28. E. Rockstrom, in *Carbohydrate Sweeteners in Foods and Nutrition* (P. Koivistoinen and L. Hyvonen, eds.), Academic Press, New York, 1980, p. 225.
29. G. von Hertzen and C. Lindquist, in *Carbohydrate Sweeteners in Foods and Nutrition* (P. Koivistoinen and L. Hyvonen, eds.), Academic Press, New York, 1980, p. 163.

# 19

# Xylitol

**Albert Bär** *Bioresco Ltd., Brussels, Belgium*

## INTRODUCTION

Xylitol is a five-carbon sugar alcohol with a sweetness similar to sucrose (1,2). It is found in small amounts in a variety of fruits and vegetables (3,4) and is formed as a normal intermediate in the human body during glucose metabolism (5). Xylitol has been shown to be of value in the prevention of dental caries because it is not an effective substrate for plaque bacteria (6–10). Due to its largely insulin-independent metabolic utilization, it may also be used as a sweetener in the diabetic diet and as an energy source in parenteral nutrition (11–13). As a sweetening agent, xylitol has been used in human food since the 1960s. Different aspects of its applications, metabolic properties, and safety evaluation have been the subject of several previous reviews (14–20). Its physicochemical properties are summarized in Table 1.

## PRODUCTION

Xylitol was first synthesized and described in 1891 by Emil Fischer and his associate (21). On a commercial scale it is produced by chemical conversion of xylan (22,23). Potential sources of xylan include birchwood, almond shells, straw, corn cobs, and wastes from the paper and pulp industries. The practical suitability of these raw materials depends, among other factors, upon their xylan content, which may vary considerably, and the presence of by-products (e.g., poly- or oligo-

**Table 1** Some Physical and Chemical Properties of Xylitol

| | |
|---|---|
| Formula | $C_5H_{12}O_5$ (molecular weight, 152.15) |
| Appearance | White, crystalline powder |
| Odor | None |
| Specific rotation | Optically inactive |
| Melting range | 92–96°C |
| Boiling point | 216°C (760 mmHg) |
| Solubility at 20°C | 169 g/100 g $H_2O$; 63 g/100 g solution; sparingly soluble in ethanol and methanol |
| pH in water (100g/liter) | 5–7 |
| Density of solution | 10%, 1.03 g/ml; 60%, 1.23 g/ml |
| Viscosity (20°C) | 10%, 1.23 cP; 60%, 20.63 cP |
| Heat of solution | +34.8 cal/g (endothermic) |
| Heat of combustion | 16.96 kJ/g |
| Refractive index (25°c) | 10%, 1.3471; 50%, 1.4132 |
| Moisture absorption (4 days, RT) | 60% RH, 0.051% $H_2O$; 92% RH, 90% $H_2O$ |
| Relative sweetness | Equal to sucrose; greater than sorbitol and mannitol |
| Stability | Stable at 120°C, no caramelization; stable also under usual conditions in food processing; caramelization occurs if heated for several minutes near the boiling point |

cP = centipoise.
RH = relative humidity.
RT = room temperature.

saccharides), which have to be removed during the production process. In principle, the commercial synthesis of xylitol involves four steps. The first step is the disintegration of natural xylan-rich material and the hydrolysis of the recovered xylan to xylose. The second step is the isolation of xylose from the hydrolysate via chromatographic processes to yield a pure xylose solution. Third, xylose is hydrogenated to xylitol in the presence of a nickel catalyst. Alternatively, hydrogenation of the impure xylose solution may be conducted first, followed by purification of the xylitol syrup. Ultimately, xylitol is crystallized in orthorhombic form (24).

The synthesis of xylitol by fermentative or enzymatic processes is possible in principle (25–28). However, such procedures have so far not been utilized on a commercial scale. Other appproaches to the synthesis of xylitol are of merely scientific interest (29,30).

## PROPERTIES AND APPLICATIONS

At present, xylitol is used as a sweetener mainly in noncariogenic confectionery (chewing gum, candies, chocolates, gum drops), and less frequently in dietetic foods (food products for diabetics), in pharmaceutical preparations (tablets, throat lozenges, multivitamin tablets, cough syrup), and in cosmetics (toothpaste) (31–37). Application in soft drinks is limited because the intake of excessive quantities of polyol-sweetened liquids would tend to exceed the threshold dose above which gastrointestinal side effects (laxation) may be experienced. In principle, the manufacture of various baked goods with xylitol is possible. However, if crust formation, caramelization, or nonenzymatic browning is required, the addition of a few percent of fructose is necessary. Since xylitol inhibits the growth and fermentative activity of yeast, it is not a suitable sweetener for products containing yeast as a leavening agent (38,39).

### Sweetness

Xylitol is the only sugar alcohol that has about the same sweetening power as sucrose (Fig. 1) (2,40–44). All other sugar alcohols are less sweet, and when they are used singly in the production of confectionery and other sweet foods, intense sweeteners must be added to bring the

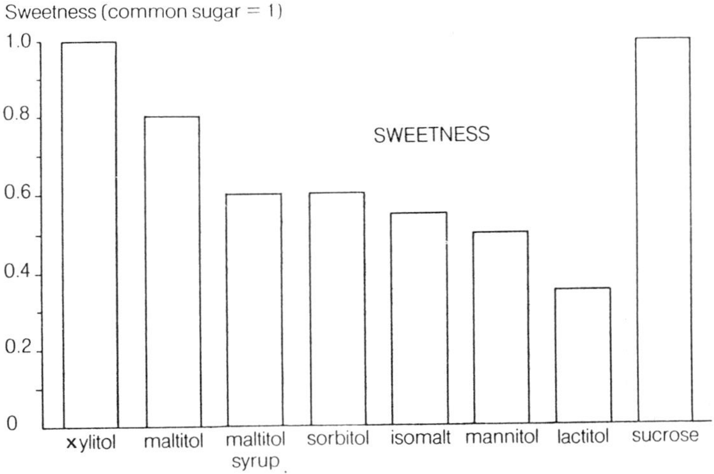


**Figure 1** Relative sweetness of sugar alcohols.

sweetness up to an acceptable standard. Another possibility is to use these less sweet sugar substitutes in combination with xylitol and thus take advantage of the synergistic effect of sweetener combinations.

## The Cooling Effect

The heats of solution of crystalline sugar alcohols and sucrose are shown in Figure 2. The loss of heat when dissolving sugar alcohols in water is much greater than with sucrose. The greatest cooling effect is obtained with crystalline xylitol. This results in a pleasant cooling sensation when xylitol is dissolved in the mouth. Use of this interesting organoleptic property is made mainly in peppermint- or fruit-flavored sweets, particularly in chewing gum, as well as in fillings of mint candies and similar products that are supposed to have a cooling effect or a particular freshness.

In chocolate, this cooling effect is perhaps somewhat unusual and not exactly typical, but in chocolate products that are flavored with peppermint oil, a cooling effect can be quite desirable.

A cooling effect is obviously not perceived from products in which xylitol is already dissolved (e.g., toothpaste, mouth rinse) or in which it exists in an amorphous form (jellies; boiled, transparent candies).

## Solubility of Sugar Alcohols

Solubility of the various sugar substitutes plays an important role in the production of, for example, fondants, jellies, and toffees. In the case of

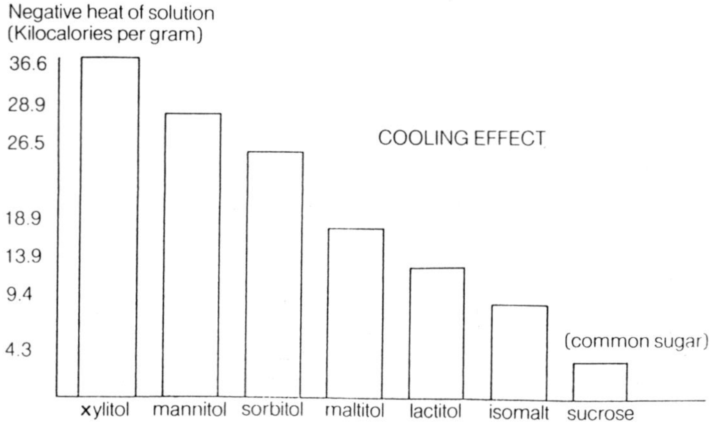


**Figure 2** Cooling effect of sugar alcohols.

**Table 2** Solubility of Sugar Alcohols in Water at 20°C

| Sweetener | Solubility (% w/w) |
|---|---|
| Sucrose | 66 |
| Xylitol | 63 |
| Sorbitol | 75 |
| Maltitol | 62 |
| Lactitol | 55 |
| Isomalt | 28 |
| Mannitol | 18 |

fondants, a controlled recrystallization should take place, whereas in the case of the other fillings, crystallization must be avoided. Table 2 shows the solubility of the most important sugar alcohols at 20°C in comparison to sucrose. With the exception of sorbitol, all presently used sugar alcohols are less soluble than sucrose.

## Other Characteristics

The viscosity of sugar alcohol solutions depends on their molecular weight. Monosaccharide alcohols such as sorbitol and xylitol have a relatively low viscosity, whereas maltitol syrups with a high content of hydrogenated oligosaccharides are fairly viscous.

As the sugar alcohols are chemically quite inert due to the lack of active carbonyl groups, they cannot participate in browning reactions. This means that there is no carmelization during heating, as is typical of sugars. This is perhaps also the reason why sugar-free chocolates appear to lack certain flavor components which in the sucrose-sweetened product are formed during the conching process.

## The Use of Xylitol in Confectionery and Chocolates

Xylitol is comercially used in many countries either alone or in combination with other sugar substitutes in the manufacture of sugarless chocolates and confectionery for diabetics and for the protection of teeth. Some of the articles produced with xylitol are hard caramels, fruit gums, licorice sweets, wafer fillings, chocolate, pastilles, and, most importantly, chewing gums.

Although it is in principle possible to produce chocolate, cream fillings, chewing gum, pastilles, and, under certain conditions, even hard caramels on the basis of xylitol alone, other sugar substitutes are normally added in order to optimize sweetness, texture, shelf life, and cost. Sugar-free pectin jellies, which are virtually indistinguishable from conventional sugar/glucose syrup jellies, can be produced with combinations of hydrogenated glucose syrup.

Boiled sweets can be produced using only xylitol or xylitol in combination with, for example, maltitol, sorbitol, lactitol, or isomalt. However, if only xylitol and/or sorbitol are used, nontransparent candies result. For the manufacturing of a glass-clear product, the addition of a dissaccharide sugar alcohol or an even more polymeric product such as Lycasin[R] is required. Boiled sweets made only with isomalt are not sweet enough and release their flavor too slowly. With the addition of a relatively small amount of xylitol (15–20%), adequate sweetness is obtained.

The production of sugarless fondants containing xylitol and sorbitol has also been reported. While sorbitol produces a somewhat waxy texture when used as the crystalline phase, a good texture and pleasant cooling taste is achieved when xylitol is applied as the crystalline phase and sorbitol as the syrup phase.

Sugar-free chewing gums usually contain mixtures of sugar alcohols and intense sweeteners in order to assure a satisfactory sweetness profile over time. While a few brands are sweetened with xylitol only, the majority of the products contain less than 25% xylitol in the sweetener mixture.

Chocolates made with only sorbitol, mannitol, or xylitol have a slightly burning and irritating aftertaste which, incidentally, is also characteristic for fructose and dextrose chocolates. Disaccharides, such as sucrose and lactose, and disaccharide alcohols, such as maltitol and isomalt, do not have this aftertaste and are mildly sweet. However, in chocolate they are not sweet enough, and intense sweeteners or a certain amount of xylitol must be added in order to augment the sweetness appropriately.

## PHARMACEUTICAL PREPARATIONS

Xylitol can be used as an excipient or as a sweetener in many pharmaceutical preparations. As in foods, the advantages are suitability for diabetics, noncariogenic properties, and nonfermentability. Cough syrups, tonics, and vitamin preparations made with xylitol can neither ferment nor mold. Because xylitol is chemically more inert than sucrose, such products may also have a longer shelf life. Xylitol-sweetened medications can be given to children at night after tooth-brushing without any harm to the teeth (36,45–49).

In tablets, xylitol can be used as a carrier and/or as a sweetener (50,51). As a carrier, xylitol has also been tested in solid dispersions because of its low melting point and stability up to 180°C. In these studies, it was found that solid dispersions of hydrochlorothiazide or *p*-aminobenzoic acid with xylitol showed a faster release than the micronized drugs (52,53). Milled xylitol may be granulated and pressed into tablets after flavoring (50). Coatings with xylitol or mixtures of xylitol and sorbitol (up to 20%) can be made by conventional pan coating as well as by sintering the surface of compressed tablets in a hot-air stream (54).

In toothpaste, xylitol may partially or completely replace sorbitol as a humectant. Because of its higher sweetness, xylitol improves the taste of the dentifrice, and in the manufacture of transparent gels, it has been said to exhibit properties slightly superior to sorbitol. In addition, there is evidence that xylitol exerts a plaque-reducing effect (55,56) and that it interferes with bacterial metabolism particularly in the presence of fluoride and zinc ions (57). An inhibitory effect on enamel demineralization has been postulated as well (58,59). Because of its overall favorable effects on dental health, xylitol has also been applied in other oral care products, such as in mouth rinses and in artificial saliva (60–63).

## METABOLISM

In principle, two different metabolic pathways are available for the utilization of xylitol: (1) direct metabolism of absorbed xylitol in the mammalian organism, mainly in the liver, or (2) indirect metabolism via fermentative degradation of unabsorbed xylitol by the intestinal flora.

### Indirect Metabolic Utilization

All polyols, including xylitol, are slowly absorbed from the digestive tract because its transport through the intestinal mucosa is not facilitated by a specific transport system. Therefore, after ingestion of large amounts, only a certain proportion of the ingested xylitol is absorbed and enters the hepatic metabolic system via the portal vein blood. A comparatively larger amount of the ingested xylitol reaches the distal parts of the gut where extensive fermentation by the intestinal flora takes place. Besides minor amounts of gas ($H_2$, $CH_4$, $CO_2$), the end-products of the bacterial metabolism of xylitol are mainly short-chain, volatile fatty acids, i.e., acetate, propionate, and butyrate (64–66). These products are subsequently absorbed from the gut and enter the mammalian metabolic pathways (67). Acetate and butyrate are efficiently taken up by the liver and utilized in mitochondria for production of acetyl-CoA. Propionate is also almost quantitatively removed by the liver, and yields propionyl-CoA (68–70).

The production of volatile fatty acids (VFA) is a normal process associated with the consumption of sugar alcohols and dietary fibers (cellulose, hemicelluloses, pectins, gums) for which hydrolyzing enzymes are lacking or poorly efficient in the small intestine. Under normal conditions, most of the generated VFA are absorbed from the gut and are further utilized via established metabolic pathways in animals and man (66,71). For the energetic utilization of slowly digestible materials, this fermentative, indirect route of metabolism plays an important role, and evidence has been presented that, even under normal dietary conditions (i.e., in the absence of sugar alcohols), the contribution of intestinal fermentations to the overall energy balance is significant (72,73).

### Direct Metabolism Via the Glucuronic Acid–Pentose Phosphate Shunt

In addition to the indirect route of utilization, a direct metabolic pathway is available for the portion of xylitol that is absorbed unchanged from the gastrointestinal tract (5,74,75).

The metabolism of xylitol and its general relationship to the carbohydrate metabolism via the pentose phosphate pathway is shown in Figure 3. This scheme illlustrates how the transformation of L-xylulose via xylitol to D-xylulose links the oxidative branch of the glucuronate pentose phosphate pathway with the nonoxidative branch, which yields glyceraldehyde-3-phosphate and fructose-6-phosphate, as well as ribose-5-phosphate for ribonucleotide biosynthesis (76–78). Thus, xylitol can be converted via the pentose phosphate pathway to intermediates of the glycolytic pathway, which may either undergo further degradation or transformation to glucose-1-phosphate, a precursor of glycogen (5,74–76). Because gluconeogenesis from exogenous xylitol is associated with the generation of NADH in the cytosol, reoxidation of cytosolic NADH is a necessary step for xylitol utilization.

The recognition of xylitol as a normal endogenous metabolite in humans originates from observations in patients with a genetic abnormality, essential pentosuria. These persons excrete considerable quantities of L-xylulose in the urine. When incubated in tissue preparations, L-xylulose was found to be metabolized to xylitol by a specific NADPH-linked enzyme, L-xylulose-reductase. This enzyme has subsequently been demonstrated to be deficient in essential pentosuria. Because pentosuric patients excrete 2–15 g of L-xylulose per day, the daily production of xylitol was estimated to be of a similar order of magnitude (5,79,80).

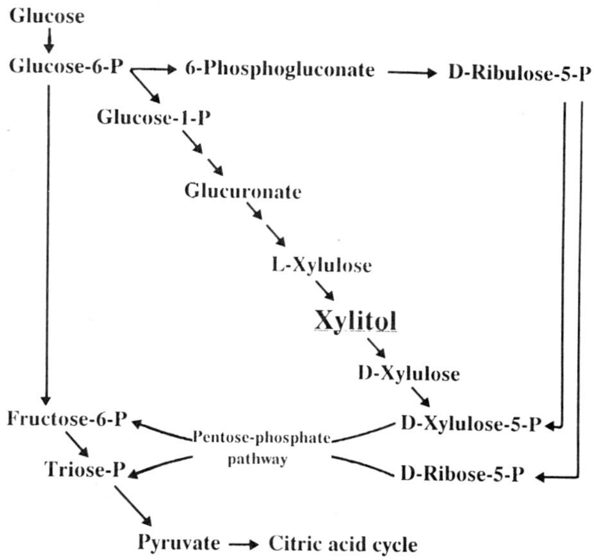


**Figure 3** Metabolism of xylitol.

## Estimation of the Caloric Value of Xylitol

If the caloric value of sugar alcohols is to be estimated on the basis of their metabolic fate, precise knowledge about the different steps of their digestion and metabolic utilization is required. In particular, it is, for example, essential to know (1) how much of an ingested dose is absorbed directly from the gut, (2) via which pathways the absorbed portion is metabolized in the human organism, (3) which proportion of an ingested dose is fermented by the gut microflora, (4) to what extent the resulting fermentation products are absorbed and metabolically utilized by the host, and (5) how much of an ingested dose leaves the intestinal tract unchanged with the feces. Although the experimental database on xylitol does not allow one to answer all these questions with sufficient precision, it is possible to obtain reasonable estimates by extrapolation from studies on other sugar alcohols that are also incompletely absorbed and are subject to the same fermentative degradation in the gut.

On the basis of a thorough assessment of numerous in vitro and in vivo experiments with xylitol and other polyols, it has been estimated that approximately one third of an ingested xylitol dose is absorbed from the gastrointestinal tract (Bär, unpublished report). This portion of xylitol, which is effectively metabolized via the glucuronate–pentose phosphate shunt, is energetically fully available and provides about 4 kcal/g. The nonabsorbed two thirds of the ingested load, however, are alsmot completely fermented by the intestinal flora. Combining a 78% retention of energy in bacterial fermentation products and a growth yield of 20 g bacteria per 100 g substrate, it has been estimated that about 42% of the energy provided with unabsorbed xylitol is consumed by bacterial metabolism and growh, while about 58% of the energy becomes available to the host after absorption of the fermentation end-products (VFA) (81). This value is well in line with an estimated 50% energy salvage proposed by a Dutch expert group (82).

Based on these estimates, a caloric value of about 2.8–2.9 kcal/g may be calculated for xylitol. This value is in line with the results of an in vivo study in which xylitol was found to be about 60% as effective as glucose in promoting growth (83).

Besides laborious and expensive energy balance trials, only indirect calorimetry allows one to determine the energetic utilization of substrates in humans. Using this noninvasive technique, the energetic utilization of xylitol has been examined in 10 healthy volunteers (84). The results of this study revealed that, during a 2.5-hour period, the cumulated increase in carbohydrate oxidation amounted to only one fourth of that caused by glucose. The total increase in metabolic rate was 52% lower after the xylitol load than after the glucose treatment. This result suggests that xylitol has a caloric value of only about 50% that of glucose. Considering, however, that the absorption (and hence metabolic utilisation) of the bacterial fermentation products may not have been complete within the 2.5-hour experimental period, it is likely that the true caloric value for xylitol is somehwat higher than the proposed 2 kcal/g.

Food regulatory authorities are increasingly taking note of the reduced caloric value of sugar alcohols, which is by now well supported by a still growing volume of scientific literature. However, for the purpose of food labeling, it would be rather impractical to allocate specific values to each one of the approved sugar substitutes. It is, therefore, not surprising that in an EEC Council Directive, a caloric value of 2.4 kcal/g has been allocated for all sugar alcohols including xylitol (Official Journal of the European Communities No. L 276/40; Oct. 6, 1990).

## Rationale for the Use of Xylitol as a Sweetener for Noncariogenic Confectionery

According to current knowledge, dental caries is caused by bacteria that accumulate in large masses, known as dental plaque, on the teeth in the absence of adequate oral hygiene. Fermentation of common dietary carbohydrates by plaque bacteria leads to the formation of acid end-products. Acid accumulation and a decrease in plaque pH will follow. A decrease in plaque pH due to bacterial fermentation of carbohydrates may lead to undersaturation of the plaque with respect to calcium and phosphate ions, to demineralization of the tooth enamel, and eventually to formation of a cavity.

Approaches aimed at the prevention or elimination of dental caries include reduction of acid dissolution of tooth mineral and stimulation of enamel remineralization by fluoride, removal of dental plaque by brushing the teeth, and reduction of the availability of fermentable carbohydrates by appropriate dietary habits (85).

For obvious reasons, fermentable sugars cannot be completely eliminated from our daily food supply. From the point of view of caries prevention, this is not even necessary, because a number of studies have indicated that eating sucrose-sweetened foods in moderation and with regular meals results in little or no caries (86). On the other hand, experiments with rats as well as some studies in humans indicate a strong relationship between the frequency of consumption of sucrose between meals and caries activity (87,88). The reason for this observation is that under conditions of frequent consumption of sugary snacks, the plaque pH drops and remains below the critical value of 5.7 for prolonged periods of time (89). As a result, periods with demineralization overwhelm the recovery phases, and dental caries may subsequently develop. This suggests that substitution of sucrose and other fermentable sugars by nonfermentable sugar substitutes has considerable impact on caries activity, even if this substitution is limited to snacks and beverages that are consumed between the main meals. Because of its noncariogenic or even cariostatic properties, xylitol is a particularly suitable substance for this purpose (90,91).

### *Acidogenicity*

There is considerable evidence documenting the fact that xylitol is not fermented by most oral microorganisms and that exposure of dental plaque to xylitol does not result in a reduction of plaque pH (6,92–97). Even after chronic exposure for 2 years, no adaptation occurred with respect to the ability of the dental plaque to ferment xylitol (98,99).

The *noncariogenicity* of xylitol was demonstrated in several rat caries studies in which xylitol was fed in the absence of other readily fermentable dietary sugars. Under these conditions, xylitol did not exhibit any cariogenic potential (6,100). The *cariostatic properties* of xylitol were investigated in numerous studies in which rats received xylitol in admixture to a cariogenic, i.e., sucrose-containing diet. The results of these studies indicate that xylitol tends to reduce the cariogenicity of such diets and that this effect is higher for moderately cariogenic diets with ≤ 25% fermentable sugars than for diets containing ≥ 50% sucrose (6,101–103). More pronounced and more consistent reductions of caries formation were observed in rat studies in which xylitol was administered alternately with sucrose-containing meals, i.e., between the cariogenic challenges (6,104–106). Under these conditions, which mimic the human meal pattern more closely, the caries scores of the xylitol-treated animals were on average 35% below those of the corresponding controls. In one study it was even found that xylitol may promote the remineralization of early caries lesions of rats (106,107). In summary, the results obtained in the rat model suggest that xylitol exhibits cariostatic properties.

### *Efficacy of Xylitol in Human Caries Studies*

The clinical benefit of xylitol for caries prevention has been demonstrated in several studies with children and adult human volunteers in which consumption of xylitol was consistently associated with a significant reduction of the caries increment. In these studies, which were conducted by independent investigators at different locations, the efficacy of xylitol was tested under carefully controlled conditions with randomized assignment of the subjects to the different treatment groups (108–112) as well as under the less standardized but "real-life"–oriented conditions of field studies involving large numbers of caries-prone children (113–115). Regardless of the differences in study design, significant reductions of the caries increment were observed in all xylitol-treated groups. Depending upon the dose (116) and frequency of the xylitol administration (117), this reduction varied between 45 and > 90%.

It is particularly noteworthy that daily fluoride brushing plus xylitol consumption was found to be more efficient than daily fluoride brushing alone (113–116). Considering the different modes of the cariostatic action of fluoride and xylitol, this is not surprising, but it is certainly relevant in terms of a most efficient combination of caries preventive measures (90).

### *Discrimination Against* Streptococcus Mutans

Microbiological investigations in caries-active humans, as well as studies in experimental animals, have demonstrated that certain types of bacteria are particularly active in initiating and promoting dental caries. Culprit number one in this respect is *Streptococcus mutans*. This microorganism has the ability to adhere strongly to the crowns of the teeth by means of sticky extracellular glucans, to grow and metabolize optimally in a relatively acidic environment, and to live under microaerobic or strictly anaerobic conditions, as met in the depth of pits and fissures (118–121).

Several studies have suggested that the predominance of *S. mutans* in cariogenic plaque is dependent upon the ability of this bacterium to remain metabolically active even in a relatively acidic environment. In fact, *S. mutans* is more active at pH 5 than at pH 7, whereas many other members of the plaque flora become metabolically inactive under such conditions. This observation suggests that the frequent ingestion of sucrose gives a competitive edge to *S. mutans* over the other plaque microbes. In this way, a circulus vitiosus for the formation of a more cariogenic plaque is formed (122–124).

In this regard, the usage of xylitol as a sucrose substitute becomes an extremely attractive means to control and prevent dental caries for two different reasons. First, no acid is formed from xylitol by the dental plaque. In fact, during and after chewing of a xylitol-sweetened gum, an elevation rather than a decrease of the plaque pH is observed (125,126). Under such conditions, however, the metabolism of *S. mutans* is, as mentioned above, not optimally active, and other bacteria may successfully compete with *S. mutans.* Second, in addition to this indirect, pH-mediated effect, xylitol appears to inhibit the growth and metabolism of *S. mutans* in a more direct way. Several experiments have shown that the addition of xylitol to a glucose-containing medium reduced the growth of *S. mutans* (127–133). This inhibition by xylitol appears to be related to the accumulation oof xylitol-5-phosphate and xylulose-5-phosphate within the cells (134–137). Probably as a result of the intracellular accumulation of these metabolites, the ability of *S. mutans* to adhere to surfaces is decreased (138,139) and disintegration of the ultrastructure of the cells may occur (140).

In line with these results, lower *S. mutans* counts were found in the plaque and/or saliva of xylitol-treated human volunteers (141–146). These observations suggest that the formation of xylitol-tolerant strains of *S. mutans* after chronic xylitol exposure does not annihilate its inhibitory effect (147).

### *General Effects on Dental Plaque*

Because dental plaque plays a crucial role in the formation of tooth decay, many investigators have examined the effects of xylitol on dental plaque. It is well established that frequent consumption of sucrose promotes the growth of a voluminous, sticky dental plaque. Xylitol is not a substrate for oral microorganisms, and it does therefore not favor plaque formation. In consequence, studies comparing the effects of sucrose and xylitol on dental plaque have demonstrated consistently that the plaque weights are lower in the xylitol-treated subjects than in the positive controls consuming sucrose (6). Although it might be expected that in this regard differences between xylitol and sorbitol would be much smaller because sorbitol is also not a preferred substrate for plaque bacteria, differences between xylitol and sorbitol in relation to plaque formation have recently been reported from two independent studies (148,149).

### Use of Xylitol as a Sweetener in Diabetic Diets

Historically, the first proposed application of xylitol concerned its use as a sugar substitute for diabetics (150). Diabetes mellitus is a chronic metabolic disorder characterized by fasting hyperglycemia and/or plasma glucose levels above defined limits during oral glucose tolerance testing. It is caused either by a total lack of insulin (Type I, insulin-dependent diabetes mellitus) or by insulin resistance in the presence of normal or even elevated plasma insulin levels, i.e. by a decreased tissue sensitivity or responsiveness to insulin (Type II, non–insulin dependent diabetes mellitus).

The major goals of dietetic and drug-based management of diabetes mellitus are to achieve normal control of glucose metabolism and glycemia, and thereby to prevent macro- and microvascular complications. The recommended treatment modalities are: dietary modification, increased physical activity, and pharmacological intervention with either an oral hypoglycemic agent or insulin.

Modification of the diet is the most important element in the therapeutic plan for diabetic patients, and for some patients with Type II diabetes, it is the only intervention needed to control the metabolic abnormalities associated with the disease.

The general aim of dietary intervention is to stabilize the blood glucose levels. Two fundamental precepts of the diabetic diet are, therefore, that smaller, more frequent meals should be taken in place of a few large meals, and that foods containing glucose, maltose, sucrose, and other carbohydrates with a high glycemic index should be avoided (151). However, sweetness is an established property of many foods, and there is evidence that, if the intake of sucrose is restricted as a part of diabetic treatment, the majority of patients feel that alternative sweeteners are a necessity. In a survey of 500 diabetics it was found, for example, that if no other sweeteners were available in the diet, 83% of those questioned would use sucrose for this purpose (152).

Even an "ideal" diet plan is worthless if patients do not adhere to it. In order to increase compliance, it seems therefore appropriate to use low- or nonglycemic sweeteners for the preparation of special diabetic products. Traditionally, one may consider using non-nutritive, intense sweeteners for this purpose. These compounds are noncaloric and have no deleterious effect upon diabetic control. Undoubtedly, they are most useful for the sweetening of beverages (soft drinks, coffee, tea). However, incorporation of nonnutritive sweeteners into solid foods causes a major technological problem. In normal products, sucrose represents a considerable portion of bulk, and its replacement by nonnutritive sweeteners in special products for diabetics requires the addition of bulking agents, i.e., typically fat and/or starch. Since diabetic patients are predisposed to macrovascular disease and are required to restrict their fat intake, products with increased fat content, although carbohydrate-modified, are not recommendable. The addition of starch, on the other hand, increases the glycemic index of the food and eliminates the glycemic and caloric advantage that one hoped to achieve by using an intense sweetener. To avoid these problems, bulk sugar substitutes like fructose, sorbitol, and xylitol may be used as an appropriate alternative. These sweeteners produce only a slight increase in blood glucose concentration and require only small amounts of insulin for their metabolism in both healthy individuals and diabetics.

The effects of xylitol on blood glucose and insulin levels and its general suitability for inclusion in foods for diabetics have been examined in several acute and subchronic studies with healthy and diabetic volunteers. When xylitol is given orally, no increase in blood glucose levels is observed even in diabetic patients (84,153–155). Similarly, plasma insulin concentrations do not rise at all (153) or only moderately (84,156) following oral xylitol application in normal and diabetic subjects. These observations indicate that the conversion of xylitol to glucose—which, in principle, is possible—is apparently too slow to raise the blood glucose concentration to a significant extent.

It has been suspected that the obvious advantage of xylitol in terms of blood glucose and insulin requirement may disappear when it is incorporated into a meal. Therefore, a study was conducted in which 30 g xylitol or sucrose was substituted for an equal amount of starch in a meal of a diabetic diet regimen. The results of this investigation demonstrate that the insulin requirement after sucrose is significantly higher than after starch or xylitol (157).

In an early subchronic study, the application of 45–60 g/day of xylitol had no adverse effects on the metabolic condition of 20 diabetic patients (150). These results were confirmed in a subsequent study in which the effect of 30 g xylitol/day on the carbohydrate and lipid metabolism of 12 well-controlled diabetics (Type II) was examined for a duration of 2–6 weeks (158). Except for transient diarrhea in a few cases, no adverse effects of xylitol administration occurred. Remarkably, however, the urinary glucose excretion disappeared with the xylitol diet at least in some of the participating patients. The good tolerance of xylitol was also established in a study with 18 diabetic children (Type I) who received 30 g/day of xylitol for 4 weeks (159).

Even at a dose of 70g/day administered over a period of 6 weeks, xylitol was well tolerated by Type I and Type II diabetics as well as by healthy nondiabetic controls. Contrary to expectations, however, no significant differences were noted in plasma glycosylated Hb, in fasting or postprandial glucose and in total urinary glucose (160). In conclusion, these investigations demonstrate that xylitol can safely be incorporated as a sweetening agent in the diabetic diet without any negative effects on their metabolic condition.

### *Use of Xylitol in Parenteral Nutrition*

The aim of parenteral nutrition is to optimize fluid, energy, nitrogen, and electrolyte balance by the infusion of adequate solutions providing energy (e.g., glucose or lipids), amino acids, and mineral salts (161,162). However, a stable metabolic condition is often difficult to reestablish in severely injured patients because the posttrauma metabolism may be seriously disturbed in various respects. In particular, the utilization of glucose, which is the body's normal energy source, is often impaired, especially in patients with shock, severe trauma, burns, sepsis, and diabetes.

Under the conditions of such a general disturbance of metabolism and of the hormone-regulated control mechanisms, insulin must be administered concomitantly with the glucose in order to assure sufficient utilization of this energy source. However, this procedure requires careful regular monitoring of the blood glucose levels in order to avoid severe complications. In addition, the infusion of glucose in excess of 0.12–0.24 g/kg/hr may be generally questioned, because higher infusion rates lead neither to a further suppression of the endogenous gluconeogenesis at the expense of physiologically important proteins, nor to an increase of the peripheral glucose utilization (163). Since with this recommended maximum rate of glucose infusion, the energy requirement of the injured or stressed organism is not completely covered, the infusion of fructose and xylitol in addition to glucose has been advocated (12,13,166,167). In several studies, the positive influence of such infusion regimes on nitrogen balance and visceral protein synthesis has been demonstrated in stressed or posttrauma patients (13,166), as well as in an animal model (167). However, these effects were not always reproducible (168,169).

More recently, it has been proposed to supplement lipid emulsions with xylitol. In this combination, xylitol is preferred over glucose because it only marginally stimulates insulin secretion and therefore does not suppress lipolysis, as demonstrated by a positive adenosine triphosphate (ATP) balance. In this way, xylitol could improve the metabolic situation (170,171).

At present, xylitol is used mainly as a glucose substitute in parenteral amino acid solutions, as well as in combination with fructose and glucose in so-called trisugar solutions. These products are frequently applied, particularly in Germany. Solutions of xylitol alone are nowadays rarely used, except in Japan.

Most clinical investigations on the use of xylitol in parenteral solutions have concentrated on the general energetic and amino acid–sparing effects. Other potentially beneficial, metabolic effects have not yet been explored in detail. Since xylitol is metabolized via the pentose phosphate pathway, it is conceivable that the levels of phosphorylated nucleotides (ATP) might be enhanced. This could be relevant in reperfused myocardial tissue, as has been suggested by studies with ribose, which is subject to the same catabolic pathway (172–174). Other potential advantageous effects include an increase of the 2,3-diphosphoglycerate levels and a NADH-mediated reduction of oxidized glutathione in erythrocytes (175–179). This latter effect may be particularly relevant in the tens of millions of people that are affected by different degrees of glucose-6-phosphate dehydrogenase deficiency, which as such causes little concern but which predisposes to radical-induced pathology under certain conditions (drug therapy, sepsis, shock). Whether a reported effective treatment of cardiac arrhythmias with xylitol is also related to a shift to a more reduced status is not known (180,181).

Although a large number of studies indicate the absence of adverse side effects with parenteral xylitol administration, some authors advise against the use of xylitol because of the possibility of increased serum lactic acid, uric acid, and bilirubin concentrations (182,183) and because of the possible deposition of calcium oxalate crystals in the kidney and in the brain (184–187). However, regarding the occurrence of renocerebral oxalosis in association with xylitol infusions, it is noteworthy that in most of these cases the recommended maximum daily dose and/or infusion rate were surpassed. If xylitol is infused at recommended rates ($\leq$ 0.25 g/kg/hr and $\leq$ 3g/kg/day) (188) and in combination with glucose and fructose, such complications are not likely to occur more frequently than in patients not treated with xylitol (189–191). In general, no adverse side effects are observed if these recommendations are observed. Contrary to fructose, sorbitol, and glucose, a metabolic intolerance to xylitol is not known.

## TOXICITY AND TOLERANCE

The results of animal tests for acute toxicity have indicated that xylitol is of very low toxicity by all routes of administration. Conventional tests for embryotoxicity and teratogenicity and for adverse effects on reproduction have given entirely negative results. Similarly, both in vitro and in vivo tests for mutagenicity and clastogenicity have given uniformly negative results (192).

Long-term studies in animals for safety evaluation have included 2-year treatment of rats, mice, and dogs. In these studies, xylitol was tested at a maximum dose level of 20% of the diet. Although the findings of these animal studies generally supported the safety of oral xylitol, some observations required further investigation. These observations were urinary tract calculi in mice and a slight increase in the incidence of adrenal medullary tumors (pheochromocytomas) in male rats (193).

The results of subsequent studies and additional data from experiments with other polyols and lactose demonstrated, however, that the adverse effects observed in mice and rats are generic in nature and lack significance for safety evaluation in humans because of the species specificity of the underlying mechanisms (19,20,194,195).

Human tolerance of high oral doses of xylitol has been investigated in numerous studies with healthy and diabetic volunteers. The results of these studies invariably demonstrate the good tolerance even of extremely high intakes (up to 200 g/day) of xylitol. Adverse changes of clinical parameters were not observed. The only side effect that was occasionally noted was transient laxation and gastrointestinal discomfort (16,192,196–199). Such effects are generally observed after consumption of high doses of sugar alcohols and slowly digestible carbohydrates (e.g., lactose). The slow absorption of these compounds from the gut and the resulting osmotic imbalance are considered to be the cause of these effects, which are readily reversible upon cessation or reduction of the amounts consumed. With continued exposure, tolerance usually develops (200,201).

## REGULATORY STATUS

On a supranational level, the Joint FAO/WHO Expert Committee on Food Additives (JECFA) has allocated an acceptable daily intake (ADI) "not specified," the most favorable ADI possible, for xylitol (202) and the Scientific Committee for Food for the European Economic Community (EEC) proposed "acceptance" of this polyol in 1984. On a national level, xylitol is approved for foods, cosmetics, and pharmaceuticals in many countries (Table 3). Claims such as "noncariogenic" or "safe for teeth" may be applied where appropriate.

**Table 3** Legal Status of Xylitol

| | Food[a] | Cosmetics[b] | Pharmaceuticals[b,c] |
|---|---|---|---|
| Argentina | + | + | + |
| Austria | + | | + |
| Australia | + | | |
| Belgium | + | + | + |
| Bolivia | + | | |
| Brazil | + | + | + |
| Canada | + | + | + |
| Chile | + | + | + |
| Colombia | + | + | + |
| Denmark | + | + | + |
| Federal Republic of Germany | + | | + |
| Finland | + | + | + |
| France | + | | |
| Greece | + | | |
| Israel | + | + | |
| Italy | + | + | + |
| Japan | | | + |
| Luxemburg | + | | |
| Mexico | + | | + |
| Netherlands | + | + | |
| New Zealand | + | + | + |
| Norway | + | + | + |
| Portugal | + | + | + |
| Russia | + | | |
| Singapore | + | + | + |
| South Africa | + | + | + |
| South Korea | | + | + |
| Spain | + | + | + |
| Sweden | + | | |
| Switzerland | + | | + |
| Taiwan | + | + | + |
| Turkey | + | + | + |
| United Kingdom | + | | |
| United States | + | + | |
| Uruguay | + | + | + |
| Venezuela | + | | |
| Yugoslavia | + | + | + |
| JECFA | ++ | | |
| EEC | ++ | | |

+ = Accepted (specific regulations on special foods, labeling, or registration are not considered in this table).
+ + = Evaluated with a recommendation of "acceptable."
[a]Approval may be limited to certain food categories.
[b]Registration of product may be required.
[c]Use as an excipient; in a few countries also for infusion solutions.

## REFERENCES


1. L. Hyvoenen, R. Kurkela, P. Koivistoinen, and P. Merimaa, Effects of temperature and concentration on the relative sweetness of fructose, glucose and xylitol, *Lebensm. Wiss.Technol., 10:* 316–320 (1977).
2. S. L. Munton and G. G. Birch, Accession of sweet stimuli to receptors. I. Absolute dominance of one molecular species in binary mixtures, *J. theor. Biol., 112:* 539–551 (1985).
3. K. K. Maekinen and E. Soederling, A quantitative study of mannitol, sorbitol, xylitol, and xylose in wild berries and commercial fruits, *J. Food Sci., 45:* 367–374 (1980).
4. J. Washuett, P. Riederer, and E. Bancher, A qualitative and quantitative study of sugar-alcohols in several foods, *J. Food Sci., 38:* 1262–1263 (1973).
5. O. Touster, The metabolism of polyols, in *Sugars in Nutrition* (H. L. Sipple and K. W. McNutt, eds.), Academic Press, New York, 1974, pp. 229–239.
6. A. Baer, Caries prevention with xylitol, a review of the scientific evidence, *Wld. Rev. Nutr. Diet., 55:* 183–209 (1988).
7. K. K. Maekinen, Sweeteners and prevention of dental caries, *Oral Health, 78:* 57–66 (1988).
8. H. A. B. Linke, Sugar alcohols and dental health, *Wld. Rev. Nutr. Diet., 47:* 134–162 (1986).
9. W. J. Loesche, The rationale for caries prevention through the use of sugar substitutes, *Int. Dent. J., 35:* 1–8 (1985).
10. I. Kleinberg, Oral effects of sugars and sweeteners, *Int. Dent. J., 35:* 180–189 (1985).
11. H. Foerster and H. Mehnert, Die orale Anwendung von Xylit als Zuckeraustauschstoff in der Diät des Diabetes mellitus, *Akt. Ernährungsmedizin, 4:* 296–314 (1979).
12. M. Georgieff, R. H. Ackermann, K. H. Baessler, and H. Lutz, Die Vorteile von Xylit gegenüber Glucose als Energiedonator in Rahmen der frühen postoperativen, parenteralen Nährstoffzufuhr, *Z. Ernährungswiss., 21:* 27–42 (1982).
13. M. Georgieff, L. L. Moldawer, B. R. Bistrian, and G. L. Blackburn. Xylitol, an energy source of intravenous nutrition after trauma, *J. Enteral Parenteral Nutr., 9:* 199–209 (1985).
14. J. N. Counsell, in *Xylitol* (J. N. Counsell, ed.), Appl. Sci. Publishers Ltd., London, 1978, p. 191.
15. L. Hyvoenen and P. Koivistoinen, Food technological evaluation of xylitol, *Adv. Food Res., 28:* 373–403 (1982).
16. K. K. Maekinen, Biochemical principles of the use of xylitol in medicine and nutrition with special consideration of dental aspects, *Experientia, 30:* 1–160 (1978).
17. R. Ylikahri, Metabolic and nutritional aspects of xylitol, *Adv. Food Res., 25:* 159–180 (1979).
18. K. K. Maekinen and A. Scheinin, Xylitol and dental caries, *Ann. Rev. Nutr., 2:* 133–150 (1982).

19. A. Baer, Safety assessment of polyol sweeteners—some aspects of toxicology, *Food Chem., 16:* 231–241 (1985).
20. A. Baer, Toxicological aspects of sugar alcohols—studies with xylitol, in *Low digestibility carbohydrates,* Proceedings of a workshop held at the TNO-CIVO Institutes Zeist, The Netherlands, November 27–28, 1986, pp. 42–50.
21. E. Fischer and R. Stahel, Zur Kenntnis der Xylose, *Berichte Dtsch. Chem. Gesellschaft, 24:* 528–539 (1891).
22. G. M. Jaffe, Xylitol,—a specialty sweetener, *Sugar y Azucar, 73:* 36–42 (1978).
23. C. Aminoff, E. Vanninen, and T. E. Doty, The occurrence, manufacture and properties of xylitol, in *Xylitol* (J. N. Counsell, ed.), Applied Science Publishers Ltd., London, 1978, pp. 1–9.
24. J. F. Carson, S. W. Waisbrot, and F. T. Jones, A new form of crystalline xylitol, *J. Am. Chem. Soc., 65:* 1777–1778 (1943).
25. H. Onishi and T. Suzuki, Microbial production of xylitol from glucose, *Appl. Microbiol., 15:* 1031–1035 (1969).
26. L. F. Chen and C. S. Gong, Fermentation of sugarcane bagasse hemicellulose hydrolysate to xylitol by a hydrolysate-acclimatized yeast, *J. Food Sci., 50:* 226–228 (1985).
27. C. S. Gong, T. A. Claypool, L. D. McCracken, C. M. Maun, P. P. Ueng, and G. T. Tsao. Conversion of pentoses by yeasts, *Biotechnol. Bioeng., 25:* 85–102 (1983).
28. V. Kitpreechavanich, N. Nishio, M. Hayashi, and S. Nagai, Regeneration and retention of NADP (H) for xylitol production in an ionized membrane reactor, *Biotechnol. Lett., 7:* 657–662 (1985).
29. D. Holland and J. F. Stoddart, A. stereoselective synthesis of xylitol, *Tetrahed. Lett., 23:* 5367–5370 (1982).
30. J. Kiss, R. D'Souza, and P. Taschner. Präparative Herstellung von 5-Desoxy-L-arabinose, Xylit und D-Ribose aus "Diacetonglucose," *Helv. Chim. Acta, 58:* 311–317 (1975).
31. T. Pepper and P. M. Olinger. Xylitol in sugar-free confections, *Food. Techn., 42:* 98–106 (1988).
32. C. Krueger, Zuckerfreie Pralinen—zuckerfreie Füllungen und Schokoladenmassen, *Süsswaren,* 11: 506–516 (1987).
33. T. Pepper, Sugar substitutes, their use in chocolate and chocolate fillings, *Manuf. Conf., 67*(6): 83 (1987).
34. C. Krueger, Zuckeraustauschstoffe, Arten, technologische, sensorische und ernährungsphysiologische Eigenschaften und Synergie-Effekte, *Zucker- und Süsswaren Wirtschaft, 41*(11); 360–365 (1988).
35. A. G. Dodson and T. Pepper, Confectionery technology and the pros and cons of using non-sucrose sweeteners, *Fd. Chem., 16:* 271–280 (1985).
36. T. Imfeld, Zahnschonende Halswehlutschtabletten, *Swiss Dent, 5:* 19–22 (1984).
37. F. Bray, Zuccheri alternativi e loro influenza nel gelato, *Ind. Aliment., 24:* 895–898 (1985).
38. A. Askar, M. G. Abd El-Fadeel, M. A. Sadek, A. M. A. El Rakaybi, and

G. A. Mostafa, Studies on the production of dietetic cake using sweeteners and sugar substitutes, *Dtsch. Lebensm. Rundschau, 83:* 389–394 (1987).

39. P. Varo, C. Westermarck-Rosendahl, L. Hyvoenen, and P. Koivistoinen, The baking behavior of different sugars and sugar alcohols as determined by high pressure liquid chromatography, *Lebensm. Wiss. Technol., 12:* 153–156 (1979).
40. H. R. Moskowitz, The sweetness and pleasantness of sugars. *Am. J. Psychol., 84:* 387–405 (1971).
41. J. Gutschmidt and G. Ordynsky. Bestimmung des Süssungsgrades von Xylit, *Dtsch. Lebensm. Rdsch., 57:* 321–324 (1961).
42. M. G. Lindley, G. G. Birch, and R. Khan. Sweetness of sucrose and xylitol, structural considerations, *J. Sci. Food Agric., 27:* 140–144 (1976).
43. S. Yamaguchi, T. Yoshikawa, S. Ikeda, and T. Ninomiya, Studies on the taste of some sweet substances, *Agric. Biol. Chem., 34:* 181–197 (1970).
44. C. K. Lee, Structural functions of taste in the sugar series: Taste properties of sugar alcohols and related compounds, *Fd. Chem.,* 2: 95–105 (1977).
45. R. J. Feigal, M. E. Jensen, and C. A. Mensing, Dental caries potential of liquid medications, *Pediatrics, 68:* 416–419 (1981).
46. P. E. Souney, D. A. Cyr, J. T. Chang, and A. E. Kaul, Sugar content of selected pharmaceuticals, *Diabetes Care, 6:* 231–240 (1983).
47. I. F. Roberts and G. J. Roberts, Relation between medicines sweetened with sucrose and dental disease, *Br. Med. J., 2:* 14–16 (1979).
48. P. J. M. Crawford, Sweetened medicine and caries, *Pharm. J., 226:* 668–669 (1981).
49. H. R. Muehlemann and A. Firestone, Drei neue "zahnschonende" Präparate, *Swiss Dent, 3:* 25–30 (1982).
50. E. Kristoffersson and S. Halme, Xylitol as an excipient in oral lozenges, *Acta Pharm. Fenn., 87:* 61–73 (1978).
51. R. Laakso, K. Sneck, and E. Kristoffersson, Xylitol and Avicel pH 102 as excipients in tablets made by compression and from granulate, *Acta Pharm. Fenn., 91:* 47–54 (1982).
52. D. W. Bloch, M. A. El Egakey, and P. P. Speiser, Verbesserung der Auflösungsgeschwindigkeit von Hydrochlorothiazid durch feste Dispersionen mit Xylitol, *Acta Pharm. Technol., 18:* 177–183 (1982).
53. I. Sirenius, V. E. Krogerus, and T. Leppaenen, Dissolution rate of p-aminobenzoates from solid xylitol dispersions, *J. Pharm. Sci., 68:* 791–792 (1979).
54. F. Voirol and R. Etter, Verfahren zur Erzeugung einer Deckschicht auf Tablettenkernen, German Patent Appl. 29 13 555 (1979).
55. K. K. Maekinen, E. Soderling, and I. Laeikkoe, Zuckeralkohole (Polyole) als "aktive" Zahnpastenbestandteile. *Oralprophylaxe, 9:* 115–120 (1987).
56. K. K. Maekinen, E. Soederling, H. Hurttia, O. P. Lehtonen, and E. Luukkala, Biochemical, microbiologic and clinical comparisons between two dentifrices that contain different mixtures of sugar alcohols, *J. Am. Dent. Assoc., 111:* 745–751 (1985).

57. A. A. Scheie, O. Fejerskov, S. Assev, and G. Roella, Ultrastructural changes in Streptococcus sobrinus induced by xylitol, NaF and $ZnCl_2$, *Caries Res., 23:* 320–327 (1989).
58. J. Arends, J. Christoffersen, J. Schuthof, and M. T. Smits, Influence of xylitol on demineralization of enamel, *Caries Res., 18:* 296–301 (1984).
59. M. T. Smits and J. Arends, Influence of extraoral xylitol and sucrose dippings on enamel demineralization in vivo, *Caries Res., 22:* 160–165 (1988).
60. J. D. B. Featherstone, T. W. Cutress, B. E. Rodgers, and P. J. Dennison, Remineralization of artificial caries-like lesions in vivo by a self-administered mouthrinse or paste, *Caries Res., 16:* 235–242 (1982).
61. A. Vissink, E. J. 's-Gravenmade, T. B. F. M. Gelhard, A. K. Panders, and M. H. Franken, Rehardening properties of mucin- or CMC-containing saliva substitutes on softened human enamel, *Caries Res., 19:* 212–218 (1985).
62. Cobanera, A. Mopasso, E. White, P. Cuevas, and R. Espinosa, Xylitol-sodium fluoride: Effect on plaque, *J. Dent. Res., 66:* Abstr. 56 (1987).
63. L. G. Petersson, M. Johansson, G. Joensson, D. Birkhed, and A. Gleerup, Caries preventive effect of toothpastes containing different concentration and mixture of fluorides and sugar alcohols, *Caries Res., 23:* 109 (1989).
64. G. Grimble, Fibre, fermentation, flora, and flatus, *Gut., 30:* 6–13 (1989).
65. A. A. Salyers and J. A. Z. Leedle, Carbohydrate metabolism in the human colon, in *Human Intestinal Flora in Health and Disease* (D. J. Hentges, ed.), Academic Press, New York, 1983, pp. 129–146.
66. J. H. Cummings, Short chain fatty acids in the human colon, *Gut, 22:* 763–779) (1981).
67. H. Ruppin, S. Bar-Meir, K. H. Soergel, C. M. Wood, and M. G. Schmitt Jr., Absorption of short-chain fatty acids by the colon, *Gastroenterol., 78:* 1500–1507 (1980).
68. J. H. Cummings, E. W. Pomare, W. J. Branch, C. P. E. Naylor, and G. T. Mac Farlane. Short chain fatty acids in human large intestine, portal, hepatic and venous blood, *Gut, 28:* 1221–1227 (1987).
69. B. M. Buckley and D. H. Williamson, Origin of blood acetate in the rat, *Biochem. J., 166:* 539–545 (1977).
70. C. L. Skutches, C. P. Holroyde, R. N. Myers, P. Paul and G. A. Reichard, Plasma acetate turnover and oxidation, *J. Clin. Invest., 64:* 708–713 (1979).
71. J. H. Cummings, H. N. Englyst, and H. S. Wiggins, The role of carbohydrates in lower gut function, *Nutr. Rev., 44:* 50–54 (1986).
72. N. I. Mc Neil, The contribution of the large intestine to energy supplies in man, *Am. J. Clin. Nutr., 39:* 338–342 (1984).
73. S. E. Fleming and D. S. Arce, Volatile fatty acids: Their production, absorption, utilization, and roles in human health, *Clin. Gastroenterol., 15:* 787–814 (1986).
74. G. E. Demetrakopoulos and H. Amos, Xylose and xylitol, *World Rev. Nutr. Diet., 32:* 96–122 (1978).
75. K. H. Baessler, Absorption, metabolism, and tolerance of polyol sugar substitutes, *Pharmacol. Ther. Dent., 3:* 85–93 (1978).

76. J. F. Williams, K. K. Arora, and J. P. Longenecker. The pentose pathway: A random harvest. *Int. J. Biochem., 19:* 749–817 (1987).
77. T. Wood, Physiological functions of the pentose phosphate pathway, *Cell Biochemistry and Function, 4:* 241–247 (1986).
78. N. Z. Baquer, J. S. Hothersall, and P. McLean, Function and regulation of the pentose phosphate pathway in brain, *Curr. Top. Cell. Requl., 29:* 265–289 (1988).
79. A. B. Lane, On the nature of L-xylulose reductase deficiency in essential pentosuria, *Biochem. Genet., 23:* 61–72 (1985).
80. K. Soyama and N. Furukawa, A Japanese case of pentosuria—case report, *J. Inher. Metab. Dis., 8:* 37 (1985).
81. A. Baer, Factorial calculation model for the estimation of the physiological caloric value of polyols. In press (1991).
82. Anonymous, De energetische waarde van suikeralcoholen, *Voeding, 48:* 357–365 (1987).
83. E. Karimzadegan, A. J. Clifford, and F. W. Hill, A rat bioassay for measuring the comparative availability of carbohydrates and its application to legume foods, pure carbohydrates and polyols, *J. Nutr., 109:* 2247–2259 (1979).
84. R. Mueller-Hess, C. A. Geser, J. P. Bonjour, E. Jequier, and J. P. Felber, Effects of oral xylitol administration on carbohydrate and lipid metabolism in normal subjects, *Infusionstherapie, 2:* 247–252 (1975).
85. J. H. Shaw, Medical progress—causes and control of dental caries, *N. Engl. J. Med., 317:* 996–1004 (1987).
86. E. Newburn, Sugar and dental caries; A review of human studies, *Science, 217:* 418–423 (1982).
87. H. G. Huxley, The effect of feeding frequency on rat caries, *J. Dent. Res., 56:* 976 (1977).
88. B. Gustaffson, C. E. Quensel, L. Lanke, C. Lundqvist, W. Grahnen, B. E. Gonow, and B. Krasse, The Vipeholm dental caries study: The effect of different levels of carbohydrate intake on caries activity in 436 individuals observed for 5 years, *Acta Odont. Scand., 11:* 232–363 (1953).
89. T. Imfeld and H. R. Muehlemann, Cariogenicity and acido-genicity of food, cofectionery and beverages, *Pharmacol. Ther. Dent., 3:* 53–68 (1978).
90. A Baer, Significance and promotion of sugar substitution for the prevention of dental caries, *Lebensm.-Wiss. u.-Technol., 22:* 46–53 (1989).
91. W. J. Loesche, The rationale for caries prevention through the use of sugar substitutes, *Int. Dent. J., 35:* 1–8 (1985).
92. H. J. Guelzow, Ueber den anaeroben Umsatz von Palatinit durch Mikroorganismen der menschlichen Mundhöhle, *Dtsch. zahnärztl. Z., 37:* 669–672 (1982).
93. B. G. Bibby and J. Fu, Changes in plaque pH in vitro by sweeteners, *J. Dent. Res., 64:* 1130–1133 (1985).
94. A. R. Firestone and J. M. Navia, In vivo measurements of sulcal plaque pH in rats after topical applications of xylitol, sorbitol, glucose, sucrose, and sucrose plus 53 mM sodium fluoride, *J. Dent. Res., 65:* 44–48 (1986).

95. M. Rekola, Acid production from xylitol products in vivo and in vitro, *Proc. Finn. Dent. Soc., 84:* 39–44 (1988).
96. T. N. Imfeld, Identification of low caries risk dietary components, *Monogr. Oral Sci., 11:* (1983).
97. Y. Maki, K. Ohta, I. Takazoe, et al., Acid production from isomaltulose, sucrose, sorbitol, and xylitol in suspensions of human dental plaque (short communication), *Caries Res., 17:* 335–339 (1983).
98. K. K. Maekinen, E. Soederling, M. Haemaelaeinen, and P. Antonen, Effect of long-term use of xylitol on dental plaque, *Proc. Finn. Dent. Soc., 81:* 28–35 (1985).
99. F. Gehring, Mikrobiologische Untersuchungen im Rahmen der "Turku Sugar Studies," *Dtsch. Zahnärztl. Z., 32:* 84–88 (1977).
100. J. S. Van der Hoeven, Cariogenicity of lactitol in program-fed rats (short communication), *Caries Res., 20:* 441–443 (1986).
101. R. Havenaar, J. H. J. Huis in't Veld, J. D. de Stoppelaar, and O. Backer Dirks, Anti-cariogenic and remineralizing properties of xylitol in combination with sucrose in rats inoculated with Streptococcus mutans, *Caries Res., 18:* 269–277 (1984).
102. R. Havenaar, The anti-cariogenic potential of xylitol in comparison with sodium fluoride in rat caries experiments, *J. Dent. Res., 63:* 120–123 (1984).
103. E. J. Karle and F. Gehring, Zur Kariogenität von Mischungen aus Kohlenhydraten und Xylit im konventionellen und gnotobiotischen Tierversuch, *Dtsch. Zahnärztl. Z., 42:* 835–840 (1987).
104. R. Havenaar, Effects of intermittent feeding of sugar and sugar substitutes on experimental caries and the colonization of Streptococcus mutans in rats, in *Sugar Substitutes and Dental Caries* (Havenaar R., ed.), Drukkerij Elinkwijk BV, Utrecht, 1984, pp. 73–79.
105. A. Hefti, Cariogenicity of topically applied sugar substitutes in rats under restricted feeding conditions, *Caries Res., 14:* 136–140 (1980).
106. S. A. Leach and R. M. Green, Effect of xylitol-supplemented diets on the progressional and regression of fissure caries in the albino rat, *Caries Res., 14:* 16–23 (1980).
107. S. A. Leach, E. A. Agalamanyi, and R. M. Green, Remineralisation of the teeth by dietary means, in *Demineralisation and remineralisation of the teeth* (S. A. Leach and W. M. Edgar, eds.), IRL Press Ltd., Oxford, 1983, pp. 51–73.
108. A. Scheinin, K. K. Maekinen, and K. Ylitalo, Turku sugar studies V. Final report on the effect of sucrose, fructose and xylitol diets on the caries incidence in man, *Acta Odontol. Scand., 33*(Suppl. 70): 67–104 (1975).
109. M. Rekola, Changes in buccal white spots during 2-year consumption of dietary sucrose or xylitol, *Acta. Odontol. Scand., 44:* 285–290 (1986).
110. M. Rekola, Approximal caries development during 2-year total substitution of dietary sucrose with xylitol, *Caries Res., 21:* 87–94 (1987).
111. A. Scheinin, K. K. Maekinen, E. Tammisalo, and M. Rekola, Turku sugar studies XVIII. Incidence of dental caries in relation to 1-year consumption

of xylitol chewing gum, *Acta Odontol. Scand., 33*(Suppl. 70): 307–316 (1975).
112. M. Rekola, A planimetric evaluation of approximal caries progression during one year of consuming sucrose and xylitol chewing gums, *Proc. Finn. Dent. Soc., 82:* 213–218 (1986).
113. A. Scheinin, J. Banoczy, J. Szoeke, I. Esztari, K. Pienihaekkinen, U. Scheinin, J. Tiekso, P. Zimmermann, and E. Hadas, Collaborative WHO xylitol field studies in Hungary. I. Three-year caries activity in institutionalized children, *Acta Odontol. Scand., 43:* 327–347 (1985).
114. D. Kandelman, A. Baer, and A. Hefti, Collaborative WHO xylitol field study in French Polynesia. I. Baseline prevalence and 32 months caries increment, *Caries Res., 22:* 55–62 (1988).
115. D. Kandelman and G. Gagnon, Clinical results after 12 months from a study of the incidence and progression of dental caries in relation to consumption of chewing-gum containing xylitol in school preventive programs, *J. Dent. Res., 66:* 1407–1411 (1987).
116. D. Kandelman and G. Gagnon, Effect on dental caries of xylitol chewing gum; two year results, *J. Dent. Res., 67:* 172 (abstr. no. 472) (1988).
117. P. Isokangas, Xylitol chewing gum in caries prevention. A longitudinal study on Finnish school children, *Proceedings of the Finnish Dental Society, 83:* 1–117 (1987).
118. S. Hamada and H. D. Slade, Biology, immunology and cariogenicity of Streptococcus mutans, *Microbiol. Rev., 44:* 331–384 (1980).
119. J. van Houte, Bacterial specificity in the etiology of dental caries, *Int. Dent. J., 30:* 305–326 (1980).
120. S. Alaluusua, M. Nystroem, L. Groenroos, and L. Peck, Caries-related microbiological findings in a group of teenagers and their parents, *Caries Res., 23:* 49–54 (1989).
121. C. Stecksen-Blicks, Lactobacilli and *Streptococcus mutans* in saliva, diet and caries increment in 8- and 13-year-old children, *Scand. J. Dent. Res., 95:* 18–26 (1988).
122. A. A. Scheie, P. Arneberg, D. Orstavik, and J. Afseth, Microbial composition, pH-depressing capacity and acidogenicity of 3-week smooth surface plaque developed on sucrose regulated diets in man, *Charies Res., 18:* 74–86 (1984).
123. D. S. Harper and W. J. Loesche, Growth and acid tolerance of human dental plaque bacteria, *Arch. Oral Biol., 29:* 843–848 (1984).
124. H. D. Sgan-Cohen, E. Newbrun, R. Huber, and M. N. Sela, The effect of low-carbohydrate diet on plaque pH response to various foods, *J. Dent. Res., 66:* 911 (Abstr. 22) (1987).
125. H. J. Maiwald, J. Banoczy, W. Tietze, Z. Toth, and A. Vegh, Die Beeinflussung des Plaque-pH durch zuckerhaltigen und zuckerfreien Kaugummi, *Zahn-, Mund-, Kieferheilk., 70:* 598–604 (1982).
126. T. M. Graber, T. P. Muller, and V. D. Bhatia, The effect of xylitol gum and

rinses on plaque acidogenesis in patients with fixed orthodontic appliances, *Swed. Dent. J., 15:* 41–55 (1982).
127. S. Assev, S. M. Waler, and G. Roella, Further studies on the growth inhibition of some oral bacteria by xylitol, *Acta Path. Microbiol. Immunol. Scand. Sect. B., 91:* 261–265 (1983).
128. C. Vadeboncoeur, L. Trahan, C. Mouton, and D. Mayrand, Effect of xylitol on the growth and glycolysis of acidogenic oral bacteria. *J. Dent. Res., 62:* 882–884 (1983).
129. S. Assev, G. Vegarud, and G. Roella, Addition of xylitol to the growth medium of *Streptococcus mutans* omz 176—effect on the synthesis of extractable glycerol-phosphate polymers, *Acta Path. Microbiol. Immunol. Scand. Sect. B, 93:* 145–149 (1985).
130. S. Assev and G. Roella, Further studies on the growth inhibition of *Streptococcus mutans* omz 176 by xylitol, *Acta Path. Microbiol. Immunol. Scand. Sect. B, 94:* 97–102 (1986).
131. S. C. Ziesenitz and G. Siebert, Nonnutritive sweeteners as inhibitors of acid formation by oral microorganisms, *Caries Res., 20:* 498–502 (1986).
132. H. J. A. Beckers, Influence of xylitol on growth, establishment, and cariogenicity of *Streptococcus mutans* in dental plaque of rats, *Caries Res., 22:* 166–173 (1988).
133. G. Siebert, P. Thim, H. P. Brenner, and S. C. Ziesenitz, Antiacidogenic effects of non-cariogenic bulk sweeteners on fermentation of sucrose by *S. mutans* NCTC 10449, *J. Dent. Res., 67:* 696 (Abstr. 110) (1988).
134. S. Assev and A. A. Scheie, Xylitol metabolism in xylitol-sensitive and xylitol-resistant strains of *Streptococci, Acta Path. Microbiol. Immunol. Scand. Sect. B., 94:* 239–243 (1986).
135. L. Trahan, M. Bareil, L. Gauthier, and C. Vadeboncoeur, Transport and phosphorylation of xylitol by a fructose phosphotransferase system in *Streptococcus mutans, Caries Res., 19:* 53–63 (1985).
136. S. Assev and G. Roella, Sorbitol increases the growth inhibition of xylitol on *Strep. mutans* OMZ 176, *Acta Path. Microbiol. Immunol. Scand. Sect. B, 94:* 231–237 (1986).
137. S. M. Waaler, S. Assev, and G. Roella, Metabolism of xylitol in dental plaque, *Scand. J. Dent. Res., 93:* 218–221 (1985).
138. E. Soederling, L. Alaraeisaenen, A. Scheinin, and K. K. Maekinen, Effect of xylitol and sorbitol on polysaccharide production by and adhesive properties of *Streptococcus mutans, Caries Res.,* 21: 109–116 (1987).
139. J. Verran and D. B. Drucker, Effects of two potential sucrose-substitute sweetening agents on deposition of an oral Streptococcus on glass in the presence of sucrose, *Archs. Oral Biol., 27:* 693–695 (1982).
140. H. Tuompo, J. H. Meurman, K. Lounatmaa, and J. Linkola, Effect of xylitol and other carbon sources on the cell wall of *Streptococcus mutans, Scand. J. Dent. Res., 91:* 17–25 (1983).
141. C. A. Overmyer, E. Soederling, P. Isokangas, J. Tenovuo, and K. K.

Maekinen, Oral biochemical status of children given xylitol gums, *J. Dent. Res., 66:* (Spec. Issue) Abstr. 1551 (1987).
142. E. Soederling, P. Isokangas, J. Tenovuo, K. K. Maekinen, S. Mustakallio, and H. Maennistoe, Habitual xylitol consumption and *Streptococcus mutans* in plaque and saliva, *J. Dent. Res., 68:* (Spec. Issue) Abstr. 953 (1989).
143. W. J. Loesche, The effect of sugar alcohols on plaque and saliva level of *Streptococcus mutans, Swed. Dent., J., 8:* 125–135 (1984).
144. K. K. Maekinen, E. Soederling, P. Isokangas, J. Tenovuo, and J. Tiekso, Oral biochemical status and depression of Streptococcus mutans in children during 24- to 36-month use of xylitol chewing gum, *Caries Res., 23:* 261–267 (1989).
145. J. Banoczy, M. Orsos, K. Pienihaekkinen, and A. Scheinin, Collaborative WHO xylitol field studies in Hungary. IV. Saliva levels of Streptococcus mutans, *Acta Odontol. Scand., 43:* 367–370 (1985).
146. F. Gehring, Mikrobiologische Untersuchungen im Rahmen der "Turku Sugar Studies," *Dtsch. Zahnärztl. Z., 32:* 84–88 (1977).
147. L. Trahan and C. Mouton, Selection for Streptococcus mutans with an altered xylitol transport capacity in chronic xylitol consumers, *J. Dent. Res., 66:* 982–988 (1987).
148. E. Soederling, K. K. Maekinen, C. Y. Chen, H. R. Pape, and P. L. Maekinen, Effect of sorbitol, xylitol or xylitol/sorbitol gum on plaque, *J. Dent. Res., 67:* 279 (Abstr. 1334) (1988).
149. P. Soparkar, P. DePaola, S. Vratsanos, and I. Mandel, The effect of xylitol chewing gums on plaque and gingivitis in humans, *J. Dent. Res., 68:* Abstr. 833 (1989).
150. C. H. Mellinghoff, Ueber die Verwendbarkeit des Xylit als Ersatzzucker bei Diabetikern, *Klin. Wochenschr., 39:* 447 (1961).
151. J. P. Felber, Diabetes mellitus (VI): The dietary use of sugar substitutes in diabetes, *Beta Release, 7:* 6–9 (1983).
152. H. Mehnert, Ueber den relativen Wert von Zuckeraustauschstoffen und Süsstoffen in der Diabetesdiät, *Wiss. Veröff. Dtsch. Ges. Ernähr., 20:* 79–84 (1971).
153. F. Manenti and L. Della Casa, Effetti dello xylitolo sull'equilibrio glicidico del diabetico, *Boll. Soc. Medico-Chirurgica di Modena, 65:* 1–8 (1965).
154. J. K. Huttunen, K. K. Maekinen, and A. Scheinin, Turku Sugar Studies XI. Effects of sucrose, fructose, and xylitol diets on glucose, lipid and urate metabolism, *Acta Odontol. Scand., 33:* (Suppl. 70): 239–245 (1975).
155. S. Yamagata, Y. Goto, A. Ohneda, M. Anzai, S. Kawashima, M. Chiba, Y. Maruhama, and Y. Yamauchi, Clinical effects of xylitol on carbohydrate and lipid metabolism in diabetes, *Lancet, ii:* 918–921 (1965).
156. W. Berger, H. Goeschke, J. Moppert, and H. Kuenzli, Insulin concentrations in portal venous and peripheral venous blood in man following administration of glucose, galactose, xylitol and tolbutamide, *Horm. Metab. Res., 5:* 4–8 (1973).
157. W. Hassinger, G. Sauer, U. Cordes, J. Beyer, and K. H. Baessler, The

effects of equal caloric amounts of xylitol, sucrose and starch on insulin requirements and blood glucose levels in insulin-dependent diabetics, *Diabetologia, 2:* 37–40 (1981).
158. S. Yamagata, Y. Goto, A. Ohneda, M. Anzai, S. Kawashima, J, Kikuchi, M. Chiba, Y. Maruhama, Y. Yamauchi, and T. Toyota, Clinical application of xylitol in diabetics, in *Pentoses and Pentitols* (B. L. Horecker, K. Lang, and Y. Takagi, eds.), Springer-Verlag, Berlin 1969, pp. 316–325.
159. H. Foerster, S. Boecker, and A. Walther, Verwendung von Xylit als Zuckeraustauschstoff bei diabetischen Kindern, *Fortschr. Med., 95:* 99–102 (1977).
160. R. A. Bonner, D. C. Laine, B. J. Hoogwerf, and F. C. Goetz, Effects of xylitol, fructose and sucrose in types I and II diabetics: A controlled, cross-over diet study on a clinical research center, *Diabetes, 31* (Suppl. 2): 80A (Abstr.) (1982).
161. L. K. R. Shanbhogue, W. J. Chwals, M. Weintraub, G. L. Blackburn, and B. R. Bistrian, Parenteral nutrition in the surgical patient, *Br. J. Surg., 74:* 172–180 (1987).
162. J. E. Schmitz, R. Doelp, K. H. Altemeyer, A. Gruenert, and F. W. Ahnefeld, Parenterale Ernährung: Stoffwechsel und Substrate, *Arzneimitteltherapie, 3:* 162–172 (1985).
163. R. R. Wolfe, F. O'Donneil Jr., M. D. Stone, D. A. Richmand, and J. F. Burke, Investigation of factors determining the optimal glucose infusion rate in total parenteral nutrition, *Metabolism, 29:* 892–900 (1980).
164. M. Georgieff, L. L. Moldawer, D. Wagner, K. Geiger, W. Fekl, G. L. Blackburn, B. R. Bistrian and H. Lutz, Stoffwechselorientierte postoperative Ernährungstherapie—Möglichkeiten und Grenzen der Anwendung von Glukose und Xylit, *Infusionstherapie, 14:* 53–64 (1987).
165. D. Loehlein, Hypokalorische parenterale Ernährung, *Beitr. Infusionstherapie klin. Ernähr., 16:* 54–63 (1986).
166. R. Doelp, A. Gruenert, E. Schmitz, and F. W. Ahnefeld, Klinische Untersuchungen zur periphervenösen parenteralen Ernährung, *Beitr. Infusionstherapie klin. Ernähr., 16:* 64–76 (1986).
167. M. Georgieff, L. L. Moldawer, B. R. Bistrian, and G. L. Blackburn, Mechanisms for the protein-sparing action of xylitol during partial parenteral feeding after trauma, *Surg. Forum, 35:* 105–108 (1984).
168. W. Behrendt, C. Minale, and G. Giani, Parenterale Ernährung nach herzchirurgischen Operationen, *Infusionstherapie, 11:* 316–322 (1984).
169. M. Semsroth and K. Steinbereithner, Stickstoffmetabolismus und renale Aminosäurenausscheidung während totaler parenteraler Ernährung hypermetaboler Patienten mit verschiedenen Kohlenhydratregimen, *Infusionstherapie, 12:* 136–148 (1985).
170. M. Georgieff, Zur Stoffwechselwirkung von intravenös verabreichter Glukose, Xylit oder Glyzerin während unterschiedlicher Dosierung und Kombination, *Beitr, Infusionstherapie klin. Ernähr., 16:* 103–119 (1986).
171. C. Dillier and A. Leutenegger, Vereinfachung der parenteralen Ernähr-

ungstherapie durch eine neue, gebrauchsfertige all-in-one-Lösung, presented at the Annual Meeting of the Schweizerische Gesellschaft für Ernährungsforschung, March 5, 1988.
172. H. G. Zimmer and E. Gerlach, Stimulation of myocardial adenine nucleotide biosynthesis by pentoses and pentitols, *Pflügers Arch., 376:* 223–227 (1978).
173. H. G. Zimmer et al., Combination of ribose with calcium antagonist and $\beta$-blocker treatment in closed-chest rats, *J. Mol. Cell. Cardiol., 19:* 635–639 (1987).
174. J. F. McAnulty, J. H. Southard, and F. O. Belzer, Improved maintenance of adenosine triphosphate in five-day perfused kidneys with adenine and ribose, *Transplantation Proceedings, 19:* 1376–1379 (1987).
175. V. Stigge, D. Strauss, H. Roigas, and H. J. Raderecht, Verbesserte Erhaltung der zellulären Konzentration von 2,3-Diphosphoglyzerat im Konservenblut durch Zusatz von Xylit, *Dtsch. Ges. Wesen, 28:* 1805–1810 (1973).
176. S. Vora, Metabolic manipulation of key glycolytic enzymes: A novel proposal for the maintenace of red cell 2,3-DPG and ATP levels during storage, *Biomed. Biochim. Acta, 46:* 285–289 (1987).
177. T. Asakura, K. Adachi, and H. Yoshikawa, Reduction of oxidized glutathione by xylitol, *J. Biochem., 67:* 731–733 (1970).
178. J. L. Pousset, B. Boum, and A. Cavé, Action antihémolytique du xylitol isolé des écorces de carica papaya, *J. Med. Plant Res., 41:* 40–47 (1981).
179. W. A. Ukab, J. Sato, Y. M. Wang, and J. van Eys, Xylitol mediated amelioration of acetylphenylhydrazine-induced hemolysis in rabbits, *Metabolism, 30:* 1053–1059 (1981).
180. T. Tanaka and H. Arimura, Validity of xylitol for management of cardiac arrhythmia during anesthesia, *J. Uoeh, 1:* 365–368 (1979).
181. N. Yoshimura, H. Yamada, and M. Haraguchi, Anti-arrhythmic effect of xylitol during anesthesia, *Jap. J. Anesthesiol. 28:* 841–848 (1979).
182. E. R. Froesch, Parenterale Ernährung: Glucose oder Glucoseersatzstoffe? *Schweiz. Med. Wochenschr., 108:* 813–815 (1978).
183. K. Korttila and M. A. K. Mattila, Increased serum concentrations of lactic, pyruvic and uric acid and bilirubin after postoperative xylitol infusion, *Acta Anesthesiol. Scand., 23:* 273–277 (1979).
184. R. A. J. Conyers, A. M. Rofe, R. Bais, H. M. James, J. B. Edwards, D. W. Thomas, and R. G. Edwards, The metabolic production of oxalate from xylitol. *Int. Z. Vitam. Ernährungsforsch. (Beih.), 28:* 9–28 (1985).
185. B. Ludwig, E. Schindler, J. Bohl, J. Pfeiffer, and G. Kremer, Reno-cerebral oxalosis induced by xylitol, *Neuroradiology, 26:* 517–521 (1984).
186. R. G. Galaske, M. Burdelski, and J. Brodehl, Primär polyurisches Nierenversagen und akute gelbe Leberdystrophie nach Infusion von Zuckeraustauschstoffen im Kindesalter, *Dtsch. med. Wschr., 111: 978–983 (1986).*
187. K. Rosenstiel, Nierenversagen nach Infusion von Zuckeraustauschstoffen, *Dtsch. med. Wschr., 111:* 1340–1341 (1986).

188. R. Schroeder, Beachting der Dosierungsgrenzen für Xylit bei der parenteralen Ernährung, *Dtsch. med. Wschr., 109:* 1047–1048 (1984).
189. J. Pfeiffer, E. Danner, and P. F. Schmidt, Oxalate-induced encephalitic reactions to polyol-containing infusions during intensive care, *Clin. Neuropathol., 3:* 76–87 (1984).
190. S. Mori and T. Beppu, Secondary renal oxalosis: A statistical analysis of its possible causes, *Acta Pathol. Jpn., 33:* 661–669 (1983).
191. A. Baer and D. Loehlein, Limited role of endogenous glycolate as oxalate precursor in xylitol-infused patients, *Contr. Nephrol., 58:* 160–163 (1987).
192. WHO/FAO, Summary of toxicological data of certain food additives, Twenty-first Report of the Joint FAO/WHO Expert Committee on Food Additives, Geneva, WHO Technical Report Series No. 617, 1977, pp. 124–147.
193. WHO/FAO, Summary of toxicological data of certain food additives and contaminants, Twenty-second Report of the Joint FAO/WHO Expert Committee on Food Additives, Geneva, WHO Technical Report Series No. 631, 1978, pp. 28–34.
194. A. Baer, Sugars and adrenomedullary proliferative lesions: The effects of lactose and various polyalcohols, *J. Am. Coll. Toxicol., 7:* 71–81 (1988).
195. Life Sciences Research Office. Health aspects of sugar alcohols and lactose. Report prepared for the Bureau of Foods, Food and Drug Administration, Washington, DC under Contract No. FDA 223-2020 by the Life Sciences Research Office, Federation of American Societies for Experimental Biology, Bethesda, MD, 1986, p. 85.
196. H. Foerster, R. Quadbeck, and U. Gottstein, Metabolic tolerance to high doses of oral xylitol in human volunteers not previously adapted to xylitol. *Int. J. Vit. Nutr. Res.* (Suppl. 20): 67–88 (1981).
197. K. K. Maekinen, R. Ylikahri, P. L. Maekinen, E. Soederling, and M. Haemaelaeinen, Turku sugar studies XXIII. Comparison of metabolic tolerance in human volunteers to high oral doses of xylitol and sucrose after long-term regular consumption of xylitol, *Int. J. Vit. Nutr. Res.* (Suppl. 22): 29–51 (1981).
198. S. J. Culbert, Y. M. Wang, H. A. Fritsche Jr., D. Carr, E. Lantin, and J. vanEys, Oral xylitol in American adults, *Nutr. Res., 6:* 913–922 (1986).
199. H. K. Akerblom, T. Koivukangas, R. Punka, and M. Mononen, The tolerance of increasing amounts of dietary xylitol in children, *Int. J. Vit. Nutr. Res.* (Suppl. 22): 53–66 (1981).
200. H. Foerster, Tolerance in the human adults and children, in *Xylitol* (J. N. Counsell, ed.), Appl. Sci. Publishers Ltd., London, 1978, pp. 43–66.
201. K. K. Maekinen and A. Scheinin, Turku sugar studies VI. The administration of the trial and the control of the dietary regimen, *Acta Odontol. Scand., 33:* 105–127 (1975).
202. WHO/FAO, Evaluation of certain food additives and contaminants, Twenty-seventh Report of the Joint FAO/WHO Expert Committee on Food Additives, Geneva, WHO Technical Report Series No. 696, 1983, pp. 23–24.

# III
# MULTIPLE INGREDIENT APPROACH

# 20

# Mixed Sweetener Functionality

**Abraham I. Bakal** *ABIC International Consultants, Inc., Fairfield, New Jersey*

Previous chapters have dealt with the properties and functionality of several currently approved sweeteners or compounds under development or consideration. The requirements for the ideal sweetener or sugar substitute, as usually defined (1), are to:

1. Have the taste and functional properties of sugar
2. Have low-calorie density on a sweetness equivalency basis
3. Be physiologically inert
4. Be nontoxic
5. Be noncariogenic
6. Compete economically with other sweeteners

An analysis of the organoleptic or functional properties of each single sweetener clearly shows that none of the currently known sugar substitutes comes close to the taste and functional properties of sucrose. Most exhibit one or more of these differences:

1. Taste properties such as sweetness lag, undesirable and lingering aftertaste, narrow taste profile, or bitterness. For example, saccharin is generally reported to have a bitter aftertaste (2); stevioside, to have a menthol aftertaste (3); and aspartame, to have a delayed sweetness (4).
2. Lack of bulking properties.
3. Stability problems during processing and storage. For example, aspartame loses its sweetness in aqueous solutions and is not stable at high temperatures (5); thaumatin reacts with tannins and loses its sweetness (6).

4. Competitive prices. For example, saccharin and cyclamate reportedly cost, on a sweetness equivalency basis, less than sucrose and other nutritive sweeteners. On the other hand, in the United States, aspartame currently costs significantly more (7).

The food industry has partially overcome the bulking limitations in some selected applications. With the expansion of polydextrose use, it is expected that other reduced-calorie products based on polydextrose and intense sweeteners will be introduced to the market. Fibers, and fibers and polydextrose have recently found expanded use as bulking agents. However, this problem is universal for all intense sweeteners; thus it will not be discussed in this chapter. Except for the bulk issue, by far the most important limitations to consumer acceptability are taste properties and cost.

The use of more than one sweetener provides the food technologist with a tool for overcoming these limitations. The advantages of combining sweeteners are many; some of the aims and goals are to:

1. Formulate products that closely imitate the taste and stability of their sugar-sweetened counterparts

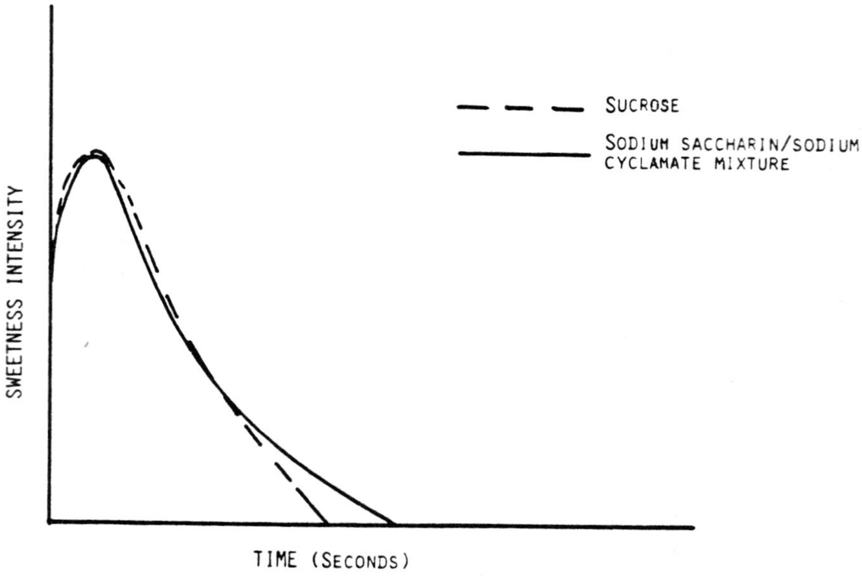


**Figure 1** Sweetness profile of aqueous solutions of sucrose versus a sodium saccharin/sodium cyclamate combination (1:10).

2. Create totally new taste experiences by utilizing sweeteners in the same manner the food industry utilizes flavors (8)
3. Meet cost restraints

Indeed, the use of combined sweeteners in food applications has been and is currently being practiced by the food industry. Before the cyclamate ban in the United States in 1970, the food industry used a combination of saccharin and cyclamate in the formulation of products. This mixture showed synergistic properties and improved taste profile as shown in Figure 1. For comparison, the taste profile of saccharin against sucrose is shown in Figure 2. With the approval of aspartame and its expanded use worldwide, several intense sweetener mixtures are being used in many foods and beverages. Figure 3 shows the taste profile of an aqueous solution sweetened with a saccharin/aspartame/cyclamate mixture in the ratio of 1:5:8, respectively. These data clearly demonstrate that combined sweeteners result in an improved sweetness profile when compared with each of the single sweeteners. In addition to the sweet-

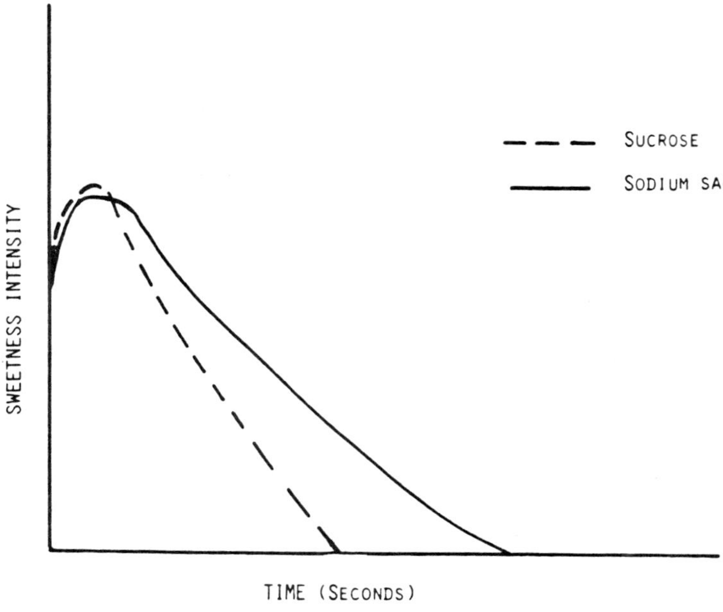


**Figure 2** Sweetness profile of aqueous solutions of sucrose versus sodium saccharin.

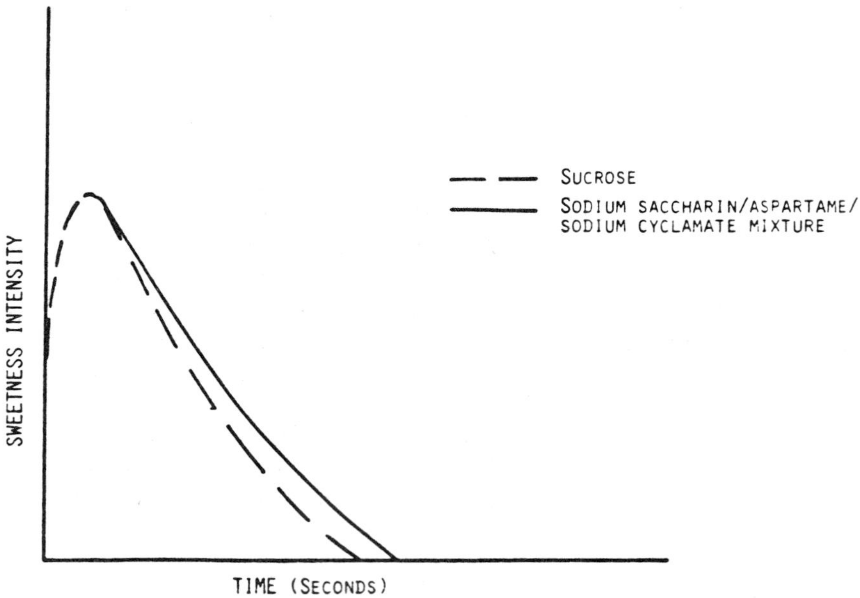


**Figure 3** Sweetness profile of aqueous solutions of sucrose versus a sodium saccharin/aspartame/sodium cyclamate combination (1:5:8).

ness profile, the quality of the taste is significantly improved, as shown in organoleptic tests conducted in our laboratories. Applications of combined sweeteners in various products are discussed in the remaining sections of this chapter.

## APPLICATIONS

### Table-Top

Table-top sweeteners are a major product category used by consumers for a variety of applications. They are commonly available as powders packed in bulk or sachets, as tablets, or as liquids. The effective amount of sweetener or sweetener combinations required to provide the sweetness equivalency of two teaspoons (10 g) of sugar to one cup (240 ml) of coffee is given in Table 1. This table clearly illustrates the reduction in the level of sweetener ingested when a combination of sweeteners is used. When a mixture of saccharin/cyclamate is used, the amount of saccharin is reduced by a factor of 5 (i.e., 40 mg alone to 8 mg in combination), and the amount of cyclamate by a factor of 2.5 (i.e., 200 mg alone to 80 mg in

**Table 1** Typical Effective Amounts of Sweeteners or Sweetener Combinations in Table-Top Applications

| Sweetener | Amount (mg)[a] |
|---|---|
| Sodium saccharin | 30–40 |
| Sodium cyclamate | 150–200 |
| Aspartame | 35–45 |
| Acesulfame-K | 50–60 |
| Sodium cyclamate/sodium saccharin | 80/8 |
| Aspartame/sodium saccharin | 5/15 |
| Acesulfame-K/aspartame | 30/3 |
| Aspartame/sodium saccharin/sodium cyclamate | 10/4/30 |

[a]Effective amount required to provide the sweetness equivalency of two teaspoons (10 g) of sugar to one cup (240 ml) of coffee.
*Source:* Refs. 1 and 9.

combination), as compared to the use of any one single sweetener. Similarly, in the case of an aspartame/saccharin mixture, the level of aspartame may be reduced by a factor of 2, and saccharin by a factor of 10.

Several patents and articles describe the synergism among aspartame, saccharin, cyclamate, acesulfame-K, and other sweeteners. Scott described the use of aspartame in combination with saccharin, cyclamate, or both to provide beverages with improved taste and consumer acceptability (9,10). In a 1973 patent, Scott described the use of aspartame with saccharin in a ratio range of 15:1 to 1:15 (10). A patent issued to General Foods discloses a sweetening composition containing aspartame, saccharin, and cyclamate (11). The inventors state that the sweetness is intense and lacks the lingering or bitter aftertaste associated with these sweeteners singly. A patent issued to E. R. Squibb proposes the use of "dipeptides" and saccharin, with the dipeptides masking the bitter aftertaste of saccharin (12–14). The saccharin to dipeptide ratio is 48:1.

Taste panels conducted in our laboratories on table-top preparations confirm the superiority of a combination of aspartame with saccharin, cyclamate, or acesulfame-K over any of these sweeteners singly. Panelists judged the taste quality of coffee sweetened with a combination sweetener, closely resembling the taste quality of sugar-sweetened coffee. Studies conducted in our laboratories suggest that aspartame/saccharin, aspartame/cyclamate, and combinations have improved stability, even in hot aqueous solutions, and also exhibit superior taste profiles. The superior taste may be associated with the flavor-enhancing properties of aspartame.

From 20 to 30% of table-top sweeteners are sold in bulk in the United States, a significant portion of which is used in cooking and baking. Consumers who cook with artificial sweeteners are usually those who must restrict their sugar intake for medical reasons. Currently, saccharin is the only sweetener approved for sale in bulk in the United States. In approving aspartame, the Food and Drug Administration (FDA) limited its sale to sachets and tablets (15). The NutraSweet Company has petitioned the FDA for the approval of aspartame in a spoon for spoon type product. As of this writing, the FDA has not acted upon this petition.

The liquid table-top sweetener market represents approximately 10% of U.S. sales. Aspartame-based products may not be feasible for this application because of their limited stability in aqueous solutions.

Recent regulatory changes in France resulted in the introduction of a number of table-top products containing a mixture of intense sweeteners. Some of these products use a combination of saccharin, aspartame, and acesulfame-K.

In summary, table-top formulations employing a single sweetener meet the current needs of the consuming public, however, each preparation has one or more serious defects, such as taste or stability. The consensus of experts and consumers is that table-top preparations combining two or more sweeteners are superior in taste and stability, and more closely imitate the sweetness of sucrose.

### Carbonated Beverages

Low-calorie soft drinks represent a significant segment of the market. According to published reports (16), sugar-free or low-calorie carbonated beverages were 17.7% of the U.S. market share in 1982. Current market share of low-calorie carbonated beverages is significantly higher and reportedly exceeds 25% of the total market. Until the approval of aspartame for use in carbonated beverages in the United States, only saccharin was available. The level of saccharin used to sweeten one fluid ounce varies between 8 and 11 mg, the actual level depending on the soft drink flavor and the product brand. To achieve similar sweetness levels with aspartame as the single source sweetener, significantly higher amounts of aspartame are required. Table 2 summarizes the levels of saccharin and aspartame used in selected soft drinks and syrups, showing that, on the average, the amount of aspartame used is about 1.5 times the amount of sodium saccharin when each of the sweeteners is used alone (17).

**Table 2** Typical Concentrations of Saccharin and Aspartame in Selected Soft Drinks and Syrups

| Flavor | Sodium saccharin[a] (mg/100 ml) | Aspartame[b] (mg/100 ml) |
|---|---|---|
| Soft drinks | | |
| Cola | 31–42 | 57.7 |
| Orange | 37–38 | 92.6 |
| Lemon-lime | 26–42 | 50.1 |
| Root beer | 27–37 | 60.5 |
| Syrups | | |
| Cola | — | 347.6 |
| Orange | — | 401.0 |
| Lemon-lime | — | 234.2 |
| Root beer | — | 355.2 |

[a]From Ref. 9.
[b]From Ref. 11.

**Table 3** Sensory Evaluation of the Sweetness Intensity of Aspartame-Sweetened Cola Beverages Stored at 80°F[a]

| Storage time (weeks) | Brand A | Brand B |
|---|---|---|
| 0 | 7.0 | 4.0 |
| 2 | 4.5 | 3.8 |
| 3 | 4.0 | 3.0 |
| 4 | 4.0 | 2.5 |
| 6 | 1.0 | 1.0 |

[a]Cola beverages were purchased in Canada. No storage history data were available. Ten expert panelists evaluated the sweetness intensity on a scale of 0 = none, to 8 = very sweet.

Aspartame-sweetened beverages have been available in Canada since 1981, when aspartame was approved. In Canada, aspartame is the only sweetener approved for this application, and typical beverages contain between 50 and 95 mg aspartame per 100 ml, the amount varying with brand and flavor. Table 3 provides typical sweetener concentrations in aspartame-based soft drinks.

Aspartame-based beverages lose sweetness as a function of storage time, temperature, and pH. Data submitted by G.D. Searle to the FDA indicate that about 50% of the initial aspartame remains in cola bever-

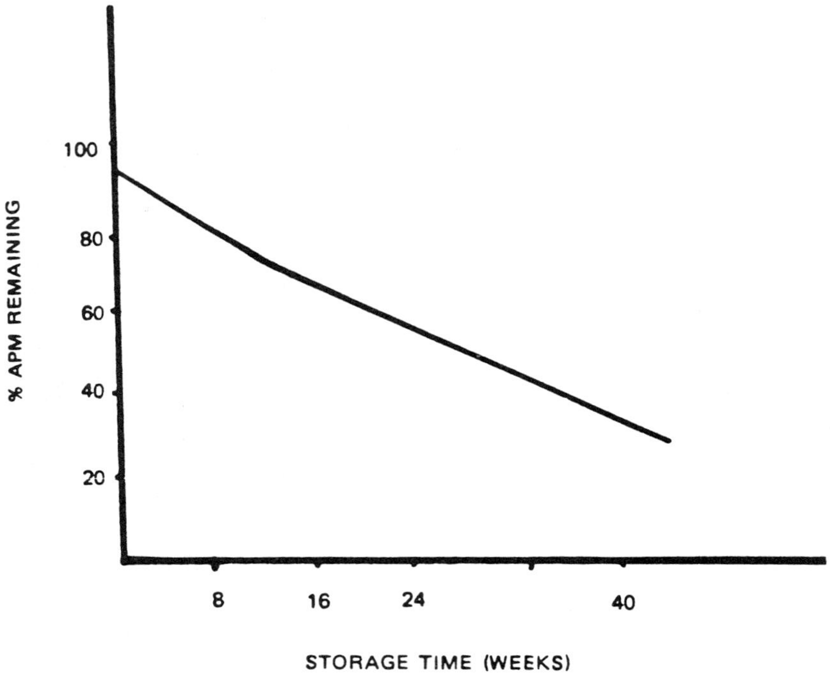


**Figure 4** Aspartame (APM) stability in carbonated cola beverages stored at 30°C. (From Ref. 17.)

ages stored at 30°C for 24 weeks (Fig. 4). In cola syrups (pH 2.4), 75% of initial aspartame remains after about 2 weeks storage at 30°C (Fig. 5).

Combination sweeteners have been used by the soft drink industry for many years. Superior products were available when cyclamate and saccharin were both approved by the FDA. Most products used 10:1 cyclamate/saccharin combinations because this ratio provided acceptable products. The ban of cyclamate in 1970 resulted in significant deterioration in the taste profile of soft drinks, which did employ saccharin as the single sweetener. The use of aspartame in combination with saccharin is described by G.D. Searle (17). Cola beverages sweetened with aspartame and saccharin at a ratio of 1:1 and stored at 20°C show significantly better sweetness stability than cola beverages sweetened with aspartame alone and stored under the same conditions. The performance of this combination of sweeteners is exceptional. The pH of the aspartame/saccharin product is lower than the pH of the aspartame-

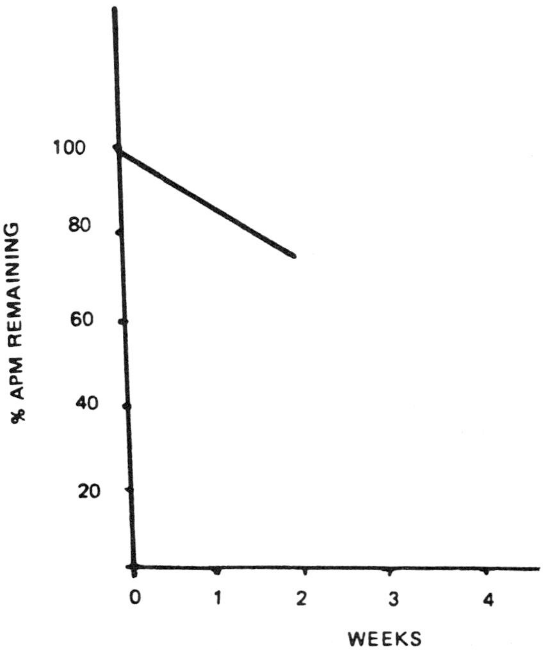


**Figure 5** Aspartame (APM) stability in cola syrup at 30°C. (From Ref. 17.)

sweetened product (pH 2.8 versus 3.05, respectively). The data are shown in Figure 6.

Following the approval of aspartame in the United States, beverage manufacturers announced their plans to introduce soft drinks containing a combination of aspartame and saccharin. One cola beverage marketed at that time contained approximately 18 mg of saccharin and 8 mg of aspartame per 100 ml.

Marketing and other considerations have led the major soft drink manufacturers to convert to 100% aspartame in the United States and in the U.K. in bottled and canned drinks. However, in other markets, products containing two or more intense sweeteners were introduced. For example, a major soft drink manufacturer markets products that contain saccharin, aspartame, and acesulfame-K in France.

Experiments conducted in our laboratory indicate that the addition of very small amounts of aspartame, in accordance with the specifications of the Lavia and Hill patent (12) improve the acceptance of saccharin-sweetened soft drinks. The cola beverages evaluated were

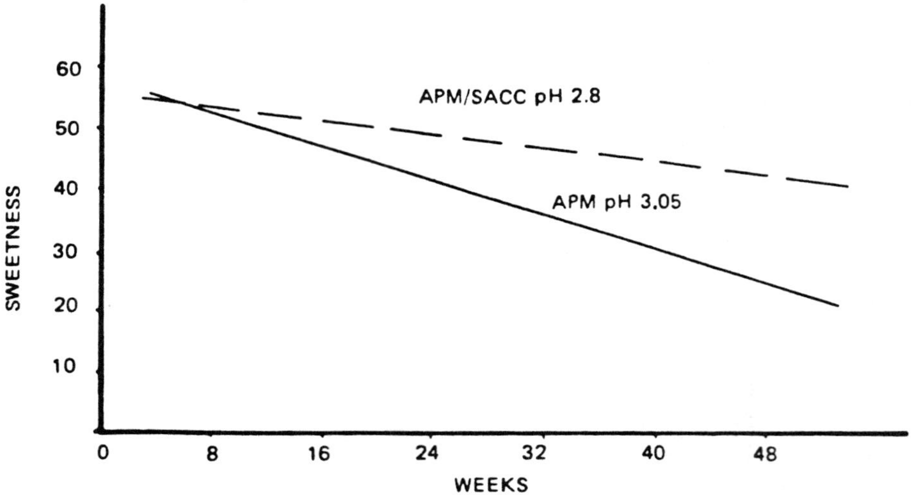


**Figure 6** Average scores for sweetness of cola carbonated beverages stored at 20°C over a period of 52 weeks. (From Ref. 17.)

sweetened with 33 mg saccharin per 100 ml. The addition of 0.7 mg aspartame to 100 ml of the saccharin-sweetened beverage (0.0007% concentration) resulted in a significant consumer preference for the beverage. It is clear that the use of aspartame at this concentration falls below the sweetness threshold of aspartame, which, according to Beck, is in the range of 0.007–0.001% (4).

Acesulfame-K and aspartame is another sweetener combination that will undoubtedly yield some superior products. These compounds were found to be synergistic in our laboratory. A sweetness enhancement of about 35% was observed. In addition to sweetness enhancement, aspartame broadens the taste profile of acesulfame-K and brings it closer to the sweetness profile of sucrose.

Aspartame and cyclamate combinations may also prove highly beneficial if cyclamates are approved in soft drinks. Acesulfame-K and cyclamate is another combination that is of interest because it yields excellent taste quality and exceptional storage stability.

Experiments conducted in our laboratories indicate that neohesperidin dihydrochalcone (NHDC), when used as a single sweetener, is inadequate for the preparation of soft drinks. However, when used in combination with saccharin, NHDC has a synergistic effect and gives improved taste perceptions. This finding is in accordance with observations described in the patent literature (18).

Stevioside was also recently evaluated as a single sweetener in cola beverages. Results were discouraging. The beverages had a licorice-like taste and were judged unacceptable. Fructose and stevioside are successfully combined in Japan to produce reduced-calorie soft drinks (50% reduction). These products enjoy good consumer acceptance in Japan (19).

Moskowitz addressed the issue of sweetness optimization in cola-flavored beverages using combination sweeteners (20) and presented a quantitative model for developing products acceptable to the consumer. Hoppe discussed the effect of various mixtures of sucrose, saccharin, and cyclamate on sweetness perception in aqueous solutions (21). Van Tornout et al. (32) evaluated the taste characteristics of mixtures of fructose with saccharin, aspartame, and acesulfame-K in soft drinks. Their data indicate that combinations of small amounts of fructose with these intense sweeteners result in soft drinks that cannot be distinguished from sucrose-sweetened beverages. All of these data indicate the benefits to the consumer that can be derived from the use of combination sweeteners in soft drinks.

## Dry Mixes

This category encompasses a variety of food products that are sold in dry form and are reconstituted by the consumer before use. They include beverage mixes, presweetened cereals, puddings, and desserts. In the United States, until the approval of aspartame, saccharin was the only intense sweetener used in this application. The approval of aspartame resulted in significant proliferation of these products in the market positioned as reduced-calorie and sugar-free. These aspartame-sweetened products have enjoyed wide consumer acceptance. To date, no dry mix products are available in the marketplace using combinations of saccharin and aspartame, although in the past, cyclamate/saccharin combinations were used. Table 4 provides a comparison of typical concentrations of saccharin and aspartame in selected products. The table clearly shows that, in these applications, the effective aspartame level is about twice the effective sodium saccharin level. Despite the recent approval of acesulfame-K for use in some of these products, no products have been introduced to the U.S. market as of this writing.

A combination of saccharin/aspartame, with a concentration of 15–20 mg sodium saccharin and 6–10 mg aspartame, in 100 g of finished product, yields a highly acceptable product and is significantly less expensive. In addition, such a combination represents a major reduction in the consumption of each of the sweeteners: almost a 50% reduction in

**Table 4** Typical Concentrations of Saccharin and Aspartame in Selected Dry-Mix Products

| Product | Sodium saccharin[a] | Aspartame[a] |
|---|---|---|
| Beverage mix | 27–34 | 40–55[b] |
| Gelatin dessert | 21–28 | 19–44 |
| Puddings | 20–25 | 50–60 |

[a]Concentration in mg/100 g of reconstituted product.
[b]Beck (5) suggests using 50–65 mg/100 g.

sodium saccharin and about a 75% reduction in aspartame level. Most important, results of consumer tests conducted in our laboratory clearly indicate the superiority of the products sweetened with a combination of the two sweeteners. Similarly, combinations of acesulfame-K and aspartame yield products with a taste profile much more closely resembling that of the sucrose-sweetened products than those sweetened with either intense sweetener singly.

### Chewing Gums

Sugarless chewing gums represent a significant segment of the chewing gum market. These gums typically contain sorbitol, mannitol, and/or xylitol (mostly in Europe) as the bulking agents to replace sugar. However, the sweetness intensity of sorbitol and mannitol is about half that of sucrose. These gums receive lower consumer acceptance because of lack of sweetness and flavor; therefore, intense sweeteners have been and are being used to improve the taste qualities of these products.

Saccharin is currently used in sugarless gums at concentrations between 0.1 and 0.2%. A typical chewing gum stick, weighing approximately 3 g, contains from 3 to 6 mg of saccharin. Several patents indicate the use of insoluble saccharin to produce chewing gums with long-lasting flavor (22,23). Taste tests comparing the sweetness quality of chewing gums with and without saccharin establish the superiority of the saccharin-sweetened products. Saccharin taste does not seem to be a limiting factor in this application because of the low saccharin concentration.

Figure 7 illustrates the sweetness-duration properties of chewing gums containing saccharin acid compared with chewing gums containing the same concentration of sodium saccharin. Figure 8, which shows the overall quality rating of the same chewing gums on a scale of 0 (dislike extremely), to 4 (neither like nor dislike), to 8 (like extremely),

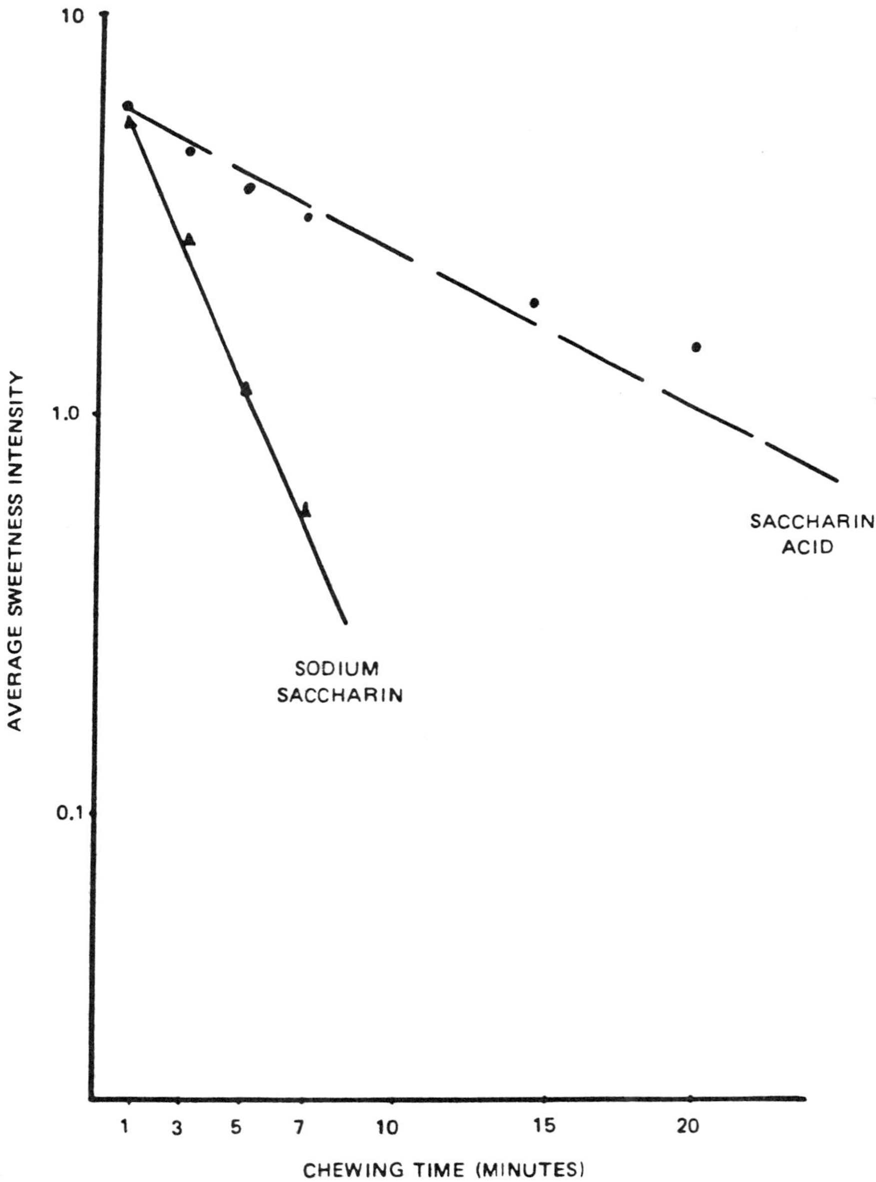


**Figure 7** Sweetness intensity of chewing gums with sodium saccharin and saccharin acid.

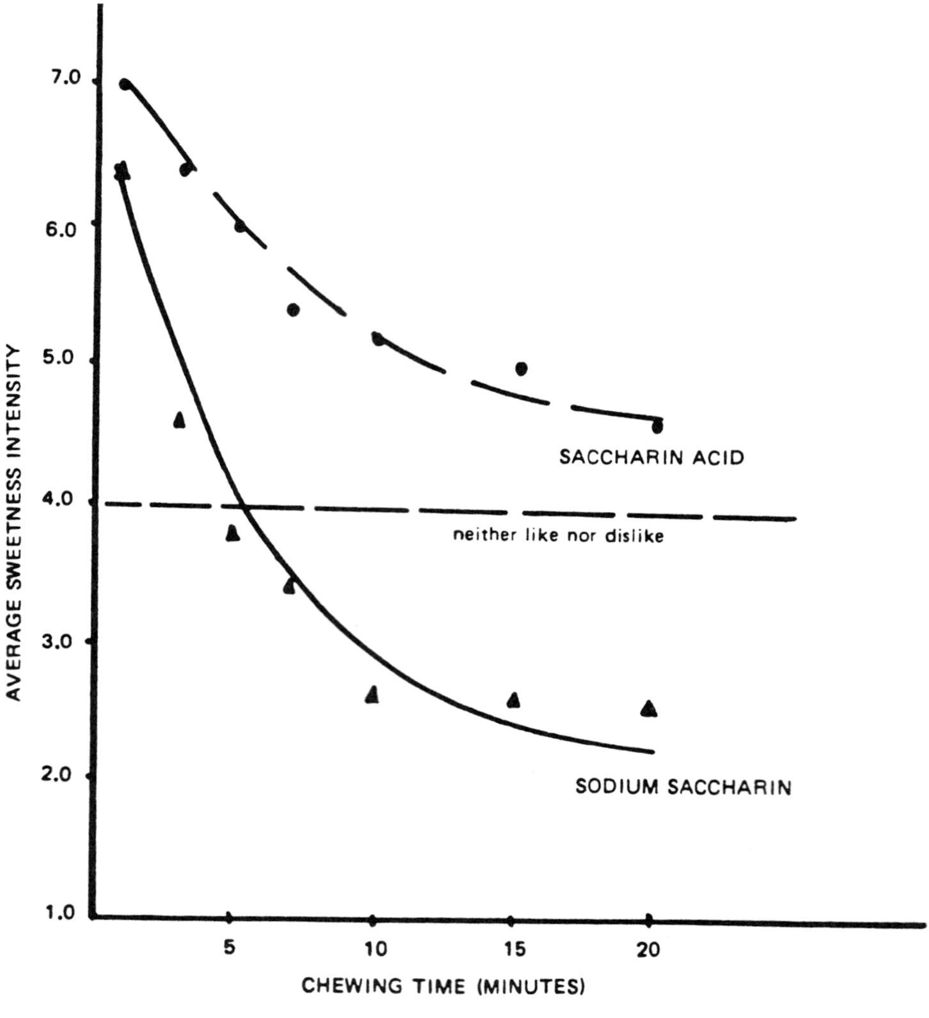


**Figure 8** Overall acceptability rating of chewing gums with sodium saccharin and saccharin acid.

clearly shows consumer preference for the chewing gum containing the insoluble saccharin. Because of its relatively low sweetness intensity, cyclamate is not used as a single sweetener by the chewing gum industry. Before the 1970 ban on cyclamates, cyclamate/saccharin combinations were commonly used at ratios between 2:1 and 10:1. Chewing gums using this combination of sweeteners were acceptable and had taste profiles similar to sugar-based products. FDA regulations allow the use of aspartame in chewing gum, both as a sweetener and flavor enhancer (15). The literature describes the use of aspartame in chewing gum at levels of 0.3% or higher (24). These gums are reported to have longer-lasting flavors.

One of the problems encountered by chewing gum manufacturers in their use of aspartame is the relative instability of aspartame in chewing gums. A patent issued to General Foods describes a method for stabilizing aspartame-containing gums by removing the calcium carbonate filler from the gum base for formulation and replacing it with talc (25). This procedure apparently alters the pH of the gum base and results in a slower rate of aspartame decomposition.

A method investigated in the industry is the application of aspartame only on the surface of the gum through dusting or glazing (26). This technique, although technically successful, presents consumer acceptance problems. Recently, encapsulation techniques have been investigated and are reportedly employed by some chewing gum manufacturers to overcome sweetness loss and to extend the sweetness and flavor of both aspartame- and saccharin-sweetened products.

Beck (5) describes the use of a combination of aspartame and saccharin in chewing gum. The ratio used was 2:1, aspartame to saccharin. This gum was preferred to gum containing any of the single sweeteners. Furthermore, although no stability data are available on this combination of sweeteners, it is believed that stability will be significantly improved. Preliminary experiments conducted in our laboratory indicate that the use of a combination of sweeteners in chewing gum presents many advantages to the manufacturer and the consumer. A major U.S. chewing gum manufacturer has recently introduced an acesulfame-K–sweetened product in the United States.

Several products containing aspartame and mixtures of aspartame and saccharin are currently under evaluation. Figure 9 provides a sweetness profile of chewing gums containing a mixture of saccharin and aspartame. The data clearly show that this product exhibits longer-lasting sweetness properties.

Several other sweeteners have been investigated. Experiments conducted with NHDC indicate that this sweetener is good in chewing gum applications. It is stable and provides an acceptable sweetness profile and flavor-enhancing properties (27). Stevioside is currently used in sugarless gums in Japan. Although the products lack the sensory qualities of saccharin- or aspartame-sweetened products, they enjoy high consumer acceptance in Japan (Maruzen Fine Chemicals, personal communication, 1982).

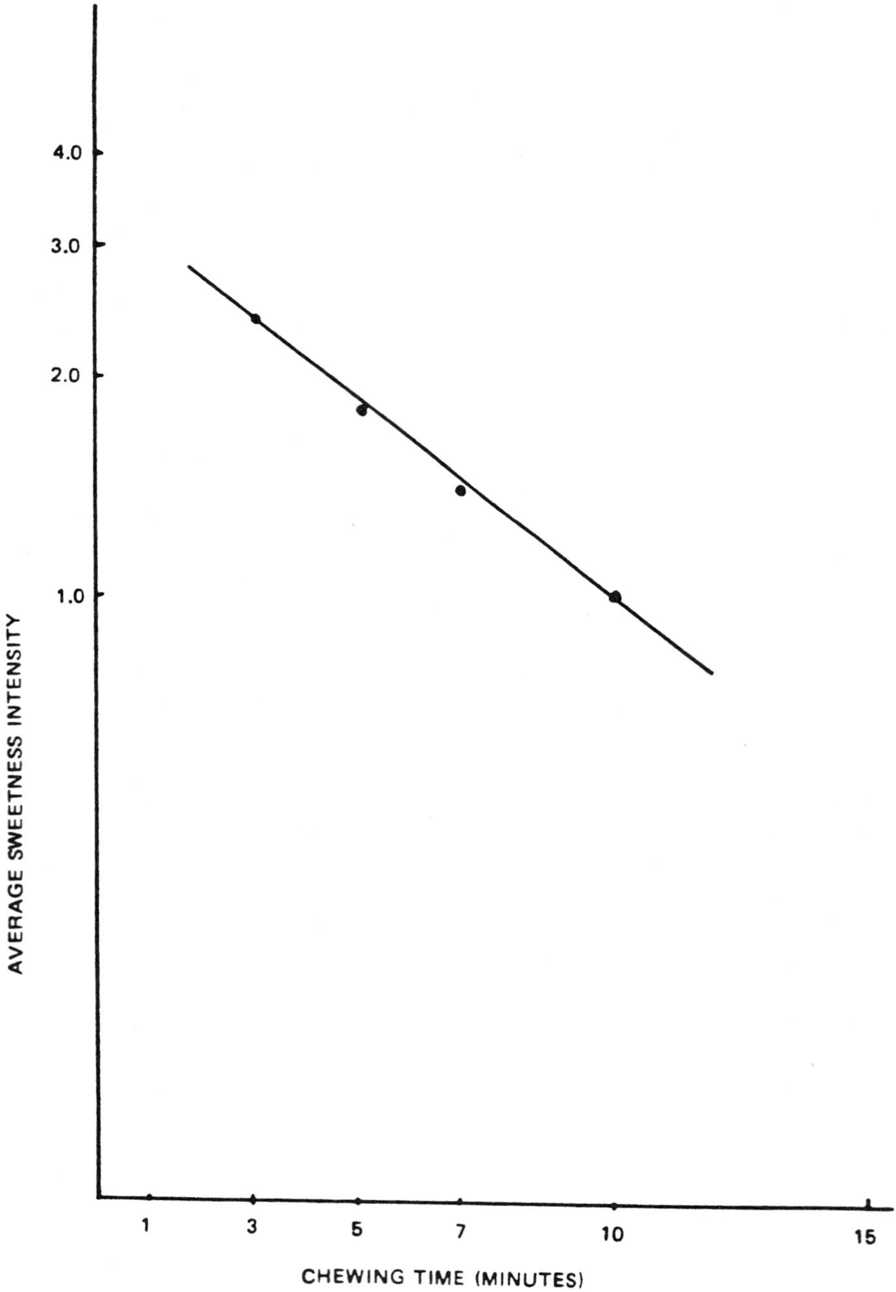


**Figure 9** Sweetness intensity of chewing gums containing a mixture of saccharin and aspartame.

### Processed Foods

This application category includes a variety of food products. Heat-processed foods, which include baked goods as well as canned dried foods, represent a variety of low-calorie products. Refrigerated and frozen products also employ low-calorie sweeteners.

Saccharin is extensively used in baked products, especially by diabetics. It is estimated that approximately 40% of the bulk table-top sweetener products marketed in the United States and Europe are used in baked goods. (M. Eisenstadt, personal communication, 1982). Recipes employing saccharin table-top sweeteners are available (28). Saccharin's heat stability makes it a good functional sweetener for this application.

With the recent approval of polydextrose, product development activities are underway to combine polydextrose (as a low-calorie bulking agent) with saccharin to produce reduced-calorie cakes. Saccharin is also used in low-calorie salad dressings, ice cream, and yogurt, but those uses are minimal because of taste limitations and limited consumer acceptance.

The instability of aspartame at high temperatures makes this application unlikely until the encapsulated form is approved. However, recipes are available for preparing baked products to which aspartame is added after baking and cooling (29). Aspartame added to baked goods at low levels improves their aroma (30).

The use of aspartame in frozen desserts and other liquid products in the United States has proliferated since approval in these applications was granted by the FDA. Beck has reported that aspartame-sweetened frozen desserts were preferred by a taste panel to ice milk (4,5). Aspartame also performs well in frozen beverages (31), no-bake cheesecake, yogurt, and other products. No combinations of sweeteners are used in these products in the United States.

Cyclamate/saccharin combinations, when available, were extensively used in baked goods because of the heat stability of this combination. Cyclamate to saccharin ratios ranged from 2:1 to 10:1 in most uses, with 10:1 being the preferred ratio when organoleptic considerations were taken into account. Similarly, this combination was used in low-calorie frozen desserts, salad dressings, jams and jellies. Acesulfame-K may find specific application in this category of product applications. However, because of its low relative sweetness intensity, its successful use may depend on its use with other sweeteners. Recently, Lim et al. (33) reported on the use of combinations of intense sweeteners in cookies.

No data are readily available on the use of other sweetener combinations in these applications. However, the success of combinations of sweeteners in table-top products, beverages, dry mixes, and chewing gum leads to the expectation that combining available sweeteners will surely improve the taste profile of artificially sweetened, heat-processed foods as well as refrigerated and frozen products.

## CONCLUSIONS

The search for the ideal noncaloric sugar substitute will continue. The ideal sweetener must not only meet the taste and functional properties of sugar, but can also undoubtedly create opportunities for new products that are not yet possible using currently available ingredients. On the basis of available evidence, the likelihood of identifying a single compound capable of meeting all requirements of all food applications is extremely small. Thus, the use of multiple sweeteners, specifically designed to meet desired taste and functional characteristics in a specific food application, is the most logical and promising route.

### REFERENCES

1. A. I. Bakal, Functionality of combined sweeteners in several food applications, *Chem. Ind.* (September 19, 1983).
2. A. Salant, Non-nutritive sweeteners, in *Handbook of Food Additives,* 2nd ed. (T. E. Furia, ed.), CRC Press, Cleveland, 1972, pp. 523–585.
3. G. E. Inglett, Natural and synthetic sweeteners, *Hortscience, 5:* 139–141 (1970).
4. C. I. Beck, Sweetness, character, and applications of aspartic acid-based sweeteners, in *ACS Sweetener Symposium* (G. E. Inglett, ed.), AVI Publishing, Westport, Conn., 1974, pp. 164–181.
5. C. I. Beck, Application potential for aspartame in low calorie and dietetic food, in *Low Calorie and Special Dietary Foods* (B. K. Dwivedi, ed.), CRC Press, West Palm Beach, Fla., 1978, pp. 59–114.
6. J. D. Higginbotham, *Talin, a New Natural Sweetener from Tate and Lyle,* Tate and Lyle, Reading, 1980.
7. G. D. Searle, personal communication, 1984.
8. A. B. Porter, Effectiveness of multiple sweeteners and other ingredients in food formulation, *Chem. Ind.* (September 19, 1983).
9. D. Scott (G. D. Searle), Use of dipeptides in beverages, German Patent 2,022,499 (September 20, 1979).
10. D. Scott (G. D. Searle), Saccharin-dipeptide sweetening compositions, British Patent 1,256,995 (1971).
11. T. P. Finucane (General Foods), Sweetening compositions containing aspartame and other adjuncts, Canadian Patent 1,043,158 (November 28, 1978).
12. A. F. Lavia and J. A. Hill (E. R. Squibb), Sweeteners with masked saccharin aftertaste, French Patent 2,087,843 (February 4, 1972).
13. A. F. Lavia and J. A. Hill (E. R. Squibb), Sweetening compositions, U.S. Patent 4,001,455.

14. J. A. Hill and A. F. Lavia (E. R. Squibb), Artificial sweeteners, U.S. Patent 3,695,898 (1978).
15. U. S. Food and Drug Administration, Aspartame, 21*CFR*172, 804 (March 19, 1982).
16. National Soft Drink Association, *Carbonated Beverages Statistics*, 1982.
17. G. D. Searle and Company, Aspartame in carbonated beverages, Vol. 1, U.S. Food Additive Petition No. 2A3661 (August 13, 1982).
18. K. Ishii et al. (Takeda Chemical Industries), Sweetening composition, U.S. Patent 3,653,923 (April 4, 1972).
19. H. Fujita and T. Edahiro, Safety and utilisation of stevia sweetener, *Shokuhin Kogyo, 22*(20): 66–72 (1979).
20. H. R. Moskowitz et al., Sweetness and acceptance optimisation in cola flavored beverages using combinations of artificial sweeteners—a psychophysical approach, *J. Food Quality, 2:* 17–26 (1978).
21. K. Hoppe, Taste interaction of citric acid with sucrose and synthetic sweeteners, *Nahrung, 25:* 11–14 (1981).
22. A. I. Bakal et al. (Life Savers), Long-lasting flavored chewing gum including sweetener dispersed in ester gums and method, U.S. Patent 4,087,557 (May 2, 1978).
23. D. A. M. Mackay et al. (Life Savers), Long-lasting flavored chewing gum, U.S. Patent 4,085,227 (April 18, 1978).
24. B. J. Bahoshy et al. (General Foods), Chewing gums of longer lasting sweetness and flavor, U.S. Patent 3,943,258 (March 9, 1976).
25. R. E. Klose and B. J. Bahoshy (General Foods), Chewing gums of improved sweetness retention, U.S. Patent 4,246,286 (January 20, 1981).
26. T. Cea and M. Glass (Warner-Lambert), Aspartame sweetened chewing gum of improved sweetness stability, S. African Patent ZA 80 06,065 (September 30, 1981).
27. E. B. Westall et al. (Nutrilite Products and Warner-Lambert), Preservation and stability of flavors in chewing gum and confectioneries, German Patent Offen. 2,155,321 (May 10, 1972); British Patent 1,310,329; U.S. Patent 3,821,417.
28. G. L. Becker, *Cooking for the Health of It*, Benjamin, Elmsford, N.Y., 1981.
29. B. Gibbons, *Slim Gourmet Sweets and Treats*, Harper and Row, New York, 1982.
30. I. Chibata et al. (Tanabe Seiyaku), Aspartyl amide sweetening agents, U.S. Patent 3,971,822 (July 27, 1976).
31. T. P. Finucane (General Foods), Methyl ester of L-aspartyl-L-phenylalanine in storage-stable beverages, Canadian Patent 1,050,812 (March 20, 1979).
32. P. Van Tornout et al., Sweetness evaluation of mixtures of fructose with saccharin, aspartame or acesulfame-K, *J. Food Science, 50:* 469–472 (1985).
33. Lim, Hyesook, et al., Sensory studies of high, potency multiple sweetener systems for shortbread cookies with and without polydextrose. *J. Food Science, 54:* 624–628 (1989).

# 21

# Polydextrose

**Frances K. Moppett** *Pfizer Inc, Groton, Connecticut*

## INTRODUCTION

The question has been asked, "how sweet is polydextrose?" Polydextrose is not sweet at all. The design of this ingredient is not to provide sweetness, but rather to be the ultimate companion ingredient to high intensity sweeteners.

Polydextrose is an ingredient designed to give the bulk, texture, mouthfeel, and functional attributes of caloric sweeteners. These are attributes that are often lost in the formulation of calorie-modified products. The key to the performance of polydextrose is its caloric value of 1 calorie per gram. This, in combination with its excellent water solubility, makes it unique in its application in reduced-calorie and low-calorie foods.

Unlike the serendipitous discovery of many unique products, polydextrose is the result of a targeted research program. The goal of the program was to fill the need for a reduced-calorie bulking agent for the reduced-calorie foods market. For such food products to be widely accepted, they must be comparable to their fully caloric counterparts. Achieving desirable body mouthfeel and texture are critical to achieving success.

Polydextrose received FDA approval for use in select food categories in 1981 (1). It was not until the mid-1980s, however, that significant commercial success was realized for this product. Proven successful formulations and a mushrooming of approved high intensity sweetener applications have resulted in a secure place for polydextrose in reduced-calorie food products.

## COMMERCIAL PRODUCTION

Polydextrose is a randomly bonded melt condensation polymer of dextrose. This unique product is a patented material invented in Pfizer Central Research Laboratories by Dr. Hans Rennhard (2). The patent describes the process for manufacture and applications of a novel carbohydrate substitute.

Polydextrose is prepared commercially by vacuum polycondensation of a molten mixture of food grade starting materials. The starting materials are dextrose, sorbitol, and citric acid in approximately an 89:10:1 mixture. The final product of this reaction is a weakly acidic water-soluble polymer that contains minor amounts of bound sorbitol and citric acid.

The theoretical chemical structure of polydextrose is illustrated in Figure 1. This structure is drawn to represent the types of bonding that can occur during polymerization. The R group may be hydrogen, glucose, sorbitol, citric acid, or a continuation of the polydextrose polymer. As evidenced by this representative structure, polydextrose is a very complex molecule, being highly branched with varied glucose linkages.

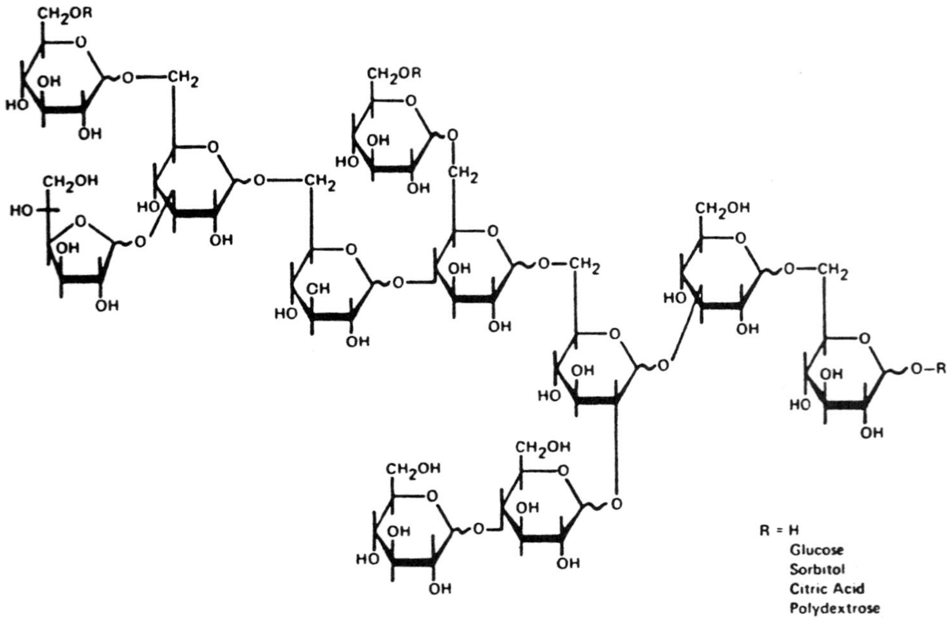


**Figure 1** Representative chemical structure of polydextrose.

**Table 1** Approximate Molecular Weight Distribution of Polydextrose

| Molecular weight range | Percent |
|---|---|
| 162–5,000 | 88.7 |
| 5,000–10,000 | 10.0 |
| 10,000–16,000 | 1.2 |
| 16,000–18,000 | 0.1 |

**Table 2** Characteristics of Polydextrose

| | |
|---|---|
| Compendial designation | Food Chemical Codex |
| Appearance | White to light tan powder |
| Odor | None |
| Polymer | > 90%[a] |
| Glucose | < 4%[a] |
| Sorbitol | < 2%[a] |
| Levoglucosan | < 4%[a] |
| Water | < 4%[a] |
| Citric acid (free) | ~ 0.1% |
| pH (10% w/v solution) | 2.5–3.5 |
| Solubility in water (25°C) | 80% |
| Optical rotation | +60 |
| Viscosity (cps, 50% solution) | 35 |

[a]Anhydrous, ash-free basis

It is these chemical parameters that result in the water-soluble, reduced-calorie nature of polydextrose.

The typical molecular weight distribution of the polymer is shown in Table 1. During the manufacturing process of polydextrose, the size of the polymer is controlled to restrict the formation of large molecular weight molecules. This control prevents the formation of insoluble material. Further discussion on this topic is presented by Allingham (3) and Beereboom (4).

As manufactured, polydextrose conforms to the typical composition described in the product description presented in Table 2. Polydextrose is supplied to the food industry in compliance with Food Chemical Codex compendial specifications.

## PRODUCT FORMS

Polydextrose is commercially available in three forms: polydextrose, polydextrose solution, and polydextrose-K. The latter two product

Table 3 Polydextrose Product Forms

| | |
|---|---|
| Polydextrose | White to light tan powder<br>Acidic pH 2.5 to 3.5 |
| Polydextrose solution | 70% Polydextrose solution<br>Neutralized to pH 5.0 to 6.0<br>Clear, colorless to straw-yellow liquid |
| Polydextrose-K | White to light tan powder<br>Blended with potassium bicarbonate to yield a pH of 5.0–6.0 in solution |

forms were developed to allow greater flexibility in the manufacture of food products. Table 3 presents a summary of the key features of each product form.

Polydextrose is an acidic, amorphous material. A 10% w/v solution of this product has a pH of 2.5 to 3.5 (5). It is a white to light tan powder. This product form can be used for most food applications, however, the slight acidity may require use of a buffering agent.

Polydextrose solution is a clear, stable 70% aqueous solution (6). The solution is colorless to pale yellow and is more viscous than a comparable concentration of sucrose. This product has been partially neutralized with potassium hydroxide to a pH of 5–6. This product form offers an ease of handling that has been particularly useful for frozen dessert and some confectionery manufacturers.

The third product form available is polydextrose-K (7,8). It is a dry blend of polydextrose with food-grade potassium bicarbonate at a level to yield a pH of 5.0–6.0 in solution. This product is particularly useful in the less acidic food products and those that are moisture sensitive. Only minor formulation changes may be necessary when using this form of polydextrose to replace significant levels of carbohydrate in a recipe.

## PROPERTIES

### Caloric Content

The key property that gives polydextrose its important role in the formulation of reduced calorie foods is the caloric value of 1 calorie per gram (9). This value is significant when polydextrose is used to replace sucrose, which has a caloric value of 4 calories per gram. The degree of caloric reduction is even greater when polydextrose is used to replace fat, which has 9 calories per gram.

The reason polydextrose has a low caloric value is that it is a large, complex, randomly bonded polymer which is not broken down by mammalian digestive enzymes. The one calorie is a result of the metabolism of microorganisms in the intestinal tract. This subject is discussed more fully in this chapter under "Metabolism."

### Water Solubility

The excellent water solubility of polydextrose differentiates it from the insoluble bulking agents such as cellulosic products. It is soluble to approximately 80% at 25°C. Polydextrose acts in a similar manner to sucrose in that it results in a clear solution without haze or turbidity. Good mechanical mixing is required to prepare concentrated solutions. The rate of solubility will depend upon the efficiency of the mixing equipment as determined by shear speed and rate of addition. Use of moderate heat and slow rate of addition aid in making these solutions. A preblend with another water-soluble ingredient is another technique that will greatly speed the rate of solution of polydextrose.

Polydextrose-K has a faster rate of solution than polydextrose. This is a result of potassium bicarbonate acting as a dispersing agent in this blend.

### Viscosity

The viscosity that polydextrose contributes to a solution is greater than that contributed by an equal amount of sucrose. Figure 2 illustrates the viscosity of 70% solutions of polydextrose, sucrose, and sorbitol at varying temperatures. In a manner similar to sucrose solutions, the viscosity of polydextrose solutions decreases with an increase in temperature.

The viscosity-enhancing effect of polydextrose plays an important role in food uses such as reduced-calorie dressings and puddings. Figure 3 shows the viscosity of 20–70% solutions of polydextrose, sucrose, and sorbitol. This illustrates the potential for using polydextrose to replace the viscosity typically provided by high sugar content.

### Humectancy

Under conditions of high relative humidity, polydextrose is fairly hygroscopic. Figure 4 illustrates the water pickup of polydextrose in storage at 75 and 52% relative humidity. In food products, polydextrose functions as a humectant and can play an important role in product quality by controlling the rate of moisture gain or loss. An example of this property is found in the baked goods area. Polydextrose can retard the loss of

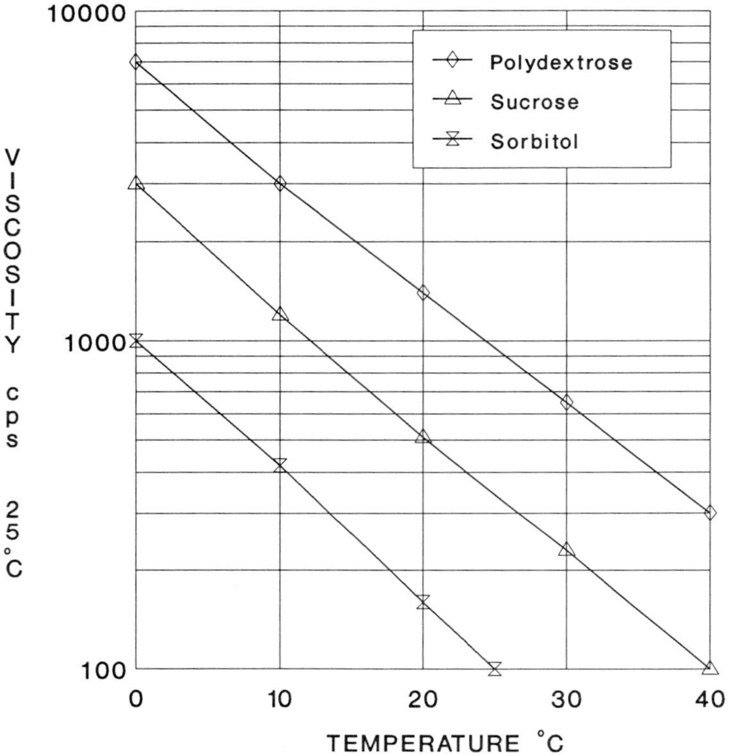


**Figure 2** Temperature/viscosity relationships for 70% solutions.

moisture, which helps to protect against staling. In this way, polydextrose serves as an important ingredient in extending shelf life.

Another very important characteristic of polydextrose in solution is the effect that it has on water activity ($a_w$). Water activity is a measure of the availability of the water to participate in chemical reactions, physical reactions, and support of microbial growth in a food product.

The water activity of various polydextrose solutions is reviewed in Table 4. At equal concentrations, polydextrose has less effect on water activity than smaller molecular weight products such as sucrose and sorbitol.

### Taste

Polydextrose is not sweet. It is used to replace the physical functionality of high calorie ingredients such as sugar, but not the sweetness. In some

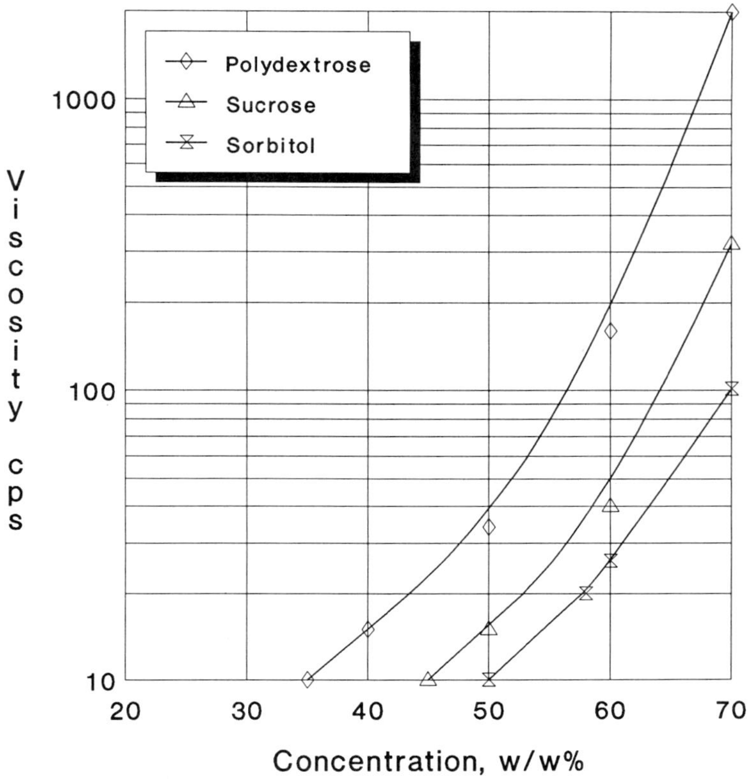


**Figure 3** Concentration/viscosity relationship.

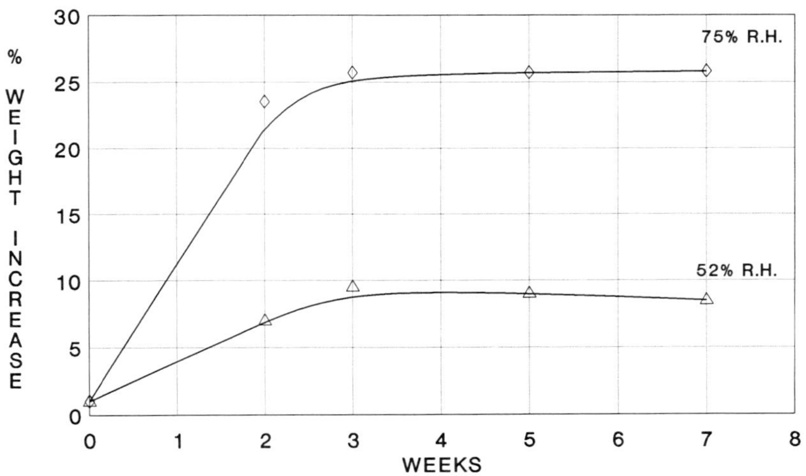


**Figure 4** Hygroscopicity of polydextrose at 25°C.

Table 4 Water Activity of Various Polydextrose Solutions

| | Solution concentration (% w/w) | |
|---|---|---|
| Sweetener | 50% | 60% |
| Sorbitol | 0.90 | 0.85 |
| Sucrose | > 0.95 | 0.91 |
| Polydextrose | > 0.95 | 0.92 |

food products, a less sweet taste may be desirable, and thus at least partial sucrose replacement can be achieved without compensating for the loss in sweetness. In most cases, it is necessary for calorie-modified foods to have the same quality and quantity of sweetness as the fully caloric counterpart. In such foods sweetness can be provided by a high intensity sweetener. It is in these applications that polydextrose is an ideal companion product for high intensity sweeteners.

At high concentrations, polydextrose can contribute a characteristic flavor. This can be overcome by formulation adjustments, flavors, and flavor modifiers.

### Noncariogenicity

Studies carried out by Professor Muhleman at the University of Zurich demonstrated that polydextrose has a very low potential for promoting dental caries (10). These tests showed polydextrose (1) to be inert to in vitro systems, (2) not to cause caries in rats, and (3) to pass the human intraproximal plaque pH test. These three tests are good indicators of the cariogenicity of foods. As a direct result of these studies, the Swiss government allows a "Safe for Teeth" labeling associated with polydextrose.

### Stability

Polydextrose is a very stable ingredient. The hygroscopic nature of the product requires good packaging and reasonable storage under conditions of low humidity to avoid moisture pickup.

Polydextrose solution is also very stable. The high solids level does not allow for the support of microbial growth. The solution can darken after prolonged storage at elevated temperatures, and, therefore, storage under cool temperatures is recommended.

## FUNCTIONS OF POLYDEXTROSE

The unique combination of properties found in polydextrose results in a variety of functional attributes that are important in the formulation of reduced-calorie, low-calorie, and sugarless products. The principal change that is made in these food products is the removal of sucrose. High potency sweeteners replace the sweetness lost when sucrose is removed. However, in most food products, a great deal more than sweetness is lost along with the sugar. Sucrose also provides the functional attributes of bulk, mouthfeel, humectancy, viscosity, freezing point depressant, and preservation (11,12). As illustrated in Table 5, polydextrose can at least partially fill all these roles except for sweetness. In a similar manner, when calorie reduction is achieved by reducing the fat level, there are functional attributes of the fat that polydextrose can replace.

### Bulk

Polydextrose is most commonly referred to as a "bulking agent." Probably the greatest single challenge when formulating reduced-calorie products is replacing the bulk of sugar. In liquid products such as beverages, this is not such an important issue. In others, however, it is critical. Probably the most dramatic examples of the importance of a bulking agent are in baked goods and confections, which rely heavily on the bulk of sugar to give the products their character. Any significant reduction in the bulk and body of these products would greatly decrease their acceptability.

Other ingredients that could be used to replace the bulk of carbohydrates and fats include such diverse ingredients as crude dietary fiber, maltodextrin, and sorbitol. However, these products are limited in their usefulness, since none of these alternative bulking agents have the combined properties of 1 calorie per gram and water solubility.

### Mouthfeel/Texture

A pleasant, satisfying mouthfeel and texture is another important contribution of sucrose and fat. Limited success has been achieved using gums

**Table 5** Functional Attributes of Sugar and Polydextrose

| | Sugar | Polydextrose |
|---|---|---|
| Bulk | Yes | Yes |
| Mouthfeel/Texture | Yes | Yes |
| Humectancy | Yes | Yes |
| Viscosity | Yes | Yes |
| Freezing point depressant | Yes | Yes |
| Preservation | Yes | Yes |
| Sweetness | Yes | No |

to enhance the texture of reduced-calorie foods. At higher levels, a slimy mouthfeel and gelled characteristics become noticeable. The use of polydextrose gives a comparable mouthfeel and general textural eating quality to the fully caloric food product. The product categories that particularly illustrate this function are puddings, frozen desserts, and salad dressings. A rich, creamy mouthfeel is particularly important and expected in these products.

### Freezing Point Depression

One of the less obvious functions of sugar is that of being a freezing point depressant. This function is very important in making creamy, palatable frozen desserts. If the freezing point of a product is too low, the texture as consumed will be too soft. If the freezing point is too high, then an unacceptably hard product results. In conventional ice cream type products, sucrose is the primary freezing point depressant, and alternative ingredients such as polydextrose are typically compared to sucrose.

The effect that an ingredient has on freezing point depression is a function of its molecular weight. Sucrose is a disaccharide. A larger molecule such as polydextrose is somewhat less effective as a freezing point depressant. A smaller molecular weight ingredient such as sorbitol would have a greater effect on freezing point depression than either sucrose or polydextrose. Ideally, a balance can be achieved with several ingredients in order to match the effect of sucrose.

The comparative effects of polydextrose, sucrose, and sorbitol on freezing point has been reported (13,14). The freezing point of a 5% solution of these three ingredients were as follows: polydextrose, −.147°C; sucrose, −.298°C; and sorbitol, −.613°C.

### Preservation/Osmotic Activity

Another function that sucrose serves, particularly at higher use levels, is that of preservation. Sugar at high concentrations reduces water activity ($a_w$). Under conditions of reduced water activity, there is less water available for the growth of microorganisms and there is also greater osmotic pressure. When the sucrose level is reduced in a recipe, the new product may have shelf life problems with respect to bacteria, yeast, or mold growth that the original product did not have. There are several ways this problem can be addressed. One approach is to replace sucrose with ingredients which have a similar effect on water activity. Polydextrose can be used to help maintain the soluble solids level while limiting the calories. In this regard, polydextrose can be used to decrease the water activity of a product as described previously. An equal weight of polydextrose will not have as great an effect on $a_w$ as sucrose.

It may be beneficial to make formulation changes in reduced-calorie products in order to maintain a comparable shelf life. One technique is to decrease the pH in order to inhibit microorganisms. Also, the use of chemical preservatives such as sodium benzoate or potassium sorbate may be particularly useful in calorie-modified products.

### Cryoprotectant

Several applications have been suggested for the cryoprotectant capability of polydextrose. Cryoprotection involves stabilizing foods against the damaging physical effects of freezing. This is a function that is well recognized for many soluble ingredients, particularly sugars. This function has been described in fish products, including surimi (15,16) and meat products (17).

### Fiber

In Japan, polydextrose conforms to their definition of a dietary fiber: "polysaccharides, related polymers and lignins, which are resistant to hydrolysis by the digestive enzymes of man" (18). The Japanese consumer has a particularly keen interest in supplementing his/her fiber intake as part of a desirable diet. In line with these factors, polydextrose has found a unique market in Japan for use in beverage products fortified with polydextrose as a soluble fiber (19). This classification is somewhat different from the concept of fiber that we have in the United States. Currently, our consensus of the definition of "fiber" seems to be limited to materials of plant origin.

## APPLICATIONS

Polydextrose has been accepted as a food additive by the U.S. Food and Drug Administration (1). The original food additive petition included specific application in eight food categories. These categories are: baked goods and mixes; chewing gum; confections and frostings; dressings; frozen dairy desserts; gelatins, puddings and fillings; hard candy; and soft candy. These food categories are among those described in the Code of Federal Regulations (21 CFR 170.30).

At the writing of this chapter, there are three additional categories being considered by the FDA. These include products such as peanut spreads, fruit spreads, syrups and toppings.

Polydextrose may be used in the approved food categories to fulfill any of four functions:

1. Bulking agent—to provide bulk or substance to a food
2. Formulation aid—to promote or produce a desired physical state
3. Humectant—to promote retention of moisture
4. Texturizer—to affect the appearance or feel of the food

### Baked Goods

This food category more specifically includes baked goods and baking mixes (restricted to fruit-, custard-, and pudding-filled pies, cakes, cookies, and similar baked products).

A one-third calorie reduction in these products would require a fat and carbohydrate reduction, which would alter the physical and organoleptic acceptability of the baked good. Therefore, polydextrose

**Table 6** Formulation for Reduced-Calorie Yellow Cake

| Ingredients | Percent |
|---|---|
| Part A | |
| Sugar, granulated | 18.29 |
| Polydextrose | 7.00 |
| Nonfat milk solids | 2.05 |
| Polyglycerol monoesters and mono- and diglycerides | 1.03 |
| Food starch modified (National Starch) | 0.82 |
| Guar gum (National Starch) | 0.19 |
| Sodium bicarbonate | 0.16 |
| Artificial butter concentrate (Consumers Flavoring Extract Co.) | 0.08 |
| Artificial vanilla flavor (610072U-PFW) | 0.08 |
| Part B | |
| Cake flour | 21.61 |
| Baking powder (Davis) | 1.83 |
| Glucono-delta-lactone (Pfizer) | 0.20 |
| Salt | 0.35 |
| Xanthan gum | 0.10 |
| Part C | |
| Whole eggs | 17.77 |
| Water | 14.22 |
| Part D | |
| Water | 14.22 |

Procedure: 1. Mix Part A. 2. Slowly add Part B and mix. 3. Add Part C. Mix, scrape down. 4. Mix until smooth. 5. Add Part D. Mix until smooth. 6. Scale 13 oz into eight-inch cake pan (greased or paper lined). 7. Bake at 350°F for 30 minutes.

**Table 7** Yellow Cake: Alitame-Sweetened, One-Third Fewer Calories, No Added Sugar, No Added Shortening

| Ingredients | % Batter |
|---|---|
| Part A | |
| Sorbitol crystalline | 14.24 |
| Polydextrose | 7.02 |
| Maltodextrin | 4.5 |
| Emulsifier blend | 4.1 |
| Sodium bicarbonate | 0.7 |
| Artificial butter concentrate | 0.17 |
| Artificial vanilla flavor | 0.08 |
| Alitame | 0.008 |
| Part B | |
| Cake flour | 21.6 |
| Cellulose | 2.0 |
| Glucono-delta-lactone | 1.3 |
| Salt | 0.34 |
| Xanthan gum | 0.1 |
| Part C | |
| Whole eggs | 21.75 |
| Water | 9.7 |
| Part D | |
| Water | 12.4 |

plays a particularly important role in the formulation of reduced-calorie baked goods (20,21). In many baked goods formulations, polydextrose can be used without a high intensity sweetener. Table 6 (22) presents such a formula for a reduced-calorie yellow cake.

The second recipe, Table 7, illustrates another formulation option. The goals achieved in this yellow cake were to have one-third fewer calories than a conventional cake with no added sugar or shortening and be as desirable as a sugar-sweetened cake. The USDA *Composition of Foods* handbook was used for establishing caloric values.

In this recipe, alitame, a dipeptide-based high intensity sweetener, was used to provide sweetness. This sweetener is awaiting approval for use in foods and is the subject of a previous chapter. Alitame has a sweetness potency 2000 times that of sucrose and was used at 80 ppm in the batter of this reduced-calorie yellow cake. A combination of polydextrose, sorbitol, maltodextrin, and cellulose powder was used to provide bulk. An alternative to adding the alitame to Part A of the batter is its addition as a water solution (approximately 1%) to Part C of the batter. The initial bake tests focused on optimizing the volume, color, symmetry, and eating quality of the cake. The resulting product is a highly acceptable reduced-calorie cake with the textural and sweetness quality demanded by consumers.

When used in chemically leavened baked goods, either the acidic or neutralized form of polydextrose may be used. If the acid form is used, the leavening acid may need to be adjusted.

Recent work reported on the good textural qualities of polydextrose shortbread cookies formulated with high intensity sweeteners (23). For sweetness, the study evaluated synergistic combinations of aspartame, cyclamate, saccharin, and acesulfame-K in the cookies.

The benefits of polydextrose in cake type products include volume, tenderness, structure and eating quality. Typical use levels in such products would be from 7 to 15% by weight.

### Chewing Gum

Polydextrose can be used to make chewing gum with good shelf life and elasticity. A high intensity sweetener would be required to provide sweetness to the gum. Several patents exist for low-calorie gum products which cite the use of polydextrose (24,25).

### Confections and Frostings

This area of application represents products where the functional properties of carbohydrates and fats are critical to the character of the food. Polydextrose is not crystalline, and therefore the crystalline/sugar texture cannot be achieved. The bulk and mouthfeel of taffy-like products is more closely matched using polydextrose. Polydextrose use level would typically be in the range of 25%.

In products such as these, the acidity of polydextrose is adjusted by the use of food-approved neutralizing agents. In lieu of this, the neutralized form, polydextrose-K, could be used.

### Salad Dressings

Many pourable salad dressings are surprisingly high in calories, containing up to 60–80 calories per tablespoon. Of their calories, approximately 80–95% come from fat (26). Therefore, making reduced-calorie dressing involves, primarily, reducing the oil level.

Polydextrose finds use in this application by replacing the functionality and mouthfeel that was contributed by the fat. Some dressings do contain a significant level of sugar. Russian, Thousand Island, and French-style dressings are highest in sugar. These dressings are good candidates for the replacement of sucrose with polydextrose and a high intensity sweetener. A number of highly palatable reduced-calorie dressings have been formulated with polydextrose and enjoy considerable success in the market place.

### Frozen Dairy Desserts

This area of use typically involves the reduction of both fat and sugar in the formulation of reduced-calorie products (27). Sugar is a multipurpose ingredient providing bulk, mouthfeel, sweetness, and freezing point depression to the final product.

The recipe in Table 8 is for an aspartame-sweetened reduced-calorie product. It contains no added sugar compared to 15–17% sugar in ice cream. This product has a sweetness and eating quality comparable to the fully caloric product.

The firmness of a frozen dessert can be manipulated by altering the freezing point. Freezing point depression is a function of molecular weight of key ingredients, and thus, a balance of polydextrose with a lower molecular weight product such as sorbitol may best match the firmness desired.

**Table 8** Aspartame-Sweetened Vanilla Frozen Dessert (7.5% Fat)

| Ingredients | Percent |
|---|---|
| Cream (36% fat) | 20.83 |
| Nonfat dry milk | 10.35 |
| Polydextrose, Type N (70% solution) | 21.43 |
| Water | 44.985 |
| Gelatin, 120 Bloom | 0.33 |
| Crystalline cellulose (Avicel RC 591) | 1.00 |
| Mono- and diglycerides (Drewmulse 700E) | 0.30 |
| Sodium citrate, dihydrate (Pfizer) | 0.20 |
| Vanilla extract, twofold | 0.50 |
| Aspartame | 0.075 |

Procedure: 1. Combine all the ingredients except the vanilla extract and the aspartame. 2. Pasteurize by rapidly heating to 160°F and maintaining that temperature for 30 minutes. 3. Homogenize at 2500 psi (first stage) and at 500 psi (second stage). 4. Rapidly cool to 40°F, add the aspartame and store under refrigeration overnight. 5. Add the vanilla extract and freeze with a 100% overrun.

Very pleasant-tasting products are possible without the use of a high intensity sweetener. However, a series of regulatory approvals in 1987 and 1988 extended the use of aspartame to include frozen dairy desserts, and this resulted in a surge of very successful reduced-calorie products (28,29). Typical use levels of polydextrose in these products range from 7 to 15%.

Frozen yogurt is a specialty product within this category that has become very popular. Frozen yogurt is lower in fat than ice cream, and thus calorie reduction is achieved primarily from sugar replacement. Here again, polydextrose serves the same varied functions of bulk, texture, and freezing point depression.

### Gelatins, Puddings, and Fillings

A 50% calorie reduction is readily achieved in an instant pudding type product when polydextrose is used with a high intensity sweetener (30). Using polydextrose results in a creamier texture with a more uniform dispersion of cocoa and starch. It also functions as a formulation aid and helps prevent lumping and nonuniformity. The chocolate instant pudding (Table 9) has a creamy, smooth mouthfeel provided by polydextrose and a satisfying, pleasant sweetness provided by aspartame.

### Hard Candy

A major portion of the sugar in hard candy can be replaced with polydextrose (31). Clarity and good bite still are achievable with a re-

**Table 9** Chocolate Instant Pudding (50% Calorie Reduction)

| Ingredients | Percent |
|---|---|
| Polydextrose | 50.09 |
| Starch (National Starch) | 26.89 |
| Cocoa, dutched (10–12% fat) | 14.15 |
| Disodium phosphate | 4.25 |
| Sodium pyrophosphate | 2.26 |
| Sodium chloride | 0.57 |
| Vanillin | 0.42 |
| Acetylated monoglyceride (Myvacet 9-45-DPI) | 0.56 |
| Color (Cocoline brown 640-Kohnstamm) | 0.17 |
| Aspartame | 0.64 |

Procedure: 1. Prepare a uniform blend of the ingredients. 2. Pour 2 cups of cold skim milk into bowl. 3. Add 70.7 grams of the pudding mix to the skim milk. 4. Mix at low speed with electric mixer until well blended (about 2 minutes). 5. Pour into serving bowls.

duced sugar level. A high intensity sweetener is needed to supply the sweetness expected in hard candy products. The acidity of polydextrose must be adjusted with a buffering agent such as sodium citrate. The use of sodium citrate functions to improve the flavor and also to minimize excess sugar inversion during processing.

### Soft Candy

The use of polydextrose as a partial sugar replacement in soft candy serves two key functions: providing bulk and humectancy (32,33). In most cases, a neutralized form of polydextrose is most suitable.

### Other Categories

At the time of this writing, there are three additional food additive petitions under review by the Food and Drug Administration. These areas of use include sweet sauces, toppings and syrups (34), peanut spreads (35,36), and fruit spreads (37).

## METABOLISM

The basis for the 1 calorie per gram caloric value of polydextrose lies in the difficulty that mammalian digestive enzymes have in attacking a large and randomly bonded polymer. Extensive feeding studies with man and animals have confirmed the caloric contribution (1,38–40).

The definitive work on the metabolism of polydextrose studied the disposition of $C^{14}$ radiolabeled polydextrose in feeding studies. The caloric value of 1 calorie per gram and the testing methodology were fully endorsed by the U.S. Food and Drug Administration (9).

A portion of the polydextrose molecule is metabolized in the intestinal tract by microorganisms. As by-products of this metabolism, these microorganisms produce volatile fatty acids and carbon dioxide. The volatile fatty acids are absorbed in the large intestines and used as an energy source. Therefore, it is the microbial function that contributes the 1 calorie per gram of ingested polydextrose. This is a phenomenon common to essentially any complex carbohydrate in the diet.

Another area of metabolism that has been closely investigated is the effect of polydextrose on insulin demand and the implications of this on use by diabetics. Studies show that the use of polydextrose does not create an insulin demand using glucose toleration techniques. The position of the American Diabetes Association concludes that "polydextrose could be of benefit to people with diabetes in that it may substitute for sugar or for fat in a number of food recipes, particularly in desserts and baked goods."

As described previously, tests show that polydextrose has a very low cariogenicity potential.

## TOLERANCE

Numerous clinical studies show that polydextrose is well tolerated when consumed in moderation as part of a normal daily diet. This has been confirmed by consumer results in the years since polydextrose has been approved.

When consumed in excessive amounts, polydextrose can have a laxative effect. Studies of adults showed a mean laxative threshold of 90 grams per day (41). This compares to a level of 70 grams of sorbitol for similar effect.

## REGULATORY STATUS

Except where specifically stated otherwise, the comments offered on regulatory issues are geared to United States regulations and practices. The reader should also be aware that regulations are constantly being changed and updated.

### FDA Approval

There were years of extensive research into the safety, properties, and applications of polydextrose prior to the approval of the food additive petition in 1981 (1,3) by the U.S. Food and Drug Administration. The regulations allow for the safe use of polydextrose as a multipurpose food additive in specific foods. It is used in accordance with good manufacturing practices as a bulking agent, formulation aid, humectant, and texturizer.

As of 1989, the use of polydextrose is specifically allowed in eight food categories. These are listed in Table 10.

**Table 10** Approved Food Categories for Polydextrose

| |
|---|
| Baked Goods and Mixes (restricted to fruit, custard and pudding-filled pies; cakes; cookies; and similar baked products) |
| Chewing gum |
| Confections and frostings |
| Dressings for salads |
| Frozen dairy desserts |
| Gelatins, puddings, and fillings |
| Hard candy |
| Soft candy |

**Table 11** Countries that Permit Polydextrose Use in Foods as of September 1989

| | |
|---|---|
| Australia | Japan |
| Austria | Netherlands |
| Belgium | New Zealand |
| Canada | Sweden |
| Denmark | Switzerland |
| Finland | United Kingdom |
| France | United States |

There is no maximum established use limit for polydextrose. Good manufacturing practices limit the quantity to the amount necessary to accomplish the intended purpose in the food.

### International

The use of polydextrose in foods has received approval in numerous countries (42). A list of countries that allow the use of polydextrose in some food applications is presented in Table 11.

Regulations governing the use of food ingredients must be closely investigated for any given country. The differences can be in both food restrictions as well as ingredient restrictions. An example is the difference in the concept of "reduced-calorie" foods. The U.S. criterion for these products is one-third reduction in calories. The definitions in other countries, however, range from a 25% to a 50% calorie reduction requirement.

## LABELING

When polydextrose is used in foods for special dietary purposes such as reduced calorie foods, it must be labeled in accordance with 21 CFR Part 105. Should a single serving of food contain more than 15 grams of polydextrose, the label must read: "Sensitive individuals may experience a laxative effect from excessive consumption of this product."

For nutritional labeling, polydextrose should be included in the carbohydrate portion, even though its caloric value is only one calorie per gram.

"Polydextrose" is the officially recognized name and should appear as such in the ingredients list. If polydextrose-K is used, the names of the blended ingredients "polydextrose" and "potassium bicarbonate" should be in the ingredient statement in their descending order of predominance.

## REFERENCES

1. 21 CFR 172.841, 46 FR 30080 (June 5, 1981), Food Additive Petition No. 9A3441.
2. H. H. Rennhard (Pfizer), Dietetic foods, U.S. Patent 3,876,794 (1975).
3. R. P. Allingham, Polydextrose—A new food ingredient: Technical aspects, in *Chemistry of Foods and Beverages: Recent Development,* Academic Press, 1982, p. 293.
4. J. J. Beereboom, Technical aspects of polydextrose, Polydextrose Trade Press Briefing, May 28, 1981.
5. Specification-C, Polydextrose F.C.C., Pfizer Chemical Specifications, Material Code S6810.
6. Specification-C, Polydextrose Solution F.C.C., Pfizer Chemical Specifications, Material Code S6811.
7. Specification-C, Polydextrose-K, Pfizer Chemical Specifications, Material Code S6814.
8. Information Sheet 2090, Polydextrose-K, Pfizer Chemical Division.
9. E. C. Coleman, U.S. Food and Drug Administration, letter dated April 21, 1981, to E. F. Bouchard.
10. H. R. Muhlemann, Polydextrose—a low-calorie sugar substitute. Dental tests. *Swiss Dent., 2*(3): 29 (1980).
11. R. E. Smiles, The functional applications of polydextrose. *Chemistry of Food and Beverages: Recent Developments,* Academic Press, 1982, pp. 305–322.
12. J. T. Liebrand, R. E. Smiles, and T. M. Freeman, Functions and applications for polydextrose in foods, *Food Technology in Australia, 37*(4): (1985).
13. R. J. Baer and K. A. Baldwin, Freezing points of bulking agents used in manufacture of low-calorie frozen desserts, *J. Dairy Science, 67:* 2860–2862 (1984).
14. R. J. Baer and K. A. Baldwin, Bulking agents can alter freezing, *Dairy Field* (Feb. 1985).
15. J. W. Park and T. C. Lanier, Combined effects of phosphates and sugar or polyols on stabilization of fish myofibrils, *J. Food Science, 52*(6): 1509–1513 (1987).
16. J. W. Park, T. C. Lanier, and D. P. Green, Cryoprotective effects of sugar, polyols, and/or phosphates on Alaska pollack surimi, *J. Food Science, 53*(1): 1–3 (1988).
17. J. W. Parks, Effects of cryoprotectants on properties of beef protein during frozen storage, dissertation, North Carolina State University, 1986.
18. The Foundation for Health and Physical Development, Ministry of Health and Welfare, Japan.
19. M. Hamanaka, Polydextrose as a water-soluble food fibre, *Food Industry, 30*(17): 73–80 (1987).
20. T. M. Freeman, Polydextrose for reduced calorie foods, *Cereal Foods World, 27*(10): 515–518 (1982).
21. Kwong-ok Kim, L. Hansen, and C. Setser, Phase transitions of wheat

starch-water systems containing polydextrose, *J. Food Science, 51*(4): 1095–1097 (1986).
22. Pfizer polydextrose for the market that's shaping up, technical brochure, Pfizer Chemical Division, 1985.
23. H. Lim, C. S. Setser, and S. S. Kim, Sensory studies of high potency multiple sweetener systems for shortbread cookies with and without polydextrose, *J. Food Science, 54*(3): 625–628 (1989).
24. R. E. Klose and R. E. Sjonvall, Low calorie, sugar-free chewing gum containing polydextrose, European Patent 0123742B1 (1987).
25. S. R. Cherukuri, F. Hriscisce, and Y. C. Weis, Low calorie chewing gum and method for its preparation, European Patent 0252874A (1988).
26. F. S. Goulart, How to read a salad dressing label, *The Nutrition & Dietary Consultant,* pp. 37–38, June 1985.
27. All the satisfaction of ice cream, but its a diet bar, *Food Engineering, 56*(9): 50–51 (1984).
28. D. H. Goff and W. K. Jordan, Aspartame and polydextrose in a calorie-reduced frozen dairy dessert, *J. Food Science, 49:* 306–307 (1984).
29. K. C. Reddy, Effect of aspartame and polydextrose levels on organoleptic, physio-chemical and economic values of ice cream and ice milk products, M.S. thesis, Mississippi State University, 1985.
30. Reduced calorie desserts made with aspartame and polydextrose (puddings, gelatins), *Food Processing,* pp. 21–22, July 1983.
31. K. J. Klacik, Continuous production of sugar-free hard candies, *The Manufacturing Confectioner,* pp. 61–67, August 1989.
32. J. T. Liebrand and R. E. Smiles, Polydextrose for reduced calorie confections, *Manufacturing Confectioner, 61*(11): 35–36 (1981).
33. A. Cridland, Developments in dietetic chocolate, *Confectionery Manufacturing and Marketing, 24*(10): (1987).
34. F.A.P. (9A4126), Use of polydextrose in sweet sauces, toppings and syrups, *Federal Register, 54*(49): (March 1989).
35. F.A.P. (7A3998), Use of polydextrose in peanut butter spread, *Federal Register, 53*(16): (January 26, 1988).
36. U.S. Patent #4,814,195, Polydextrose in peanut butter.
37. F.A.P. 8A4068, Use of polydextrose in fruit spreads, *Federal Register, 53*(102): (May 1988).
38. S. K. Figdor and H. H. Rennhard, Calorie utilization and disposition of [$^{14}$C] polydextrose in the rat, *J. Agric. Food Chem., 29:* 1181–1189 (1981).
39. J. S. White, C. M. Parsons, and D. H. Baker, An in-vitro digestibility assay for prediction of the metabolizable energy of low-calorie dextrose polymeric bulking agents, *J. Food Science, 53*(4): 1204–1207 (1988).
40. S. K. Figdor and J. R. Bianchine, Caloric utilization and disposition of [$^{14}$C] polydextrose in man, *J. of Agricultural and Food Chemistry, 31:* 389 (1983).
41. Raphan Study (IV), Pfizer communication.
42. Bulking agent moves to international commercialization, *Food Processing, 47*(8): (1986).

# 22

# Other Low-Calorie Ingredients: Fat and Oil Substitutes

**Robert G. LaBarge** *Dow Chemical Co., Midland, Michigan*

## INTRODUCTION

When people think of weight control, the two things that come to mind are dieting and use of high intensity sweeteners. Although control of calories is generally accepted as effective, most do not want to sacrifice any of their gustatory pleasures. Even though it has not yet been recognized by the general population, calorie control could well turn out to be the fad of the nineties.

In order to effectively control calories, the food formulator must consider a number of contributors. Sugar is not the only, or in many cases even the main, source of calories. The American caloric diet comes 35–40% from fat (1). Therefore, fat replacement or diminution in many cases becomes at least as and sometimes more important than sugar replacement.

The high caloric density of fat, i.e., 9 or more kcal/g versus the 4 from carbohydrates and proteins, makes fat the substance of choice for attaining the most efficiency in caloric reduction. Replacement of 1 g of fat is more efficient than replacement of 2 g of sugar. An excellent summary of the contribution of fat to the diet may be found in the Scientific Status Study by the Institute of Food Technologists' Expert Panel on Food Safety and Nutrition (2).

Fats and oils are found in a wide range of foods, but at highly variable levels. Amounts range from a low of a few percent in bread and rolls to 80% or more in mayonnaise and margarines. Cookies will contain 10–15% fat, icings around 25%, cakes 10–20%, and salad dressings and potato chips anywhere from 30 to 45%. Fat provides many properties: these include giving the desired lubricity to the product to afford a sense of flavor richness and tenderness, carrying of fat-soluble nutrients, and a means of providing essential fatty acids (3).

Today, many of us are becoming interested in getting the fat out. Whether it involves the trimming of excess fat from meat, eating low-calorie salad dressing, margarine, or mayonnaise, or consuming products like Lean Cuisine, consumers seem serious about lowering their fat intake. We may already be seeing the seeds of a change. With the increasing evidence of the link of calories and cancer (4–6), calorie control may be the next move—or it may simply be a further indictment of fat because of its high caloric density. Portion control may come back into vogue as the answer. Whatever happens, there will continue to be great interest in finding good fat substitutes. This chapter will review fat digestion and a wide variety of chemistries and ways people have tried to develop fat substitutes or mimic its properties. Even though fibers are often used to reduce calories, this review will make no attempt to cover this subject.

A number of review articles have dealt with various aspects of this issue over the last several years (2,7–12). Many of these considered fat replacers and fat substitutes to be the same. However, subtle differences exist. In this review, fat replacers will be defined as materials that decrease the caloric value of fat, but whose use involves modification of the base food. In contrast, fat substitutes are ingredients designed to decrease caloric value without the necessity of major alterations in the base food. These will inherently possess physical properties closer to those of common triglycerides.

Fat replacers differ significantly both in how they serve to replace the fat and in their caloric value. Most make use of water in some manner, either by increasing viscosity sufficiently to simulate some of the sensory properties of fat or by binding additional water, thereby effectively decreasing calories. These are usually proteins or carbohydrates and contribute in the range of 4 kcal/g. Examples include gums, modified starches or proteins, and water-soluble compounds.

In contrast, fat substitutes can serve as a direct substitute for the original fat or oil. Examples include carbohydrate fatty acid esters, polycarboxylic acid esters, hindered esters, medium-chain triglycerides, polyglycerine esters, polysiloxanes, and substituted ethers.

The objective of the formulator is to either reduce calories or remove fat. Therefore, in commercial products, it should be recognized that many systems could work via various combinations of these techniques.

## FAT DIGESTION

Fats are relatively straightforward molecules, being simply an esterified glycerine. Glycerine, a straight-chain, three-carbon substance, has three hydroxyl groups: two primary hydroxyls on the one and three carbon atoms, and a secondary hydroxyl group at C2. Each of the hydroxyls is combined with a long-chain acid to form a triester of glycerine. They are known as triglycerides, biological amphiphiles having glycerine as the small polar backbone, with the fatty acids being the large nonpolar segment.

Variations in physical properties are brought about by the myriad choices and combinations of acid groups. Many different acid moieties such as oleic, stearic, palmitic, linoleic, and myristic are found. Differences are caused by chain length, degree of saturation, and varying combinations of these fatty acids on the glycerine backbone. It is this nonpolar portion that makes fats so insoluble in water.

Humans consume an average of 100 g of exogenous fat per day. In addition to exogenous fats, the body produces over 50 g/day of endogenous lipids, the majority of which are biliary in nature. About 30 g of these are bile salts, 10–15 g phospholipids, and 1–2 g cholesterol. In addition to the biliary lipids, the body produces 2–6 g/day of lipids from desequamated cells sloughed off the lining of the small intestine and 10 g/day of lipids from dead bacteria. Of this 150 g, we excrete only about 4 g/day of fecal fat. The rest is either used, stored, or recycled.

Most believe that triglycerides, once consumed, pass virtually unchanged through the first portion of the gastrointestinal tract, including the highly acidic conditions encountered within the stomach. This is only partially true, but suffices for most cases. Once through the stomach and into the small intestine, fats are not so lucky. Lipase, a catalytic enzyme secreted by the pancreas, rapidly cleaves the triglyceride at the ester linkage forming free glycerine and fatty acids. These can then pass through the intestinal membrane and, after a series of steps, are ultimately either catabolized or stored within the body. This approach is actually rather simplistic, but accurate in a broad sense.

There are actually a number of lipases that act upon exogenous lipids. The first of these is human milk lipase. It is present in two places—human and gorilla mother's milk. This is why babies, both human and gorilla, are able to utilize triglycerides so efficiently early in life, even though they consume 4–5 g of triglyceride/kg/day. Without these lipases they would suffer steatorrhea. Human milk lipase is stable to the acid pHs found in the stomach and is activated by the bile salts in the small intestine.

There are two other lipases present which act in the acidic environment of the stomach: tongue and gastric lipase. These function at a pH of about 2–4, although 4 is optimum. A fasting stomach will exist at a pH of 1–1.5, but, as food enters, the pH is raised to about 4–5. Interestingly, tongue lipase works about four times as fast at the $S_N3$ position (or 3rd carbon atom of glycerine), as it does at the $S_N1$ position. Perhaps this is because the SN 3 position of milk triglycerides is always a short- or medium-chain fatty acid. Subsequent reaction of the diglyceride is much slower. Tongue lipase works in either oil/water or water/oil emulsions. This is important because of the poor emulsification occurring in the stomach. Here about 30–40% of the triglycerides, predominantly low- and medium-chain moieties from milk and certain vegetable oils, are absorbed through the stomach wall and transported via the portal system to the liver.

Next we get to the small intestine. There are four pancreatic lipases that function in this environment at a pH of 6–7. The first and most important is the classic pancreatic lipase. However, there is also a bile salt–dependent lipase. These are accompanied by phospholipase $A_2$ and cholesterol esterase.

Action of the pancreatic lipase is actually rather complex. The shear forces existing in the pyloric valve are responsible for reducing fat globules from the 20–40 $\mu$m size in the stomach to 0.1 $\mu$m in the duodenum. Initial attack by the lipase, which functions only at the oil/water interface, is ameliorated by the bile salts, which actually push the lipase off the interface. Colipase, which has a high surface activity, is necessary so it may bind to the interfacial surface in the presence of bile salts. The other end of the colipase is capable of binding strongly with the lipase, pulling it down to the surface so hydrolysis may begin.

The glycerine and fatty acids resulting from this action pass through the intestinal membrane, are resynthesized, and further transported within the body. This is why, although we consume over 100 g/day of exogenous fats and produce a large additional amount endogenously, only about 4 g/day is excreted in the feces—about 1 g each from the diet, bile, cells, and bacteria.

## EARLY EFFORTS AT FAT REPLACEMENT

Many strategies have been tried in attempts to either fool the enzymes, interfere with bile acid emulsification, or in some way modify or replace a triglyceride so the body can either not recognize it or does not know how to deal with it when it does see it.

Early efforts were aimed at finding ways to substitute for or reduce the caloric value of oil-like substances in food. Perceived health risks as well as economic reasons seem to be the main driving forces.

Mieth and his coworkers (13) have described efforts to modify the high energy properties of fats as taking place in four areas:

1. Combination of water and surface-active lipids or nonlipidic additives with lower energy values. These may include special proteins or polysaccharides having emulsifying or high swell gel properties.
2. Compounds such acetoglycerides and medium-chain triglycerides having a low heat of combustion
3. Acaloric compounds whose structure differs significantly from triglycerides, such as silicones and paraffins
4. Ester modification—acaloric compounds having fat-like properties close to those of ordinary lipids, but whose ester bonds have been modified or converted to ethers. These would include glycerol ethers, pseudofats, and carbohydrate fatty acid esters.

Efforts have been going on for a long time to find the perfect fat substitute. In 1948, Harry Deuel published a paper on the digestibility of rapeseed oil in the rat (14). Rapeseed oil is predominantly erucic acid, a long-chain fatty acid. He was interested because Boer reported (15) that rats grew less favorably on a diet containing rapeseed oil than on one containing butter, a finding that Deuel was able to confirm in his laboratory. This also corroborated his 1940 finding that rapeseed oil was very slowly absorbed (16). In two separate rat-feeding studies he found that trierucin was hydrolyzed in the gastrointestinal tract and excreted as a soap. He concluded that the low digestibility of 77% was associated with poor absorption of the erucic acid portion. After hydrolysis, the erucic acid formed was thought to be both of too high a molecular weight and too saturated to be readily absorbed.

## AQUEOUS-BASED FORMULATIONS

Initially, researchers tried to design around the fat by use of a simple, cheap, safe ingredient known as water. As one would expect, this worked to a limited degree. Water, in combination with other substances such as gums, used for their thickening power, is sometimes effective, but often leads to deterioration of one or more desirable properties. So, what is the formulator to do if he or she wants to reduce the high fat level in a product?

One approach is that of simple substitution, illustrated by the work of Morley and Ashton (17). They taught how to prepare a soft serve ice cream with very low levels of fat and high levels (57%) of water. Formulation with 100–200% overruns didn't hurt caloric value. Essentially, they found that the right combination of protein, water, sugar, and emulsifiers and stabilizers allowed them to get these large overruns and end up with a low-fat, low-calorie soft serve ice cream.

Another interesting approach is taught by Moran and Sharp (18). They teach preparation of a low-fat spread as an oil-in-water-in-oil emulsion, where the continuous phase is an oil and the dispersed phase a cream containing whey protein. Use of whey allows reduction of fat from 80 to under 50%. In this case gelatin is used to provide the creamy mouthfeel the inventors were looking for.

Many have used a variety of other gums and thickeners to afford the proper mouthfeel and/or viscosity profile. The most commonly used gums are algin and xanthan, generally in the range of 0.1–0.5%. They are used at such low levels because they dramatically increase the viscosity, thus leading to emulsion stability. Representative end uses include such diverse areas as whipped toppings, candy, mayonnaise, cream cheese, and table syrup.

## MINERAL OIL

A number of efforts have been made to use substances like paraffins and mineral oils as fat substitutes. Mineral oil is a hydrocarbon having limited absorbability. It is used both to reduce the strain of constipation with a dosage of 15–45 ml at bedtime and, in an emulsified form, to reduce seepage through the anal sphincter.

The question, "Why not use mineral oil?" is always the first people ask. Mention of negative side effects such as diarrhea and other toxicological objections usually satisifes them. Diarrhea is undoubtedly a problem, but the toxic effects were considered more questionable for some time.

The big question is whether mineral oil is a carcinogen. A government study has looked at carcinogenicity data and reported mixed results, but it is probably not conclusive enough to do anything but raise a caution sign. Most studies seem to give it a clean bill of health, but contaminants may pose a problem. These have been indicated far more often than purified mineral oil.

However, three serious objections have been put forth. First, it interferes with vitamin absorption. Second, it is particularly susceptible to absorption when it occurs in the emulsified form and third, it has been implicated in production of lipid pneumonia.

A great deal of work has been done to determine the influence of mineral oil on fat-soluble vitamins A, D, E, and K. The evidence is often conflicting, but generally seems to indicate that an effect will occur. Specifically, these fat-soluble vitamins may, in some cases, have their absorption reduced because of solubility in the unabsorbed mineral oil.

The next question, that of absorbability, is really a question of degree. Mineral oil is essentially but not completely nonabsorbed. Analysis is difficult, so quantified results are problematic. Although early studies were done in 1884, the first good ones were not run until 1929. In these (19), mineral oil was recovered from feces and tissue examined for evidence of absorption. Rats were fed 5% oil for 5 weeks and a pig 100 ml/day for 54 days. Some absorption was found to occur with storage taking place in the liver.

An extensive study was done (20) feeding rabbits, rats and pigs for periods of 13–19 months. In this study, Strycker found little change up to 6 months, but then an increasing amount seemed to be absorbed with significant changes occurring in the lymph nodes. Later it was shown (21) that a combination of bile salts, oleic acid, and glyceryl monostearate is able to emulsify mineral oil to the 0.5 $\mu$m particle size necessary for intestinal wall absorption, although several workers later questioned these results.

Recently, the Food Additives Branch of the British Ministry of Agriculture, Fisheries and Food indicated it would be banning use of mineral oil as both a direct food additive and in food processing due to a lack of toxicological information on the wide range of materials being used. Recent studies have indicated toxicity at low doses due to cumulative accumulation in organs and tissues (22). In the United States, mineral oil is permitted under U.S. Code of Federal Regulations, Title 21, 172.878 for a variety of uses from a release agent in capsules to a dust control agent for sorghum.

## WATER-SOLUBLE FAT REPLACERS

A number of water-soluble compounds, in addition to the gums mentioned earlier, have been promoted as partial or complete fat replacers. Among those most often mentioned are Pfizer Inc.'s polydextrose and National Starch & Chemical Corporation's N-Oil.

Polydextrose is prepared by high temperature polymerization of glucose under catalytic conditions. It is a water-soluble, amorphous solid having a caloric value of 1 cal/g. However, studies of Cooley and Livesey (23) indicate this value is closer to 3–4 kcal/g. Pfizer disputes these results (24). They maintain the radiolabel method is the method of choice, and this indicates a value of 1 kcal/g in both animals and humans. Although the literature contains a great deal of discussion as to which is the best method for determining caloric value, the formulator is primarily interested in what he or she can claim. FDA has agreed with Pfizer Inc.'s claim that polydextrose has 1 kcal/g.

Polydextrose is being marketed primarily as a bulking agent, but is also described as having some utility as a partial substitute for fats. It serves reasonably well if utilized in the proper combination with other ingredients as part of a designed program for partial fat replacement. Like a number of other low-calorie products, it has a laxative effect due to creation of an osmotic load in the lower intestine. The mean laxative effect was found to be 90 g/day or 15 g/serving, a number sufficiently high to allow significant use in food before gastrointestinal problems begin to occur.

N-Oil™ is marketed to both replace fat and enhance fat properties. Four parts of oil may be replaced by one part N-Oil and three parts water. It is especially recommended for ice cream and both pourable and spoonable salad dressings at a 25% concentration. N-Oil actually provides 4 cal/g but when used at this 25% level only provides 1/9 the calories of the fat it replaces.

N-Oil is a bland, low-viscosity tapioca dextrin which allows one to retain the mouthfeel of oil without the caloric content. It is also available in an instant form having cold water stability for use in frozen desserts.

National Starch (25) obtained a patent for a converted starch that can form a water gel. Specifically, starches such as tapioca, corn, or potato having a dextrose equivalent (DE) of under 5 are covered. The preferred embodiment is that of tapioca dextrins with a DE of 2 or less and hot flow viscosity and gel strength of 20–100 sec and 65–930 g at 25–35% solids.

Tapioca dextrins reportedly overcome the major disadvantages of undegraded cereal starches, i.e., low temperature stability under long-term storage conditions. When used, they have a tendency to irreversibly change when frozen and, when defrosted, form a water insoluble mass.

N-Oil is marketed both in a dry and instant form. The former must be heated, while the latter can be used in a dry mix. They are reported to have been used to replace fat both in sausage and in tofu-containing frozen desserts.

Creme de tapioca (Tipiak) is a starch which comes from cassava flour. It is capable of also acting as a partial fat replacer by absorbing water and improving moisture and texture. Specifically, it is recommended for use in a wide range of products such as luncheon meats, pastas, and confections (26).

Grain Processing Corporation sells a number of maltodextrins, especially Maltrin M040, a low DE maltodextrin containing almost 98% penta and higher saccharides. It is made from corn starch, spray-dried, and is capable of being used as a partial fat substitute, although it too is water-soluble. Company literature claims it provides a creamy, fat-like texture while only contributing 4 kcal/gram. Perhaps a combination of good film-forming characteristics and low hygroscopicity contributes to these properties.

Maltodextrins are low DE corn starches prepared by enzyme or acid hydrolysis. These oligosaccharides are composed of fewer than 20 $\alpha(1\rightarrow4)$-linked glucose units. They are white powders marketed in three classes based on DE ranges: $< 8$, 9–13, and 14–19. They are bland, nonhygroscopic, readily digestible powders of high water solubility, which build a high viscosity and form a heat-reversible gel. Their low osmolality is beneficial in high-energy diets as this property allows gastric emptying to occur at a more normal rate.

Although they contribute 4 kcal/g, they are often used in reduced-fat products. In combination with water, the gel that is formed has a texture close to that of hydrogenated oils. Incorporation into salad dressings, frozen desserts, table spreads, sour cream, and imitation peanut butter has been described (27).

Another water-soluble compound marketed as a general binding agent and filler in foods is Paselli SA2, from AVEBE. They claim it to be a unique potato starch capable of replacing over 50% of the oil or fat in a wide variety of products such as dips, dressings, ice creams, dairy products, and baked goods. Specifically, gels with a DE less than 5 will exhibit fat-like properties by enhancing creaminess and providing a stable gel, while reducing calories by one half.

SA2 is made via a limited enzymatic hydrolysis of the amylopectin and amylose fraction of potato starch. Hydration will occur at room temperature without lumping if high agitation is used, but full gellation requires an additional 2–3 hours at concentrations above 20%.

## SIMPLESSE

The first hint most people had regarding the Nutrasweet Company's announcement of its new low-calorie fat was in *Crain's Chicago Business* (28) on January 18, 1988, followed by a column in the *Wall Street Journal* (29) of January 19 indicating that Nutrasweet was holding a press conference in New York on January 27.

At the media event, Simplesse was described as a substance made from milk and egg white proteins through a process which changed its size, shape, and texture. It was claimed to have 1 1/3 calories per gram and designed to be used in ice cream, margarine, yogurt, processed cheese, salad dressing, and a range of other products, but not to be used for frying, baking, or to be exposed to a low pH (30). As it was made from natural proteins and had no artificial ingredients, Nutrasweet felt it would be Generally Recognized as Safe and not need FDA approval.

In a meeting with FDA Commissioner Frank E. Young on January 28, Nutrasweet was given three options: submission to the FDA of a (1) food additive petition or (2) GRAS review, or (3) market it and risk FDA seizure (31). Nutrasweet, being an astute organization, decided to prepare a GRAS petition.

In the interim, numerous references to Simplesse, telling its story, began to appear in print. It seems that Norman Singer, in 1979, when working at John LaBatt Limited, was brought a substance from whey which, when tasted, resembled cream cheese. When observed microscopically, the small spheres of protein appeared to roll constantly. He hypothesized that this was what gave it a creamy texture. In 1984, this technology was licensed to Nutrasweet.

A U.S. patent issued to Singer, Yamamoto, and Latella (32) disclosing a proteinacious, water-dispersible macrocolloid of dairy whey protein. Particle size was in the range of 0.1–2.0 $\mu$m and spheroidal in character. Hydration gave a smooth, emulsion-like organoleptic property. At the same time, a microparticulation technique was disclosed whereby undenatured whey protein was submitted to high shear in an aqueous environment at high temperature and low pH.

The resultant product can be hydrated with twice its weight of water, reducing the 4 calories of the protein to 1 1/3. Substitution on an equal weight basis for fat would thus result in a significant caloric reduction. All this assumes that the taste remains bland and nothing causes the particle to coagulate and lose its fat-like texture. If this were to happen, shelf life and marketability would be reduced significantly.

A petition to affirm Generally Recognized as Safe (GRAS) status was filed by Nutrasweet in September and accepted for filing on November 23, 1988 (33). Although it took longer than expected, the FDA finally granted GRAS status for use in reduced calorie frozen desserts. Trailblazer, derived from an old General Foods patent, is expected to be a competitive product when approved.

About the same time, Nutrasweet formed a business group specifically to develop and market Simplesse and announced formation of Desserve, the Sampler Cafe in Mt. Prospect, Illinois, to serve as a test marketer for low-calorie desserts. Overall, Nutrasweet feels Simplesse could replace about 15% of dietary fat and 4% of dietary cholesterol. Maximum use levels in food categories applied for in the GRAS petition would reduce fat by 4% in the diet. If Simplesse were used in all ice cream consumed, the 90th percentile would consume 23.5 g/day or 1.6 g of protein (34). This can be compared to the 17 g/day claimed for olestra in all its petitioned-for uses.

In a February 17, 1989, letter from Nutrasweet to the FDA, it was noted that the United Kingdom and West German representatives confirmed that Simplesse "constitutes a known food substance or ingredient, not a food additive or novel food." West Germany concluded that Simplesse was not a food additive under current regulation (35). With this and the aforementioned FDA approval, marketing of Simple Pleasures. Now, all will have a chance to see whether it has any shortcomings and if they can be overcome. The ultimate test will be how successful it is in the marketplace.

Phillip Morris's Kraft General Foods is also active in this area. They are looking for GRAS affirmation on their own milk and egg white protein combination, which sounds very similar to Simplesse (36). General Foods was issued a patent in 1981 (37) with remarkably similar technology to that of Simplesse. It talked of an oil-replacement composition prepared by hydrating and whipping a protein and a cellulose gum and mixing with an acid-stable modified starch. Whether this is the same technology as described by Freedman will become clear over time, but it is obvious work other than Simplesse has been going on for some time in this general area.

Reach Associates has announced a similar series of products—Nutrifat C, PC, and Instant PC—aimed at the fat-replacement market. Nutrifat C is a combination of enzyme, dextrin, and fiber, which is claimed can replace up to 50% of edible oils. It, like Simplesse, combines with water to reduce calories. The PC mixes C with 30% protein and the Instant is a pregelatinized version of PC.

Nutrasweet has also entered the battle in another way with their hemicellulose technology. They promote the use of hemicellulose A and B and modified hemicellulose A and B for use as a water-soluble bulking agent to replace functional properties of carbohydrates or fats (38). Hemicellulose A is the fraction precipitating when an alkaline extraction mixture is acidified, and B is the ethanol precipitate after A is removed. Modified hemicellulose refers to material remaining after acid or enzymatic digestion, i.e., a low molecule weight fraction. Although teaching mainly involves sugar replacement, claims are made for both sugar and fat replacement.

## CARBOHYDRATE FATTY ACID ESTERS

Although many efforts to develop fat substitutes have involved use of water-soluble compounds, most involve modifications of triglycerides. One interesting technique for circumventing the body's hydrolysis of triglycerides is to modify the molecule to render it incapable of being hydrolyzed by lipase.

Many have suggested use of carbohydrate fatty acid esters. Such compounds are relatively easy to synthesize in the lab via transesterification, acid chlorides, anhydrides, or straight fatty acid esterification—tools commonly available to most competent organic chemists. With the wide variety of fatty acids and carbohydrate backbones available, appropriate physical properties can be designed into the molecules. Many are reported to be excellent thickeners, but this property is unsuitable in an oil substitute.

The first efforts in this area appear to have begun with synthesis of polyhydric alcohol esters at Heyden Chemical Corporation (39). This was followed by a patent granted to Heyden Newport Chemical for diet products containing polyhydric alcohols, which had been completely esterified with fatty acids (40). A long string of modifications upon this basic concept was thus born.

Most of the effort seems to have focused on sugar rather than starch esters as their melting point is easier to control. Sugar esters have also been reported to caramelize in a manner similar to sucrose itself. Analytical techniques are difficult due to the large and often irregular mass of many of these substances, but good labs are capable of getting around many of the problems.

When one thinks of sugar esters, sucrose polyester, the much-touted noncaloric oil candidate "olestra" from the Procter & Gamble Company immediately comes to mind. In the late 1950s and early 1960s, Mattson and Volpenhein were doing basic research to determine the effect of pancreatic lipase on a number of alcohol esters and, in the course of screening, found that sucrose polyester (SPE) seemed to be unhydrolyzed under test conditions. Eventually this translated into the idea that this might be a useful phenomenon—if lipase did not hydrolyze the material, yet it had oil-like properties, it might be able to function as a O- rather than a 9-calorie oil in food.

It was found (41) that, as the number of ester groups were increased sequentially from the three in triolein to the eight in SPE, the rate of lipolysis decreased dramatically. This is corroborated by the work of Weiss and Mieth (42), where a similar phenomenon was reported. Putting this together with evidence that absorption is decreased markedly if hydrolysis does not occur, it seemed likely that SPE might serve well as a fat substitute.

A process patent for making SPE soon followed. Rizzi and Taylor (43,44) taught a multistep process in which a 3-to-1 ratio of soybean oil methyl ester to sucrose is reacted in the presence of potassium salt of the soybean oil ester at 95°C under reduced pressure for 75 minutes. The mixture is allowed to cool and sodium hydride in mineral oil is added and heated at 145°C for 2 hours. After cooling, more NaH was added, the process repeated two more times, and purification started. The process seemed to work, and yields of 90% were claimed, but these yields have seldom been reproduced by others.

A lot of work has gone on through the years, both in the United States and abroad, attempting to come up with a better and more economical process. Patents were subsequently issued (45) in which NaH was replaced by potassium carbonate, with use of high soap to polyol ratios (46), both serving to clean up the process considerably. The following year a U.S. patent similar to P&G's, but utilizing molecular distillation for purification, was issued to Dai-Ichi (47).

Many experiments have been carried out with SPE, and many interesting things learned. Radiolabeled studies have confirmed that SPE is not found in the blood, and thus it is reasoned it is not absorbed. Inability to detect presence of SPE in the liver of chronically fed rats also contributes to this conclusion.

Reports have appeared about potential vitamin deficiencies arising after consumption of SPE. Like with mineral oil, some of the fat-soluble vitamins are reported to be swept out of the GI tract. However, a U.S. patent (48) notes use of vitamin-fortified mixtures to compensate for this effect. Specifically, olestra will be supplemented with vitamin E to maintain nutritional adequacy.

On the other side, it seems that SPE can also serve to remove potentially toxic substances such as DDT from the body (49). It has also been claimed to be useful for treatment of duodenal ulcers (50). Procter & Gamble also has several use patents, including margarines, beverages, and a European application for low-calorie baked goods (51–54).

Perhaps the greatest obstacle to wide-scale use of SPE involves reports of a significant amount of anal leakage when too much SPE is consumed. If one stops to think about it, if one is putting an oil into the mouth and forcing it to go through the alimentary canal unchanged, one should expect it to come out the other end in substantially the same form. Now, if in addition it is a slippery, oily substance when it goes in, why shouldn't it be the same when it comes out? If it has sufficient lubricity, the sphincter will be incapable of holding it back and it will run right out the other end.

However, Procter & Gamble had a solution. A patent (55) claimed "sufficient anti-anal leakage agent to prevent leakage of said liquid polyester through the anal sphincter." Such agents were taught to be edible C18 and higher saturated fatty acids and their salts. Later feeding studies showed that this solution worked reasonably well in rats, but not nearly as well in humans. In order to reduce the incidence and severity of anal leakage, Procter & Gamble patented a new type of SPE.

This type is explained in the recent patent literature and differs from the definition of SPE that appears in the original Food Additive Petition (56). Procter & Gamble claimed that viscosity control, along with a yield point eliminated, or at least diminished, the problem. For reducing leakage, it appeared that varying the fatty acid content to give a higher melting SPE, having a higher viscosity (2.5 poise at shear rate of 800 reciprocal seconds), a yield point of at least 2500 dynes/sq.cm., a thixotropic area of at least 200,000 dynes/sq.cm.sec. and a liquid/solid stability of at least 50% afforded a substance which diminished this effect in humans (57). This material reduced leakage in both adults and children, but subjects still exhibited increased levels of flatulence, urgency, and soft stools.

Many clinical studies have been reported in the literature; almost all deal with another key property uncovered at Procter & Gamble—that of cholesterol lowering. SPE seems to increase cholesterol excretion, while not affecting bile acid excretion. A number of studies have shown that SPE has the ability to reduce plasma cholesterol by 10–20% (58–61). On a very positive note, these studies have also shown that both total and LDL cholesterol are dramatically reduced, with little effect on HDL. Highest reduction was found to occur in hypercholesterolemic people on hypocaloric diets. Many felt this would prove to be the greatest benefit of all to the use of noncaloric oils. The fact that they recently dropped their NDA (62) indicates that Procter & Gamble is unlikely to pursue the drug route. At the same time, however, they are considering the possibility of cholesterol-lowering foods. This could prove to be an interesting marketing direction.

Consumption levels of olestra are considered by Procter & Gamble to be 5.7 g/day for chronic average exposure and 8.0 g/day for a person-day estimate in the food categories mentioned in the FDA petition. The 90th percentile would consume an average of 9.7 g/day and have person/day/acute exposure of 17.4 g/day in the United States (63). These numbers might well be higher in many European countries and are not entirely consistent with consumption levels estimated by Frito-Lay (64) and Nutrasweet (34). Although categories differ, the latter seems to indicate a higher intake level relative to the stated category.

These estimates are used as a starting base for clinical trials. A conundrum arises as one looks at such numbers. It revolves around the relative values of caloric savings vs. clinical acceptance. If one is to save significant calories, high levels of olestra must be consumed. If too high a level is consumed, one risks side effects such as anal leakage. Appropriate balance is vital, particularly once the consumer has access. It is possible a self-limiting factor may exist. Validity of these numbers also affects safety factors for long-term feeding studies.

The FDA has indicated that they will look at a preponderance of the evidence and are developing a new set of rules for such compounds (65). Postmarket surveillance and exposure assessment will probably be necessary (66), and side effects will be examined closely (67). They will also have to consider the many objections raised by the Center for Science in the Public Interest (68,69), the National Renderers Association (70–72), Medlantic (73), and acamedicians (74). Procter & Gamble has a number of studies underway aimed at answering the questions raised (75,76).

Procter & Gamble has reported on unpublished data concerning the effect of chronic exposure to olestra on vitamin K. A double-blind study was run consisting of 100 people consuming 20 g/day of olestra. The control group of 100 consumed triglyceride. Overall conclusions indicated olestra would not effect vitamin K levels when consumed at this 20 g/day level (63).

Numerous U.S. patents have been issued in the last few years in the olestra area. Even more prolific however, have been European patent applications. A patent relating to foods containing both sucrose polyesters and vegetable proteins describes a dietary method to reduce blood plasma cholesterol (77). If at least 1 g per serving of the food is sucrose polyester and 1.5 g is vegetable protein, total cholesterol and plasma triglyceride levels are reduced. This is described as being a means to reduce cholesterol levels without use of drugs or making a major change in eating habits.

Another patent is for plastic shortenings containing SPE having short-chain fatty acid radicals of 2–10 carbon atoms being combined with long-chain fatty acid radicals containing 20–24 carbons (78). These combine to form a crystal matrix which functions to decrease anal leakage. Ratios between 9:1 and 3:1 of liquid to solid are recommended using caprylates and behenates. Solid shortenings are prepared in which the noncaloric solid fat portion can hold large volumes of oil within its crystal matrix. The shortening may be used in a wide range of foods such as baked goods and margarines.

Another application discloses shortenings which have a SFI curve and penetration adjusted to increase smoothness and decrease graininess (79). These shortenings are composed of a soft oil, hardstock fat, and intermediate melting triglycerides. The soft oil is a liquid oil acting as a softener, having a higher iodine value. Oleic and linoleic esters are preferred. The hardstock fat contains more saturated fatty acid chains, which tend to be of longer length. The intermediate melting triglycerides contribute to the crystal structure and increase oxidative stability, having an iodine value between 25 and 60.

Low-calorie chocolate confections can be prepared in which SPE is a replacement for natural fats and intense sweeteners. Bulking agents can replace sugar (80). A 5:3 ratio of myristate to laurate in the ester portion of the SPE affords a compound which has a melting point between 30 and 36°F and an SFI of 66, at least 6.6 degrees below its clear melting point. Combination with cocoa gives a product having a melting profile close to chocolate confectionery fat.

In a similar application (81), Procter & Gamble disclosed a cocoa butter substitute made from SPE. The fatty acid composition involves predominantly palmitic and lauric acids in one case and capric and stearic in the other. The substitute is claimed to be noncaloric and less expensive than natural cocoa butter, but has nearly identical melting properties.

SPE improves the life of frying mediums and food fried in these mediums (82). A method of frying of foods in compositions giving better stability and frylife, as well as the same fried fat-containing compositions, is claimed. SPE balance can vary from 35 to 100%, balanced by ordinary triglycerides, in these media. This becomes important because of the significant level of oil absorbed by the food under normal frying conditions.

A food or beverage having an altered flavor display may be prepared containing a fat phase composed of triglyceride and SPE (83). Such a composition will have both a viscous rheology and a high liquid solid stability. These will have lower caloric values than standard beverages of this type. Reportedly, SPE will change the flavor display exhibited by standard triglycerides in beverages. Specific components may be heightened, others may be suppressed, flavor ordering of the profile is changed, and both rate and overall flavor intensity is modified. Many examples are given combining both intermediate and hardened blends of SPEs.

Procter & Gamble does not have a monopoly on interest in SPE. Unilever has recently filed an application for a new process for preparation of polyol fatty acid esters and their use in foodstuffs and pharmaceuticals (84). They claim to have overcome the low rate of conversion to ester by using a shorter-chain fatty acid ( $< 15$ C atoms) soap at levels greater than 8%. If successful, this would significantly help the economics of SPE.

Mitsubishi has an effort underway in Japan to extend their partially esterified sucrose to a more SPE-like product. Dai-Ichi Kogyo Seiyaku has made limited samples of their material available. Other published literature indicates that several U.S. and European firms, predominantly Unilever (85,86), have done work with SPE type materials. An obvious interest exists, and many seem to be preparing to someday enter the market. Procter & Gamble's basic patent ran out in 1989, leaving the SPE market more open. Whatever the ultimate commercial outcome, SPE appears poised to be the groundbreaker in the fat replacement area, and for that Procter & Gamble deserves a lot of credit.

For additional data, two excellent reviews of SPE have been written (87,88). These provide more specific details concerning what has been done.

Although, as indicated earlier, most of the work in this area has been done with sugar, Pfizer has patented nondigestible food additives made by heating starch with food-safe di- or tricarboxylic acids or anhydrides (89). They are claimed to be resistant to action of amylolytic enzymes, and hence not absorbed.

## POLYCARBOXYLIC ACID ESTERS

Besides SPE, many other routes have been investigated to make specific esters that are not susceptible to lipase hydrolysis. One of the most obvious is synthesis of esters of polycarboxylic acids (retrofats). Many have been synthesized over the years, beginning in the early 1940s. Reactions between 1,3-diolein and 1,3-distearin with dicarboxylic acids from fumaric, succinic, and adipic acids have been reported (90). Again, proper selection of moieties can give one some control over physical properties obtained. However, the majority of such compounds afforded waxy, plastic-like materials.

Polycarboxylic acids are probably best exemplified by the work reported by Hamm at Best Foods (91). Trialkoxytricarballylate (TATCA) and trialkoxycitrate (TAC) were synthesized and evaluated in a number of food models such as mayonnaise and margarine. In the case of TATCA a 72% yield of $> 99\%$ pure material was obtained. Crude TAC was made from oleyl alcohol and citric acid. Unfortunately, yields were very low (20%) because of thermal decomposition of citric acid.

Hamm also reported synthesis of trialkoxyglyceryl ether, SPE, and use of jojoba oil. All oils exhibited viscosities similar to vegetable oil and generally behaved as Newtonian fluids. Surface and interfacial tensions were generally similar as well, with the exception of SPE.

Margarines were prepared from both TATCA and jojoba. The TATCA product was similar in appearance and texture to the control, but was softer and melted more quickly. It fared well, however, in both egg frying and cake baking. In contrast, jojoba did not perform physically as well and, although it fried an egg well, it performed poorly in cake baking—affording large air pockets and poor volume.

Hamm concluded that TATCA could be used in any typical vegetable oil application, "e.g. shortening, hard butters, salad oil, industrial frying oil, provided food safety had been demonstrated and its consumption could be restricted to low levels in the consumer diet in order to prevent the appearance of unwanted anal leakage problems."

Both jojoba and TATCA were fed to rats in attempts to determine their relative caloric values. Quantitative results were not obtained because, even at low doses of 0.5%, oily coats were obtained as a result of anal leakage. Weight gain data did, however, seem to indicate a low caloric value for both substances. At 2% levels in the diet, all animals consuming either jojoba or TATCA showed evidence of depression. Ten percent of the animals consuming jojoba died, as did 20% of the animals eating TATCA. Again, as seen earlier in the case of SPE, anal leakage had become a major negative issue.

Subsequent to this study, a U.S. patent was issued which claimed use of esterified tricarballylic acid as a substantial portion of edible oil to reduce calories in food (92). Recently (93) additional feeding studies were reported for TATCA. Using 10% corn oil as control, levels of 0.15, 0.75, 1.5, 3, 6, and 9% TATCA were substituted and fed to male rats for 2 weeks. Food consumption increased at the 9% level only and an average of 97% of the TATCA was recovered in the feces with HPLC showing absence of hydrolysis products. At the 9% level, all rats exhibited anal leakage, but at 6% only 2/10 did.

Nabisco (94) has filed a European application claiming numerous variations on the retrofat theme for use as reduced calorie oils. Much of their work seems to be related to or an extension of that of Best Foods, Frito-Lay, and others. Nabisco reports partial hydrolysis of their compounds to give 50–90% caloric reduction.

## HINDERED ESTERS

Another interesting, yet similar approach to reduced calories is via design of steric hindrance into the triglyceride. A patent claims use of alpha-branched esters of triglycerides (95). Whyte notes that absorbability of oils can be lessened by altering either the fatty acid or alcohol portion of the molecule. The latter can be done using neopentyl alcohol, but the thrust of his patent claims use of modified ester groups. Specifically, he states that substitution of alkyl groups alpha to the ester carbonyl will decrease absorbability, while at the same time maintaining both desired palatability and physical properties. Such compounds give a coefficient of absorbability of 0–50 vs. 90–100 for triolein. This probably occurs because branching interferes with enzymatic hydrolysis by pancreatic lipase. These alpha-substituted glycerides are recommended for use in salad or cooking oils, plastic shortenings used in frying, cake making and bread baking, as well as mayonnaise, margarine, and dairy products.

Another patent claims use of acylated glycerides in low-calorie foods (96). This appears related to Whyte's work on alpha-alkyl substitution. Again, these compounds were claimed to have desirable physical and palatable properties while having a lower effective caloric value because of both lower digestibility and absorbability. Pancreatic lipase hydrolysis decreases as one increases the length of the side chain, as one would expect.

Significant work has been reported by Frito-Lay on their efforts to develop a thermally stable oil substitute. At this time, they seem to be more advanced in their studies than anyone other than Procter & Gamble. Frito-Lay's material was first discussed in a patent entitled "synthetic cooking oils containing dicarboxylic acid esters" (97). This was followed by a patent for their use in fried food products (98).

In the first, they claimed diesters of substituted malonic acid where at least one of the labile hydrogens is substituted by an alkyl group of 1–20 carbon atoms. Preferred compounds were the dioleyl esters of hexadecyl or dihexadecyl malonic acid. The oleyl esters were preferred to lower molecular weight esters because of a higher boiling point. However, it is clear that a mixture of fatty acid esters is being used to produce a fat substitute that has the correct physical properties. These esters are reported to resist hydrolysis by pancreatic lipase and other digestive juices present in the stomach and small intestine, and are thus highly unabsorbed.

A paper at the American Oil Chemists Society annual meeting (64) reviewed ongoing research. Dioleyl dihexadecylmalonates (DDM) were characterized as being thermally stable and proposed for frying applications. Their melting point is slightly below body temperature, but they do not impart a waxy mouthfeel. Shelf life for fried snack foods was 40–50 days, standard for the industry.

Sensory evaluations for potato chip products prepared from a 60:40 blend with soybean oil found overall acceptance to be equal to products prepared with either neat soybean or cottonseed oils or blends. The only significant difference was in a perception of lower oiliness, a positive attribute. Tortilla chips also showed equivalent characteristics to control.

Substituted malonates were found to be resistant to porcine pancreatic lipase and other digestive enzymes. Recovery of fecal oil was high relative to ordinary triglycerides. In fact, it was comparable to mineral oil, i.e., 85–95%. Radiolabel studies showed less than 0.1% of the dose in the carcass, but this low level of absorption is believed to be real. This contrasts with reports of Procter & Gamble, who claim absorption of olestra to be below limits of detection (62), despite evidence of Aust to the contrary (99–101) for sucrose polyester. Mineral oil (102), silicone oil (103), and microcrystalline cellulose (104) have also been found to be absorbed at low levels.

Anal leakage was found to occur in both male and female Fischer 344 rats at approximately 3.2 g/kg/day in the diet after 7–10 days. Pathology indicated no gross lesions or effects on body organs. Similarly, no dose-related changes in clinical chemistry were found other than a mild leukocytopenia, which was believed to be secondary to nutritional effects.

A 60:40 blend with soybean oil would result in about a one-third calorie reduction. This translates to consumption of 6 g of oil substitute per ounce of potato chips, 4 g for corn chips, and 6–7 g for fried chicken. Mean consumption in salty snacks for teenagers would thus be 1.9 g/day for teenagers, with heavy users consuming 9.6 to more than 21 g/day. A typical consumption estimate, if used in a variety of foods, would be 43 g/day. This is much higher than for the indicated consumption of olestra of 5.7 g/day (63).

ARCO Chemical, in an effort to utilize propylene oxide, has been actively pursuing use of esters of the glycol as a low-calorie fat substitute. A European patent application (105) was published claiming epoxide-extended polyol esters of fatty acids and a method of preparation to form products which are resistant to lipase hydrolysis and digestive tract absorption. Initiator can have 3–8 hydroxyl moieties, the backbone can be epoxide varying from C-3 to C-6 and has a minimum number of units corresponding to the number of primary hydroxyls in the initiator, with ester being a fatty acid acyl moiety.

Initiators include sugars, glycerides, or saccharides. Sugars might include mannose, arabinose, glucose, galactose, sorbitose, and amylose, but glycerol is preferred. Epoxides, in addition to propylene, could also include butylene, pentene, and isobutylene. Ester can be varied, depending on specific physical properties desired. In fact, all this is fairly basic polyglycol chemistry, well known in the industry for many years. Choice of glycerine as an initiator and use in the name appears to be an excellent choice to make the product appear more like a triglyceride.

Relative digestibility, as measured by lipase activity, using olive oil as control, varies from 76.2% for nonepoxylated material to 46.2% with one, 18.9% with 2.2, and 0 with 5 or 8 epoxy units. All results are an average of four determinations. Three-week feeding studies at the 2.5% level in Sprague-Dawley rats showed typical results. Body weight increased similarly in both test and control animals with no outward signs of toxicity.

## MEDIUM-CHAIN TRIGLYCERIDES

Medium-chain triglycerides have been used clinically for many years, but their success in consumer items has been limited. Coconut oil, having a large percentage of C6–C10 fatty acids, serves as a good source. Glycerine esterification affords a product smaller in size than standard triglycerides and having greater water solubility. As such, they are more readily absorbed via the portal system than the lymphatic one and therefore proceed more directly to the liver where they are rapidly oxidized and metabolized. They are actually metabolized as fast as glucose, not more slowly as corn oil would be. Thus, they have a much lower desire to end up as depot fat than their longer-chain counterparts and may lower total cholesterol level as well.

This is in contrast to much of what appears in the popular literature relative to saturated fatty acids (SFA). If differing SFAs are absorbed by different pathways, coconut oil may not have the same effect on cholesterol as their longer-chain counterparts. As a result, although it will still have a high caloric density, coconut oil may not always act as has been reported.

Although medium-chain fatty acids have been mentioned as possible weight control substances, evidence is mixed as to their utility. Still, Babayan claims that medium chain fatty acid esters of polyglycerines give a feeling of fullness while, at the same time, reducing both lipid level and calories (106).

## POLYGLYCERINE ESTERS

Polyglycerine esters have also been recommended as tailor-made fats. They both taste and look like fats, but are lower in caloric value, while having a positive effect on palatability.

A paper has been published dealing with nutritional studies of polyglycerol esters (107). A number of esters were prepared from cottonseed and peanut oil fatty acids. After feeding weanling male rats at a 1 g/day level for 3 weeks, ester level was increased to 8%. Comparison to lard showed equivalent weight gains and no abnormalities. Diarrhea was, however, observed in test rats.

Procter & Gamble published a metabolic study on tri- and decaglycerines both in vivo in the rat and in vitro using pancreatic enzymes (108). Labeled fatty acids allowed them to determine, independently, the fate of the polyglycerines. They concluded that the ester bonds were hydrolyzed prior to absorption and free fatty acids absorbed via the thoracic duct pathway. Michael and Coots (108) hypothesized the polyglycerines were not absorbed as rapidly as the free fatty acids and, when absorbed, entered via the portal venous blood. They also determined that, following fatty acid catabolism, polyglycerines were excreted unchanged in the urine and not stored in the carcass. In excess of 90% of the triglycerine and 40% of the decaglycerine were absorbed following hydrolysis. Less than 4% of the tagged backbone was excreted in respiratory carbon dioxide. This was about equivalent to glycerine contaminant in starting material.

Hydrolysis of ester greatly decreases potential as a low-calorie fat. Even if backbone is not utilized, the fatty acid portion will contribute significantly to caloric value. If one minimizes fatty acid content, water solubility is high and oil-like characteristics are lost. If one increases fatty acid content to obtain an oily product, caloric value increases rapidly.

In a related study, glycerine was esterified with a saturated long chain fatty acid fraction from coconut oil (89% stearate and 11% palmitic) and fed to humans (109). They found the substance to be poorly absorbed vs. corn oil (31–39% vs. 98%). Results were ascribed to its high melting point, impaired emulsification, and micellar solubilization in the small intestine.

The authors proposed that combination of these long-chain saturated fatty acids with polyglycerines would lead to a poorly absorbed fat. They speculated that incorporation of these into the diet to provide the satisfaction of eating "rich" creamy foods merited further investigation. Several researchers made use of this hypothesis, claiming use of C16–C18 saturated fatty acid esters of low DP polyglycerols to produce margarine and low-fat spreads having less graininess and more butter-like properties (110).

## ETHERS

Another approach receiving a fair amount of investigation involves conversion of the ester linkages of triglycerides to ethers (111). Synthesis of the 1,3-diether is relatively straightforward, but substitution of the secondary position is far more difficult. The melting point will increase with increasing chain length and can be somewhat tailored to properties desired. Most triethers are readily oil soluble and have little or no taste, odor, or color. They are, however, susceptible to oxidation because of the alpha ether hydrogens.

Ethers are somewhat susceptible to hydrolysis in the body, although at a far slower rate than esters. Absorption appears to be directly related to the chain length of the alkyl portion—dioctadecyl being absorbed by less than 10%, while dioctyl is absorbed at a 50% level. Studies with labeled trialkyl ether have shown that 99.8% of the radioactivity was excreted in the feces. However, the small amount absorbed was detected in the liver and spleen. In monkeys, 97.7% of the triether was found in the feces, with 1.8% being recovered in the blood. Rat-feeding studies have shown little to be absorbed, no bad side effects, and some weight loss (112,113).

Hamm reported the synthesis of a trialkoxyglyceryl ether (91). In addition to synthetic difficulties, purification proved to be both time consuming and expensive. Preparative HPLC gave an acceptable product with less than 0.01% diethers. Analysis showed it to be mainly the trioleylether. As synthesis and purification were so difficult, no further evaluations were conducted. Hamm concluded this would be too great an impediment to permit commercial viability.

Monsanto has a patent for preparation of sucrose ethers for use in the detergent area (114). These might also be capable of functioning as low-calorie substitutes in food.

## POLYSILOXANES

Polymers of silicones have been proposed as fat substitutes. A European Patent Application (115) has been filed claiming use of polyorganosiloxanes having at least 15% by weight organic carbon content and a molecular weight of at least 500 in food compositions. The limitation of 500 daltons was imposed to minimize possibility of absorption in the stomach or intestines. Uses include frying oils, mayonnaise, baked goods, cereals, peanut butter, dairy products, and sandwich spreads. Caloric reductions ranged from 59% in pie crusts and 55% in ice cream to 27% in blueberry muffins.

Bracco (116) has reported on the effect of phenylmethylpolysiloxane (PS) on the body composition of obese Zucker rats. Sixteen obese female Zucker rats were fed both experimental (22% PS) and control diets for 5 weeks. Control rats gained 8.2% of initial weights, while PS-fed rats lost

11.7%. No significant differences existed in food intake. In the PS-fed rats both lipid and protein levels were reduced significantly, but PS-fed rats had an overall higher proportion of body protein. Stools were firm and well formed in both groups, but levels of cellulose in all diets were high.

In a second experiment, the low-fat group was fed a cellulose-diluted, isocaloric diet (33.6% vs. 11.7%). PS-fed rats lost 7.5% of initial weight, while cellulose diluted gained 4.6% with the same total daily intake. In this case body fat decreased significantly in the PS group, but protein levels did not.

These results are interesting because of the lack of compensation for reduction of caloric density. It should be noted that this can be strain dependent. In fact, use of Sprague-Dawley rats resulted in no loss of body weight when fed a 10% level of polydimethylsiloxane (117).

In a further study (118) polydimethylsiloxane was fed to humans as part of a 40% fat diet. No leakage was detected at 2% (30 g/day), a slight amount at 3%, and significant leakage at 4%. In addition, other signs of gastrointestinal stress were observed at 3% levels in the diet.

## AVICEL AND JOJOBA OIL

Many other approaches have been suggested for replacement or reduction of fat content in foods. Avicel or microcrystalline cellulose is one. Studies commissioned by FMC showed that low oil emulsions containing microcrystalline cellulose exhibited properties similar to those having larger amounts of fat. This allowed them to remove approximately two thirds of the oil and a corresponding number of calories without significantly changing the physical properties.

Jojoba oil, a linear mixture of long-chain fatty acid esters, has been promoted as a substitute for ordinary animal and vegetable oils (119). Digestion is only about 40% (120). Rat studies indicate that body fat can be decreased significantly (121). As Hamm has shown, however, in the food area its value is highly dubious. Claims of low caloric value have not yet been satisfactorily proven, and toxicity questions are still high on everyone's list.

Work has been done to incorporate it into food (122). The inventors taught how to make improved coating compositions for foodstuffs like chocolate and dried fruit. Hydrogenated material is especially good for coating chocolate because of its hardness and similarity to beeswax. The liquid product is claimed to have very good oxidative stability.

It does not have food approval, but many farmers have started large jojoba plantations in the southwest in hopes of making large profits.

From reading the claims, it appears that advocates believe jojoba might be the panacea of the twenty-first century. It is claimed to be good for everything from curing gout and constipation to serving as a motor oil for your car.

## CONCLUSIONS

In his 1986 patent for the use of acylated esters of glycerine, Volpenhein makes an interesting statement: "the combination of desirable fat-like properties with decreased absorbability or digestibility cannot be predicted with any degree of accuracy. Accordingly, selection of non-digestible fats suitable for food compositions is still a 'hit or miss' proposition." This simple statement highlights one of the most troublesome and confusing issues to those involved in research aimed at development of low-calorie fats. Screening techniques are inadequate at best, and it is difficult to predict what is really going to happen once a substance enters the human body. A large number of hypotheses can be generated and appear beautiful on the blackboard, but translation into synthesis of the appropriate compound and testing to prove or disprove the hypothesis is difficult, time consuming, and fraught with error and frustration.

A Unilever spokesman cogently assessed the major questions as: (1) Will it work as a fat substitute? (2) Is it safe to use without deleterious side effects? and (3) Will the cost be prohibitive? In the interim, much basic research is in progress. Eventually someone will either develop and get approval for a true fat substitute or formulators will learn how to use fat replacers effectively. Meanwhile, the consumer will be only partially satisfied until more questions have been answered.

## REFERENCES


1. J. E. Kinsella, *Food Technology, 42*(10): 128 (1988).
2. A. M. Altschul, *Food Technology, 43*(4): 113 (1989).
3. G. A. Miller and H. I. Frier, *Food Technology, 43*(4): 136 (1989).
4. D. Kritchevsky, J. Am. Oil Chemists Society Annual Meeting, Cincinnati, OH, 1989.
5. M. W. Pariza, *Nutrition Reviews, 45*(1): 1 (1987).
6. R. W. Hart, *Food Chemical News, 31*(18): 52 (1989).
7. T. H. Donnelly, American Chemical Society Annual Meeting, Miami Beach, FL, 1985.
8. A. Gillis, *J. Am. Oil Chemist Society, 65*(11): 1705 (1988).
9. K. A. Harrigan and W. M. Breene, *Cereal Foods World, 34*(3): 261 (1989).
10. B. F. Haumann, *J. Am. Oil Chemist Society, 63:* 278 (1986).
11. R. G. LaBarge, *Food Technology, 42*(1): 84 (1988).

12. Norman O. V. Sonntag, Am. Oil Chemists Society Annual Meeting, Cincinnati, OH, 1989.
13. G. Mieth, *Die Nahrung, 27*(9): 853 (1983).
14. H. J. Deuel, *J. Nutr., 35:* 295 (1948).
15. J. B. Boer, B.C.P. Jansen, and A. Kentic, *J. Nutr., 33:* 339 (1947).
16. H. J. Deuel, Jr., L. Hallman, and A. Leonard, *J. Nutr., 20:* 215 (1940).
17. R. Morley and W. Ashton, U.S. Patent 4,400,405 (1983).
18. D. Moran and D. Sharp, U.S. Patent 4,515,825 (1985).
19. H. J. Channon and G. A. Collinson, *J. Biochem., 23:* 676, (1929).
20. W. A. Strucker, *Arch. Pathol., 31:* 670 (1941).
21. A. C. Frazer, *Physiol. Rev., 26:* 103 (1946).
22. *Food Chemical News, 31*(2): 21 (1989).
23. S. Cooley and G. Livesey, *Brit. J. Nutr., 57:* 235 (1987).
24. D. C. Hobbs, personal communication, February 16, 1988.
25. J. M. Lenchin, P. C. Trublano, and S. Hoffman, U.S. Patent 4,510,166 (1985).
26. J. Rice, *Food Processing, (6):* 118 (1989).
27. A. L. Morehouse, American Chemical Society Annual Meeting, Los Angeles, CA, 1988.
28. D. Snyder, *Crain's Chicago Business, 11*(3): (1988).
29. W. Wall, *Wall Street Journal,* January 19, 1988.
30. W. Wall, *Wall Street Journal,* January 28, 1988.
31. *Food Chemical News, 29*(52): 16 (1988).
32. N. S. Singer, S. Yamamoto and J. Latella, U.S. Patent 4,734,287 (1988).
33. *Federal Register* (53FR47580), November 23, 1989.
34. J. Hjelle, American Health Foundation Meeting, New York, March 15, 1989.
35. *Food Chemical News, 31*(9): 15 (1989).
36. A. M. Freedman, *Wall Street Journal,* June 30, 1989.
37. J. M. Rispoli, J. P. Sabhlok, A. S. Ho, B. G. Scherer, and C. Giorliano, U.S. Patent 4,308,294 (1981).
38. E. R. Barnett et al., Eur. Patent Public. No. 0301440 (1989).
39. R. H. Barth, R. Park, and H. Burrell, U.S. Patent 2,356,745 (1944).
40. A. Minich, U.S. Patent 2,962,419 (1960).
41. F. H. Mattson and R. A. Volpenhein, *J. Lipid Res., 13:* 325 (1972).
42. A. Weiss and G. Mieth, *Die Nahrung, 27:* K13 (1983).
43. G. P. Rizzi and H. M. Taylor, U.S. Patent 3,963,699 (1976).
44. G. P. Rizzi and H. M. Taylor, *J. Am. Oil Chem. Soc., 55:* 398 (1978).
45. R. A. Volpenhein, U.S. Patent 4,517,360 (1985).
46. R. A. Volpenhein, U.S. Patent 4,518,772 (1985).
47. T. Yamamoto and K. Kinami, U.S. Patent 2,161,806 (1986).
48. F. H. Mattson, U.S. Patent 4,034,083 (1977).
49. R. A. Volpenhein and R. J. Jandacek, U.S. Patent 4,241,054 (1980).
50. F. E. Wood, Jr., Eur. Patent Public. No. 0290216 (1988).
51. B. A. Roberts, U.S. Patent 4,446,165 (1984).
52. P. D. Orphanos, C. A. Bernhardt, B. L. Madison, P. Seiden, A. M. Ehrman, J. D. Baginski, and T. B. Guffey, European Patent Public. No. 0290421 (1988).
53. E. J. Hollenbach and N. B. Howard, U.S. Patent 4,368,213 (1983).

54. M. D. Robbins and S. S. Rodriguez, U.S. Patent 4,461,782 (1984).
55. R. J. Jandacek, U.S. Patent 4,005,195 (1977).
56. *Fed. Reg., 52:* 23606 (1987).
57. C. A. Bernhardt, Eur. Patent Public. No. 0236288 (1987).
58. F. H. Mattson, R. J. Jandacek, and M. R. Webb, *J. Nutr., 106:* 747 (1976).
59. C. J. Glueck, F. H. Mattson, and R. J. Jandacek, *Amer. J. Clinc. Nutr., 32*(8): 1636 (1979).
60. M. J. Mellies, C. Vitale, R. J. Jandacek, G. E. Lamkin, and C. J. Glueck, *Amer. J. Clin. Nutr., 41*(1): 1 (1985).
61. G. A. Nolen, F. E. Wood, Jr., and T. A. Dierckman, *Food Chem. Toxic., 25*(1): 1 (1987).
62. C. M. Bergholz, Am. Oil Chem. Soc. Annual Meeting, Cincinnati, OH, 1989.
63. J. Spinner, American Health Foundation Symposium, New York, March 15, 1989.
64. M. E. Spearman and J. G. Fulcher, Amer. Oil Chem. Soc. Ann. Meeting, Cincinnati, OH, 1989.
65. J. Vanderveen, American Health Foundation Meeting, New York, March 15, 1989.
66. J. Vanderveen, Northeast AOCS Section Symposium on Dietary Lipids, Amherst, MA, 1989.
67. A. Forbes, Nat'l Acad. Sciences/Nat'l Heart, Lung and Blood Inst. Meeting, Washington, D.C., 1988.
68. M. F. Jacobson and E. Blume, letter to FDA, December 2, 1987.
69. *Food Chemical News, 29*(40): 22 (1987).
70. *Food Chemical News, 31*(5): 23 (1989).
71. *Food Chemical News, 30*(16): 15 (1988).
72. *Foodstuffs, 61*(9): 1 (1989).
73. *Food Chemical News, 30*(29): 35 (1988).
74. J. M. Karkeck, *J. Am. Diet. Assoc., 88*(9): 1039 (1988).
75. C. M. Bergholz and K. C. Triebwasser, personal communication, March 20, 1989.
76. *Food Chemical News, 31*(6): 2 (1989).
77. F. H. Seligson, J. E. Hunter, and A. H. St. Clair, U.S. Patent 4,789,664 (1988).
78. R. J. Jandacek and J. C. Letter, U.S. Patent 4,797,300 (1989).
79. T. B. Guffey, S. R. Talkington, S. S. Abe, and M. D. Mijac, Eur. Patent Applic. 0290420 (1988).
80. J. L. Y. Kong-Chan, U.S. Patent 4,810,516 (1989).
81. S. A. McCoy, P. M. Self, and B. L. Madison, Eur. Patent Public. No. 0271951 (1988).
82. J. R. Baginski and D. J. Weisgerber, Eur. Patent Public. No. 0287157 (1988).
83. T. B. Guffey, S. R. Takington, S. S. Abe, and M. D. Mijac, Eur. Pat. Public. No. 0290065 (1988).
84. P. Van der Plank, Eur. Pat. Public. No. 0301634 (1989).
85. F. W. Cain, Eur. Patent Public. No. 0304130 (1989).
86. F. W. Cain and L. H. Wesdorp, Eur. Patent Public. No. 0304131 (1989).

87. R. J. Jandacek, *Intl. J. Obesity, 8*(S1): 13 (1984).
88. R. W. Boggs, *Fette Seifen Anstrichmittel, 88:* 154 (1986).
89. R. Carrington and G. Halek, U.S. Patent 4,247,568 (1981).
90. R. O. Feuge and T. L. Ward, *J. Am. Chem. Soc., 80:* 6338 (1958).
91. D. J. Hamm, *J. Food Sci., 49:* 419 (1984).
92. D. J. Hamm, U.S. Patent 4,508,746 (1985).
93. M. Bieber, Am. Oil Chem. Soc. Annual Meeting, Cincinnati, OH, 1989.
94. L. P. Klemann and J. W. Finley, Eur. Patent Public. No. 0303523 (1988).
95. D. D. Whyte, U.S. Patent 3,579,548 (1971).
96. R. A. Volpenhein, U.S. Patent 4,582,715 (1986).
97. J. Fulcher, U.S. Patent 4,582,927 (1986).
98. J. Fulcher, U.S. Patent 4,673,581 (1987).
99. L. Aust, J. Bruckner, J. Proll, R. Noack, and G. Mieth, *Ann. Nutr. Metab., 25:* 255 (1981).
100. L. Aust, G. Mieth, J. Bruckner, A. Weiss, and R. Noack, *Die Nahrung, 30:* 453 (1986).
101. L. Aust, H. Schliemann, J. Proll, and G. Mieth, *Die Nahrung, 32:* 305 (1988).
102. H. P. Dinesoy, R. E. Weesner, and J. MacGee, *Am. J. Clin. Pathol., 78:* 35 (1982).
103. J. Paul and W. F. R. Poyer, *Arch. Biochem. Biophys., 87:* 312 (1960).
104. S. Nicklin and K. Miller, *Food Chem. Toxicol., 22:* 615 (1984).
105. J. F. White and M. R. Pollard, Eur. Patent Public. No. 0254547 (1988).
106. V. K. Babayan, Amer. Oil Chemists Society Meeting, Philadelphia, PA, 1985.
107. V. K. Babayan, H. Kaunitz, and C. A. Sianetz, *J. Am. Oil. Chem. Soc., 41:* 434 (1964).
108. W. R. Michael and R. H. Coots, *Toxical. Appl. Pharmacol., 20:* 334 (1971).
109. S. A. Hashim and V. K. Babayan, *Am. J. Clin. Nutr., 31:* S273 (1978).
110. J. Van Heteren, C. Poot, F. Reckweg, and M. F. Stewart, U.S. Patent 4,456,627 (1984).
111. V. W. Trost, Can. Patent 1,106,681 (1981).
112. F. Paltauf, *Biochime Biophys. Acta, 176:* 818 (1969).
113. F. Spener, F. Paltauf, and A. Holasek, *Biochime Biophys. Acta, 152:* 368 (1968).
114. V. R. Graetner, U.S. Patent 3,170,915 (1965).
115. C. L. Frye, Eur. Patent Public. No. 205273 (1986).
116. E. F. Bracco, N. Baba, and S. A. Hashin, *Am. J. Clin. Nutr., 46:* 784 (1987).
117. S. A. Hashim, American Health Foundation Meeting, New York, March 15, 1989.
118. S. A. Hashim and E. F. Bracco, Am. Oil Chemist Society Ann. Meeting, Cincinnati, OH, 1989.
119. J. H. Brown, *HAPPI* (10): 38 (1984).
120. V. Anantharaman, *J. Am. Oil Chem. Soc., 65*(1): 35 (1988).
121. G. S. Ranhotra, J. A. Gelroth, F. A. Novak, and F. Bohannon, *Cereal Chemistry, 63*(5): 459 (1986).
122. M. T. Devitt and J. B. Rossell, U.S. Patent 4,356,197 (1982).

# Index